AF684460

Food Contact Materials Analysis

Mass Spectrometry Techniques

Food Chemistry, Function and Analysis

Series editors:
Gary Williamson, *University of Leeds, UK*
Alejandro G. Marangoni, *University of Guelph, Canada*
Juliet A. Gerrard, *University of Auckland, New Zealand*

Titles in the series:

1: Food Biosensors
2: Sensing Techniques for Food Safety and Quality Control
3: Edible Oil Structuring: Concepts, Methods and Applications
4: Food Irradiation Technologies: Concepts, Applications and Outcomes
5: Non-extractable Polyphenols and Carotenoids: Importance in Human Nutrition and Health
6: Cereal Grain-based Functional Foods: Carbohydrate and Phytochemical Components
7: Steviol Glycosides: Cultivation, Processing, Analysis and Applications in Food
8: Legumes: Nutritional Quality, Processing and Potential Health Benefits
9: Tomato Chemistry, Industrial Processing and Product Development
10: Food Contact Materials Analysis: Mass Spectrometry Techniques

How to obtain future titles on publication:
A standing order plan is available for this series. A standing order will bring delivery of each new volume immediately on publication.

For further information please contact:
Book Sales Department, Royal Society of Chemistry, Thomas Graham House, Science Park, Milton Road, Cambridge, CB4 0WF, UK
Telephone: +44 (0)1223 420066, Fax: +44 (0)1223 420247,
Email: booksales@rsc.org
Visit our website at www.rsc.org/books

Food Contact Materials Analysis

Mass Spectrometry Techniques

Edited by

Michele Suman
Barilla Advanced Laboratory Research, *Parma, Italy*
Email: michele.suman@barilla.com

ROYAL SOCIETY OF CHEMISTRY

Food Chemistry, Function and Analysis No. 10

Print ISBN: 978-1-78801-124-2
PDF ISBN: 978-1-78801-297-3
EPUB ISBN: 978-1-78801-719-0
Print ISSN: 2398-0656
Electronic ISSN: 2398-0664

A catalogue record for this book is available from the British Library

© The Royal Society of Chemistry 2019

All rights reserved

Apart from fair dealing for the purposes of research for non-commercial purposes or for private study, criticism or review, as permitted under the Copyright, Designs and Patents Act 1988 and the Copyright and Related Rights Regulations 2003, this publication may not be reproduced, stored or transmitted, in any form or by any means, without the prior permission in writing of The Royal Society of Chemistry or the copyright owner, or in the case of reproduction in accordance with the terms of licences issued by the Copyright Licensing Agency in the UK, or in accordance with the terms of the licences issued by the appropriate Reproduction Rights Organization outside the UK. Enquiries concerning reproduction outside the terms stated here should be sent to The Royal Society of Chemistry at the address printed on this page.

Whilst this material has been produced with all due care, The Royal Society of Chemistry cannot be held responsible or liable for its accuracy and completeness, nor for any consequences arising from any errors or the use of the information contained in this publication. The publication of advertisements does not constitute any endorsement by The Royal Society of Chemistry or Authors of any products advertised. The views and opinions advanced by contributors do not necessarily reflect those of The Royal Society of Chemistry which shall not be liable for any resulting loss or damage arising as a result of reliance upon this material.

The Royal Society of Chemistry is a charity, registered in England and Wales, Number 207890, and a company incorporated in England by Royal Charter (Registered No. RC000524), registered office: Burlington House, Piccadilly, London W1J 0BA, UK, Telephone: +44 (0) 20 7437 8656.

Visit our website at www.rsc.org/books

Printed in the United Kingdom by CPI Group (UK) Ltd, Croydon, CR0 4YY, UK

Preface

Evolution of Mass Spectrometry Methods for Food Contact Materials Applications

Food packaging is surely a topic of worldwide relevance and importance. In recent decades, mass spectrometry (MS) has made incredible progress; in parallel, packaging solutions devoted to food have increased their number and varieties in many ways. The handling of issues with food contact materials (FCMs) requires a deep contemporaneous knowledge of both the materials and the food worlds.

Consumers' needs and expectations have been growing for many years in combination with true globalization of the markets; at the same time, the legal requirements and harmonized international regulations for FCMs are still far from being achieved.

Food safety, quality and traceability topics imply more and more continuous analytical improvements and efforts, and these find most of their solutions and effectiveness in the MS technique. In particular, this book, in addition to highlighting the mentioned topics, places special emphasis on the following three aspects:

- *demonstrates how in the last 20 years MS has become more and more available and diffused in many analytical laboratories and industrial facilities;*
- *combines different approaches utilizing MS as a tool for solving many different food packaging problems;*
- *gives concrete and practical MS protocols to stakeholders that testify to the great flexibility of MS.*

The book is therefore aimed at a wide potential audience such as postgraduates, academics and industrialists.

Food Chemistry, Function and Analysis No.10
Food Contact Materials Analysis: Mass Spectrometry Techniques
Edited by Michele Suman
© The Royal Society of Chemistry 2019
Published by the Royal Society of Chemistry, www.rsc.org

The first two chapters start to set the scene about risk assessment for both paper/board-based and plastic-based FCMs. In fact, the law requires the determination of contaminants or specific substances used for different purposes in the production of FCMs. A combination of analytical protocols with very high sensitivity and resolution allows a thorough description of all the species that must be monitored, and many of the standard methods used to determine these specific substances involve the use of MS.

Chemical substances intentionally added to FCMs to enhance the physicochemical properties and extend the product shelf life can be released during the various recycling steps and recovery processes. Starting compounds and additives (fillers, absorbers, antioxidants, photoinitiators, ...) are commonly used in the manufacture of FCMs and may remain unreacted in the matrix and consequently migrate into food, leading to a potential health issue: the third chapter illustrates the chemical nature and properties of these additives explored through MS.

Chapter 4 then reviews MS studies devoted to the complex world of adhesives (employed in packaging to manufacture laminates or to build up the final form of a packaging) and their various formulations, which involve a large number of volatile and semivolatile compounds. In this context, headspace solid-phase extraction (HS-SPE), headspace solid-phase microextraction (HS-SPME) and gas chromatography–olfactometry (GC-O) are usually combined with MS detection.

Indeed, GC-O is a suitable technique for the determination of off-odours resulting from the presence of odour-active compounds. This aspect is discussed in Chapter 5, where key parameters that have an impact on mass transfer in the migration of volatile substances from/to FCMs are considered, introducing migration and permeation tests in combination with proper sampling methods and analytical solutions, with GC-FID (flame ionization detection) and GC-MS in the foreground. In this chapter, the typical instrumentation applied for tackling different migration analytical scenarios is discussed in detail, by showcasing also the screening potential of ambient desorption and direct injection MS.

Every time migration phenomena occur within food packaging materials, they can potentially lead to sensory changes, playing a key role in how packaged food is received by consumers. Chapter 6 considers the interactions between packaging, food and the environment surrounding food. Sometimes it is necessary to avoid them as much as possible, but on other occasions, such as with so-called "active packaging," there is an intentional need to favour them. Understanding these interactions implies the application of multidisciplinary knowledge, ranging from molecular chemistry to cognitive psychology, where, again, MS maintains a place of primary importance.

Let us now move on "opening the black box" of non-intentionally added substances (NIASs): these are chemical compounds that can migrate from FCMs but were not added on purpose. Their origin can be related to, for example, degradation of the material's components, impurities of raw materials used in processing, *etc.* Is MS the right tool also for investigating these "not added originally"/not present in the positive list of authorized starting substances and additives? Chapter 7 says yes!

The identification of unknown migrants from FCMs is a really complicated issue since neither chemical and spectral databases nor analytical standards are available. Here a particularly strategic approach is the implementation of liquid chromatography coupled with high-resolution mass spectrometry (LC-HRMS) in combination with proper statistical data treatment. Chapter 8 addresses the most relevant non-target/HRMS data acquisition modes and also the data handling used for the identification/characterization of food packaging contaminants.

The application of engineered nanomaterials in FCMs represents a new direction for interesting new functions, but the toxicological implications are still a partially unknown menace. Chapter 9 comments constructively on how information collected especially by inductively coupled plasma mass spectrometry (ICP-MS) can be combined with that derived from other microscopy/fractionation techniques in order to detect the size and concentration of nanoparticles and better understand the mechanisms of the migration of nanoparticles from FCMs to food.

A final question remains after having considered such an intricate scenario: what are the best analytical solutions that can be efficiently and commonly applied among industries and/or outsourced to expert laboratories? Chapter 10 gives the answer to this delicate question, illustrating how these analytical solutions are really based on MS with a clear tendency towards HRMS, focusing on multi-methods to cover several analytes in parallel, screening out of simulants and opening the door in the future to the more comprehensive full screening of food itself.

This is the first book to be exclusively dedicated to the topic of the interconnections of mass spectrometry and FCMs and its facets. It will provide the interested reader with an excellent overview of the topic and with detailed insights into the rapidly developing field of these analytical approaches.

This book will take the reader to a final overall conclusion: "*The great effectiveness of mass spectrometry in solving food contact materials issues*". Enjoy!

Michele Suman

Contents

Food Chemistry, Function and Analysis No.10
Food Contact Materials Analysis: Mass Spectrometry Techniques
Edited by Michele Suman
© The Royal Society of Chemistry 2019
Published by the Royal Society of Chemistry, www.rsc.org

CHAPTER 1

Risk Assessment of Plastic-based Food Contact Materials: Focus on Polyolefins

FABIO TESTONI* AND INES MINGOZZI*

Basell Poliolefine Italia Srl, Italy, R&D PT&C (Process Testing and Characterization), Piazzale Donegani 12, 44122 Ferrara, Italy
*E-mail: Fabio.testoni@lyondellbasell.com, Ines.mingozzi@lyondellbasell.com

1.1 Introduction

Polyolefins is a term indicating a class of polymers commonly used for, among other applications, food packaging.[1] They are produced starting from one or more monomers, polymerization aids (catalysts, peroxides, *etc.*) and polymer production aids (*i.e.* in-process antistatic agents). Solvents can also be present. The resulting base polymer is then formulated with additives. Polymers in their final form (generally pellets, granules or flakes) are sold to converters to produce articles.

Monomer(s), additives, polymerization aids (PAs) and polymer production aids (PPAs) are considered intentionally added substances (IASs). Impurities within starting raw materials and reaction products from the manufacturing process (*i.e.* oligomers, additive by-products) are defined as non-intentionally added substances (NIASs) and are subjected to risk assessment, in addition to PAs and PPAs if not listed in EU or National Food Contact Regulations.

Food Chemistry, Function and Analysis No.10
Food Contact Materials Analysis: Mass Spectrometry Techniques
Edited by Michele Suman
© The Royal Society of Chemistry 2019
Published by the Royal Society of Chemistry, www.rsc.org

Regulation (EC) No. 1935/2004 of the European Parliament and of the Council of 27 October 2004 provides general principles of safety and inertness for all food contact materials (FCMs). In addition, EU Regulation No. 10/2011 as amended sets out rules for plastic materials in contact with food. EU Regulation No. 10/2011 describes requirements for migration tests. There are two types of limits: OML (overall migration limit) and SML (specific migration limit). OML is based on the total mass of substances that can migrate (gravimetric analysis), whereas SML is based only on quantification of one specific substance that can migrate into food.

EU Regulation No. 10/2011 is based on the principle of a "positive list" and Annex 1 contains the list of authorized chemicals (monomers, additives and certain PPAs). Some listed substances can have an SML. Other substances that are not required to be listed, such as PAs and NIASs, need to be risk assessed.

When a polymer is used for food contact applications, there are important material features that must be considered before approaching any further analytical challenge related to risk assessment:

- **ODOUR**: According to current regulation:[2] "... any material or article intended to come into contact directly or indirectly with food must be sufficiently inert to preclude substances from being transferred to food in quantities large enough to endanger human health or to bring about an unacceptable change in the composition of the food or a deterioration in its organoleptic properties." This means that it is necessary to monitor packaging material components that could migrate into food, deteriorating safety and quality, starting from a low molecular weight.
- **NIASs**: According to EFSA (European Food Safety Association),[3] reaction products and impurities (NIASs) that are not included in the EU list of authorized substances in EU Regulation No. 10/2011 are required to be risk assessed, irrespective of their source or intended function. Oligomers are also considered as NIASs. Their identification and quantification are required for risk assessment by Art. 19 of EU Food Contact Regulations 10/2011/EC, specifically for those oligomers with molecular weight (MW) <1000 dalton (Da), as potential migrants of concern which, after transfer (migration) into food, may be physiologically resorbed after ingestion.

1.2 Evaluation of Analytical Approaches

Polymer characterization usually consists of fractionation/extraction or thermal desorption. Fractionation/extraction can be carried out using different solvents and techniques. Aiming to investigate all the species present in an FCM, a combination of analytical methods is used. With polypropylene, a commonly used approach is the separation of the insoluble crystalline part (X_i) from the soluble atactic fraction and oils.[4,5] Methylene chloride (CH_2Cl_2) is another solvent that is commonly used. The fractionated or extracted part

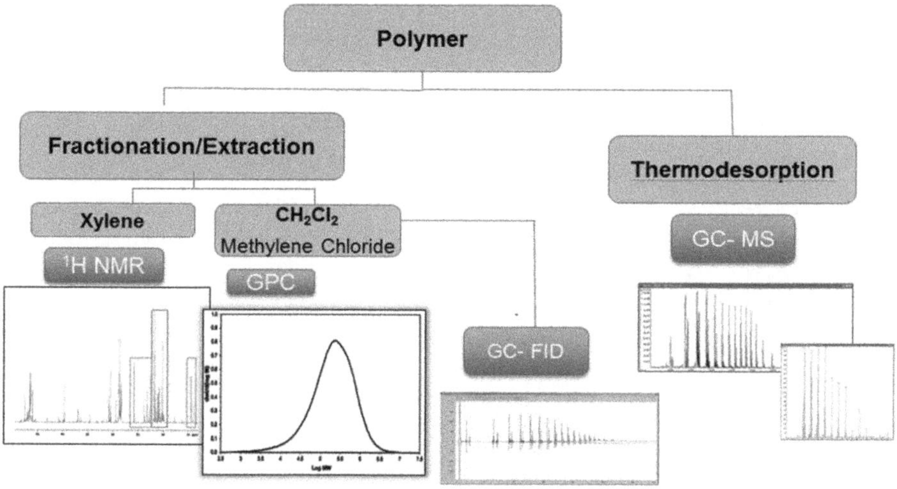

Figure 1.1 Polymer analytical approaches. PP case: fractionation, extraction, thermal desorption. Techniques involved are NMR, GPC and capillary GC.

can be analysed using different instrumental techniques: NMR (nuclear magnetic resonance), GPC (gel permeation chromatography), GC-FID (gas chromatography coupled with flame ionization detection), GC-MS (gas chromatography coupled with mass spectrometry), HPLC-MS or HPLC-DAD (high-performance liquid chromatography coupled with mass spectrometry or diode-array detection, respectively). The polymer as such can also be thermally desorbed and studied by GC-MS and GC-FID (see Figure 1.1).

What emerges is a very complex matrix, with species covering a very wide MW range, starting from a few daltons (residual monomer) up to millions of daltons (high-MW polymers with high crystallinity), as shown in Figure 1.2. This scheme is characteristic and different for each individual polymer.

Considering a very common FCM, *i.e.* a printed film to wrap food, a deeper investigation reveals a very complicated situation (see Figure 1.3). At least four different polymers were used, laminated with a printed surface and glue.

Analysis of a final item is very difficult and the result is a sum of the contributions from each individual layer component, including the printed layer (with or without paper, when present) and glue, and heating effects during extrusion/lamination. Moreover, it is worth remembering that each film layer can be formed by more than a single polymer.

Very complex matrices, due to different plastic materials and printing and lamination processes, are usually challenging to analyse if the whole production chain is unknown. The final item is analysed to characterize NIASs, but it would often be desirable to obtain more information about the origin of the NIASs.[6]

For the above reasons, to achieve a full characterization and understanding of FCMs, each individual polymer should be analysed as such, together with the laminated final film with and without the printed and glued layer.

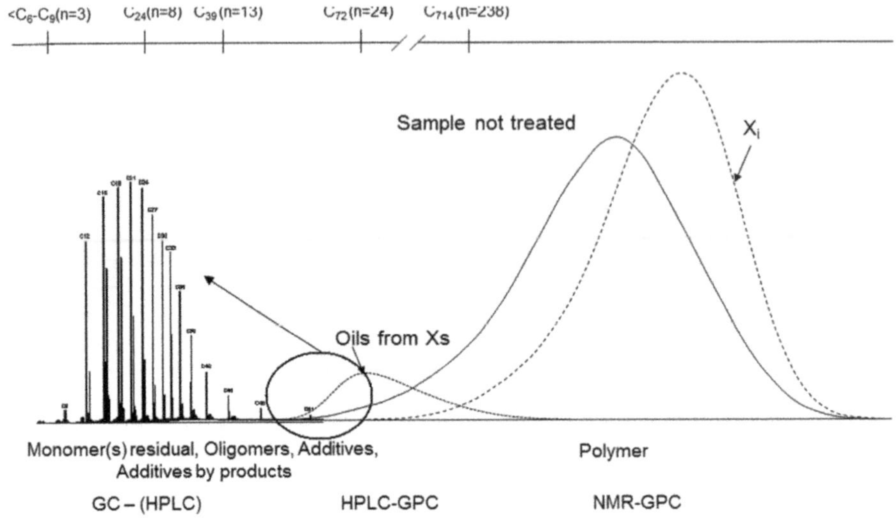

Figure 1.2 Scheme of PP polymeric matrix profile, starting from monomer residual, passing through additives and oligomers, arriving at the crystalline fraction. Polymer oligomers are zoomed (circled zone) *via* GC. The upper scale roughly indicates the region for hydrocarbon clusters; *n* in parentheses indicates the degree of polymerization when propylene is concerned.

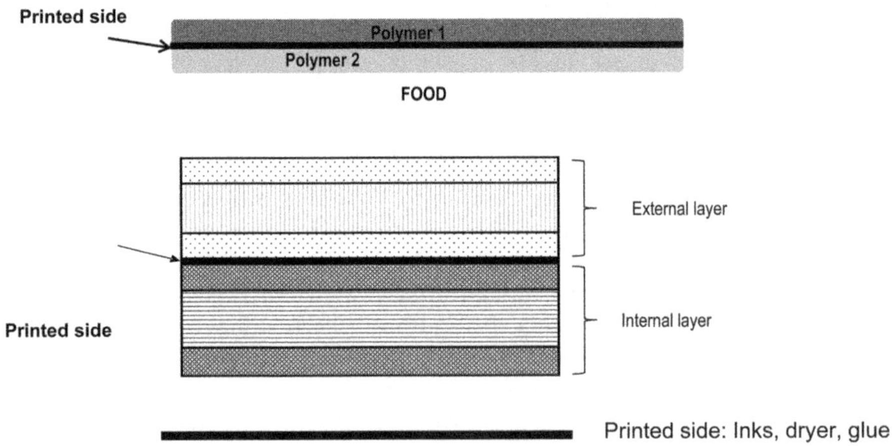

Figure 1.3 High end use film for food packaging: multi-layer laminated film with printed side and glue.

1.3 GPC: First Approach

Polymer analysis is usually carried out by GPC, in particular to describe finally the MW distribution. It could also be considered as a first approach to characterizing oligomers with MW <1000 Da. The separation path is based on the different retention times of molecules of a polymer in solution. Chemical interaction between the particles and the stationary phase can be considered negligible. In addition, no information about the chemical nature of compounds is obtained.

Starting from polypropylene homopolymers (PP HOMOs), Figure 1.4 shows three different materials with the same average MW but different MW distributions, due to different synthesis conditions.

The resulting GPC traces are shown at top left in Figure 1.4 and the NIAS–oligomers region is the grey circled zone. Even if clear differences in the MW distributions for the three samples can be measured in terms of average MW (M_n, M_w, M_z), no accurate speculations are feasible for the NIAS region *via* GPC. To overcome the poor resolution and separating power of GPC, GC must be considered.

1.4 GC-MS and HPLC-MS Approaches

For comprehensive and full characterization of ODOUR, NIAS (or IAS) substances, including oligomers, a deep and more sophisticated analytical approach is needed. Generally, an extraction step of the potential migrants from the matrices is applied and then GC or LC separation is performed. Thermal desorption (static or dynamic) is applied for volatile compounds, whereas solvent extraction is needed for the characterization of less volatile species. Capillary GC and HPLC provide the required separation, selectivity and efficiency.

As for the detector to be used, not only should it give as much information as possible for each separated organic compound, but it should also have good sensitivity. A mass spectrometer coupled with capillary GC or HPLC fits all these requirements.

Capillary GC, coupled with FID, is also widely used in many studies as it provides a relatively easy and universal quantitative approach. The FID response for hydrocarbons can be considered as weight percent response, even without any use of external standard (ESTD) or internal standard (ISTD) calibration.

Within capillary GC, the main technique used for volatile organic compounds (VOCs), three different sample analyses are necessary, starting from static headspace (SHS) for the C_3–C_{24} range, passing through a dynamic headspace–thermal desorption system (TDS) for C_9–C_{48}, and arriving at solvent extraction for C_9–C_{60} or heavier compounds. Hydrocarbons up to C_{100} can also be eluted, if high-temperature-resistant capillary columns are used. C_n roughly indicates the number of carbon atoms in the possible eluted species.

To detect additives (IASs) or some high-MW additive by-products (NIASs), HPLC-DAD-MS is necessary. HPLC cannot be used to quantify oligomers because, owing to their chemical nature, they are not detectable with either MS or DAD.

APCI (atmospheric pressure chemical ionization), ESI (electrospray ionization) or APPI (atmospheric pressure photoionization) sources are not

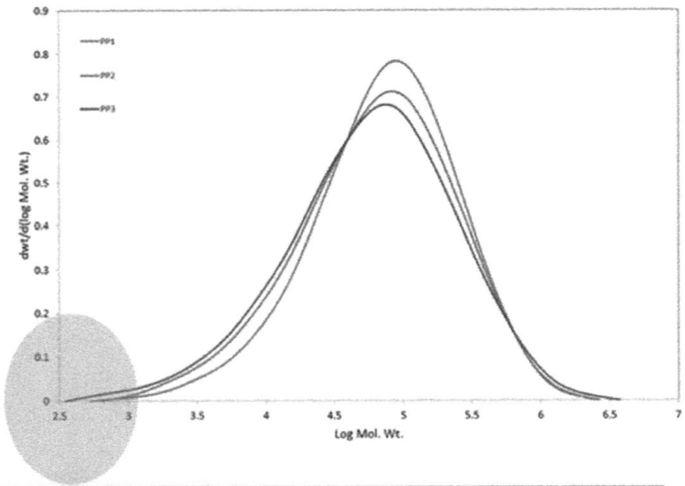

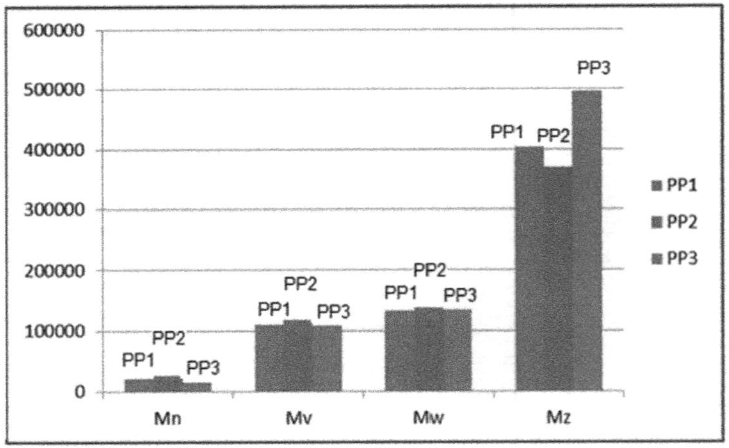

	Mn (Kg/mol)	Mw (Kg/mol)	Mz (Kg/mol)	Mw/Mn	Mz/Mw
PP1	22	133	403	6.1	18.3
PP2	28	138	370	5.0	13.5
PP3	15	135	496	8.9	32.8

Figure 1.4 Different HOMO PP with the same average MW show different MW distributions. Oligomers are in the grey circled zone.

effective in aliphatic hydrocarbon ionization. Different extraction methods have been described,[7] but essentially they are based on UASE (ultrasonic-assisted solvent extraction), OSM (one-step microwave-assisted extraction) or TSM (two-step microwave-assisted extraction).[8–11]

In order to detect NIASs or IASs with SML using HPLC-MS, a quadrupole instrument is not sufficient. APCI, ESI and APPI are considered soft ionization techniques,[12] so they do not produce EI (electron ionization)-like spectra, with fragmentation patterns of analytes, but only a few ions for each compound (*i.e.* molecular ion and one or two fragments). These few ions and their relative abundances (fragmentation pattern) depend on the solvent and temperature. Consequently, these fragmentation patterns differ considerably from those available in the database currently used for species recognition analysis and they are not searchable in commercially available mass spectral libraries. Only by using the exact mass (2 ppm or better resolution) it is possible to establish the empirical formula of the analyte. If CAD (collisionally activated dissociation) is used both in single-stage MS and multi-stage mass spectrometry (MS^n), it is possible to obtain a fragmentation pattern.[13]

Since DAD is not a destructive detection method, it can be used before MS with very good results. Coupling DAD with MS (TOF), identification of some IASs (see Figure 1.5) or NIASs (see Figures 1.6 and 1.7) can be carried out using exact mass calculation. The UV spectrum, when reported in the literature, can also be helpful.

1.5 Samples

Samples suitable for GC-MS or GC-FID investigation can be spheres or particles (polymer before extrusion), pellets, films, items or food as such. Usually, the sample amount is 1 g, for the SHS method a few grams (2–5 g) for solvent extraction and less than 1 g for TDS. If solvent extraction or SHS is used, spheres and pellets require grinding before treatment. An FCM (final item) requires cutting or grinding. TDS can be used with a few spheres or pellets, but manufactured items and films need to be cut into small pieces before thermal desorption.

Any sample manipulation should be carried out wearing cotton gloves, to avoid sample contamination. Plastic gloves should not be used, to avoid "ghost peaks." Contamination due to ubiquitous substances or interfering compounds should be considered in solvents used for HPLC and/or extraction.

1.6 Characterization: SHS

SHS is wrongly considered a relatively easy technique. Reality is different, and the results should be carefully evaluated before drawing any conclusion. Together with TDS-GC-MS, it is the main technique used to investigate odour.

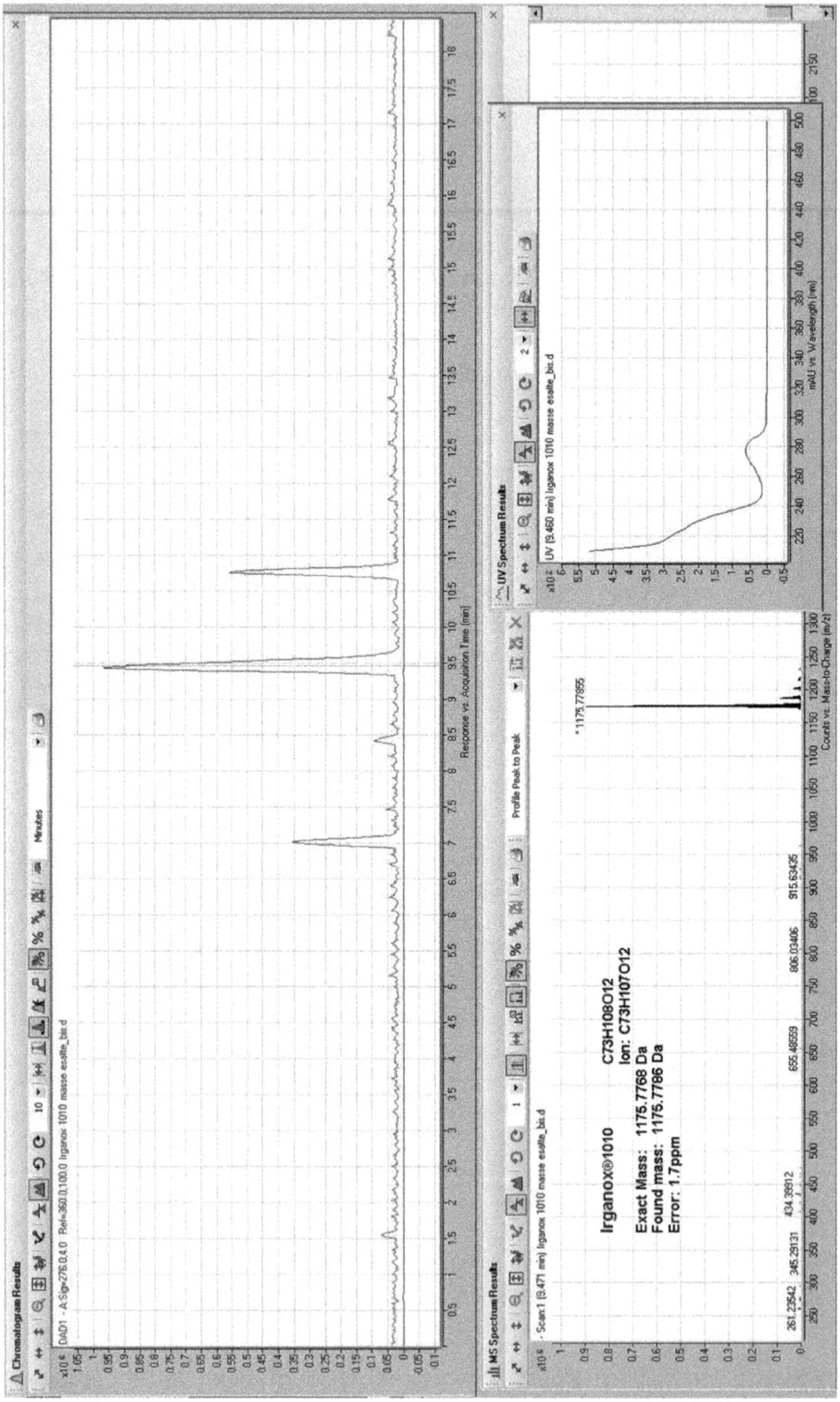

Figure 1.5 Irganox®1010 eluted in HPLC-DAD-MS(TOF). Bottom left: relative mass spectrum with exact mass calculation (ppm error: 1.7). Bottom right: Irganox®1010 UV spectrum.

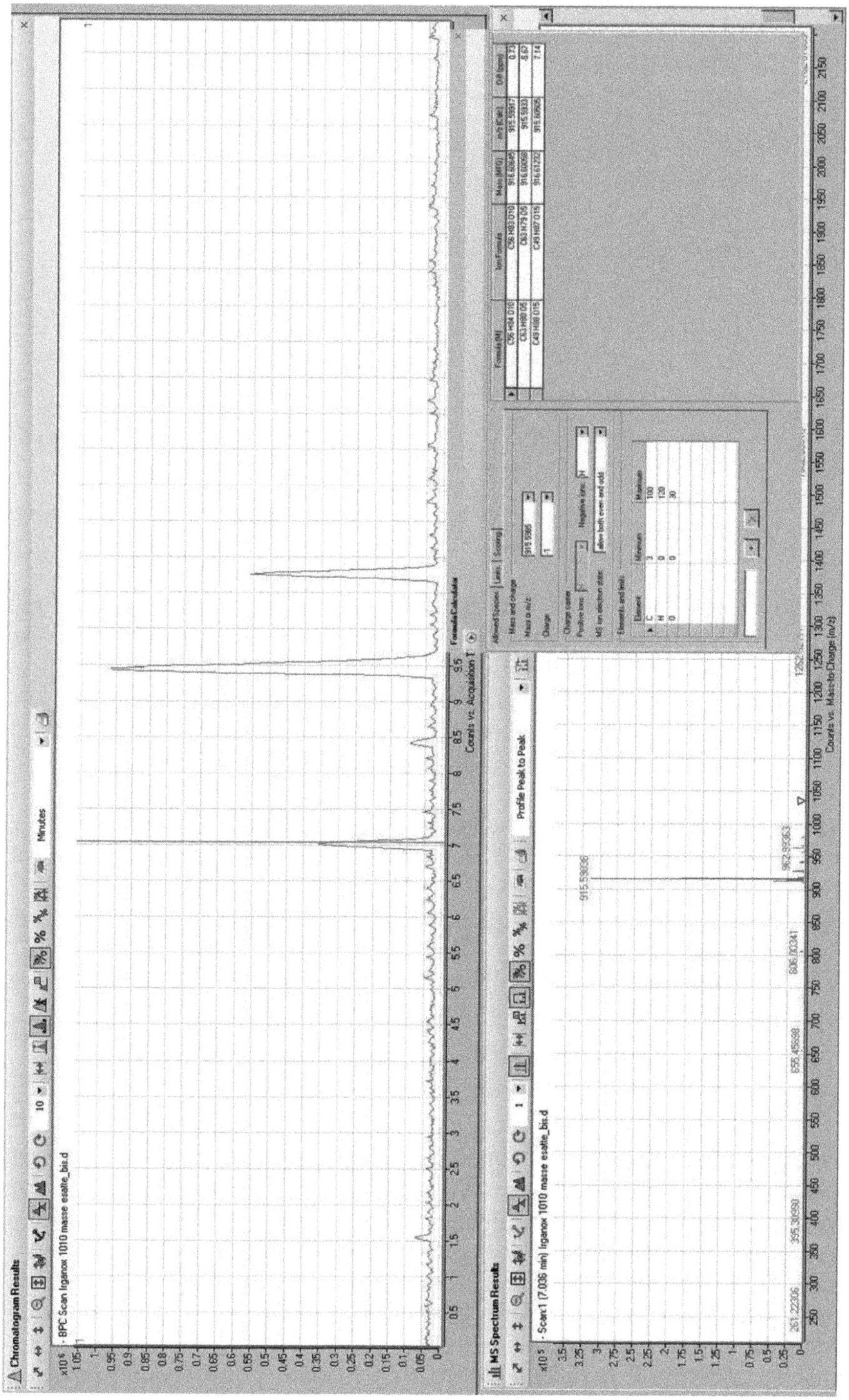

Figure 1.6 Irganox®1010 impurity (retention time = 7 min), relative mass spectrum with exact mass calculation (ppm error: 0.7).

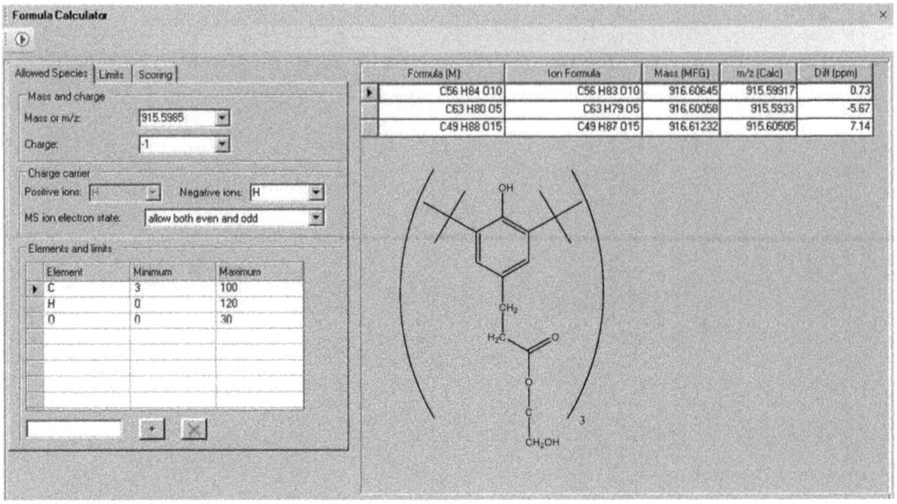

Figure 1.7 Irganox®1010 impurity, with exact mass calculation (ppm error: 0.7) and its structural formula.

Very exhaustive discussions about SHS, the amount sample and its treatment, relative vial pressure, *etc.*, can be found in the literature.[14] In the following, a few practical considerations are discussed.

Mechanisms of odour formation and its transport,[15] together with the odour threshold,[16] have been described in the literature. If food as such is analysed, useful references can be found in a flavour reference handbook.[17]

SHS does not require any particular equipment except a glass vial that can be closed with an inert septum. Vials of size 20 mL are the most commonly used, sealed with silicone or a butyl rubber internally PTFE-coated septum. Sample heating, vial gas-phase injection and chromatogram evaluation complete the analysis (see Figure 1.8).

The sample amount should not be greater than 1–2 g.[14] The maximum and analyte quantitation are described by the equation

$$\text{peak area} \propto C_G = \frac{C_0}{K + \beta} \tag{1.1}$$

where C_G = concentration of the analyte in the headspace, C_0 = original sample concentration of the analyte, K (partition coefficient) = C_S/C_G, C_S = concentration of analyte remaining in sample, β (phase ratio of the vial) = V_G/V_S, V_G = volume of the gas phase and V_S = volume of the liquid–solid sample.

If peak identification only is required, a search of mass spectral libraries is the first necessary step. If quantitation is also necessary, it is very important not to forget some fundamental concepts.

For each individual component, starting from residual monomer(s), up to oxygen-based or odorous substances, passing through hydrocarbons

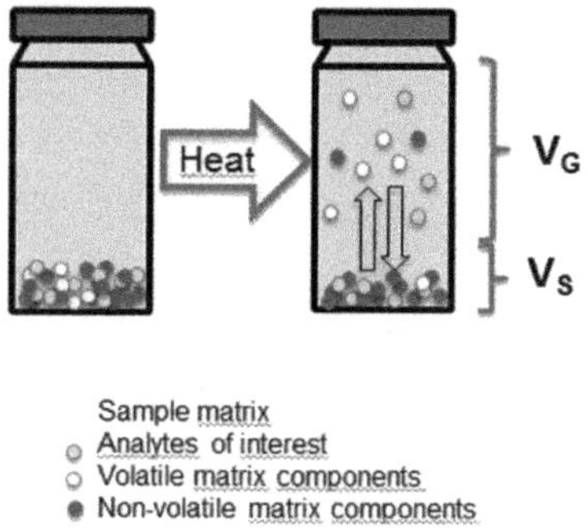

$$\textbf{Peak area} \propto C_G = \frac{C_0}{(K+\beta)}$$

where:
C_G = concentration of the analyte in the headspace
C_0 = original sample concentration of the analyte
K (partition coefficient) = C_S/C_G
ß (phase ratio of the vial) = V_G/V_S
V_G = Volume of the gas phase
V_S = Volume of the liquid - solid sample

Figure 1.8 SHS principle and relative scheme: volatile compounds are distributed (thermodynamic equilibrium) between gas and solid (or liquid) phases.

(C_2–C_{24}), the partition coefficient *K* must be considered in order to avoid incorrect analytical results.

The first approach in material investigation can be carried out by heating the closed vial containing the sample without any grinding. GC-MS and interpretation of the mass spectra give a rough indication of the compounds present in the gas phase. Chromatographic separation can be achieved with apolar columns with a thick film phase (VOC-like) or polar columns (*e.g.* Carbowax™-like).

Of course, the gas-phase concentration and *K* differ not only according to the heating temperature but also to the nature of the matrix (different polymers give different *K* values for the same substance), matrix particle size, nature of the analyte [polar, non-polar, its boiling point (bp), *etc.*] and equilibration time. The heating temperature and *K* are the limiting factors.

If an internal standard (ISTD) is used, it must be selected to be as similar as possible to the substance or chemical clusters to be quantitated (*i.e.* comparable *K* or similar bp). An example, obtained using FID, is shown in Figure 1.9.

If toluene is used, hydrocarbons with similar bp (C_9) can be quantified. If hexadecane is used, C_{12}–C_{15} hydrocarbons can be studied. Of course, a single ISTD must not be used to quantify the whole chromatogram.

To determine the effect of different post-polymerization treatments on the VOC population, SHS coupled with GC-MS can be used. An effective comparison can be obtained by heating the same sample type and amount at the same temperature for the same time (see Figure 1.10).

Grinding the cooled polymer before applying the SHS technique is strongly suggested. For some analytes, too long a period of time would be necessary to reach equilibrium with the gas phase if the sample has not been milled. Acetone and TBA (*tert*-butyl alcohol), typical by-products of some peroxides used for "visbreaking" high-MW distribution polymers, are good examples (see Figure 1.11). Visbreaking is a process commonly used in polymer

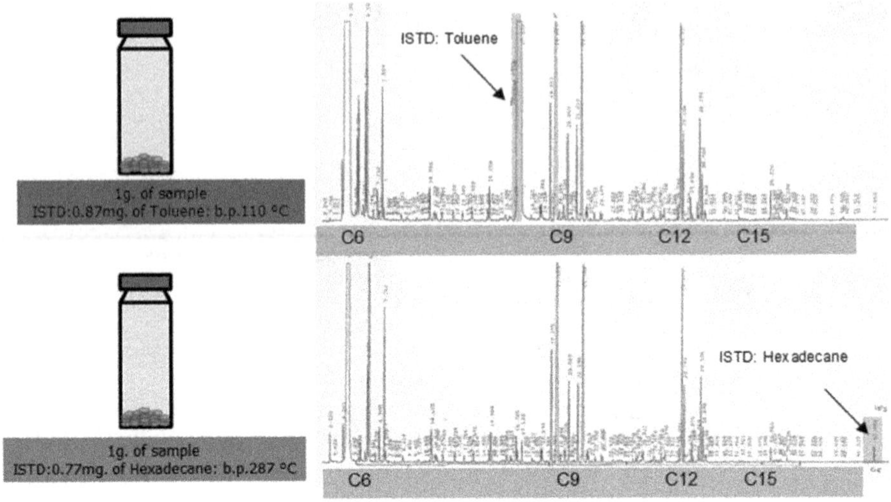

Figure 1.9 Different quantitation of the whole chromatogram is obtained if different ISTDs with different bps are used, even if their amounts are very similar.

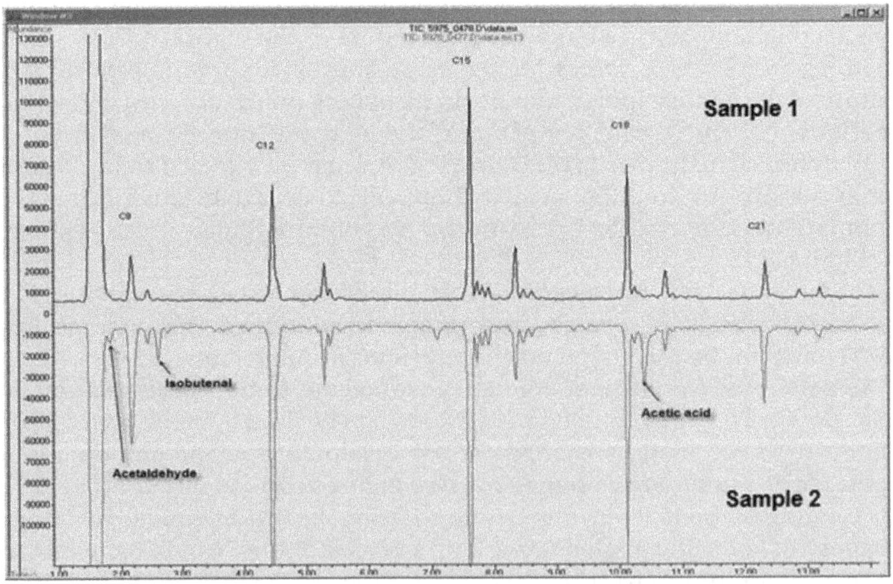

Figure 1.10 GC-MS comparison between the same polymer obtained with two different post-polymerization treatments. Sample 2 contains oxygen-based compounds: probable odour issue.

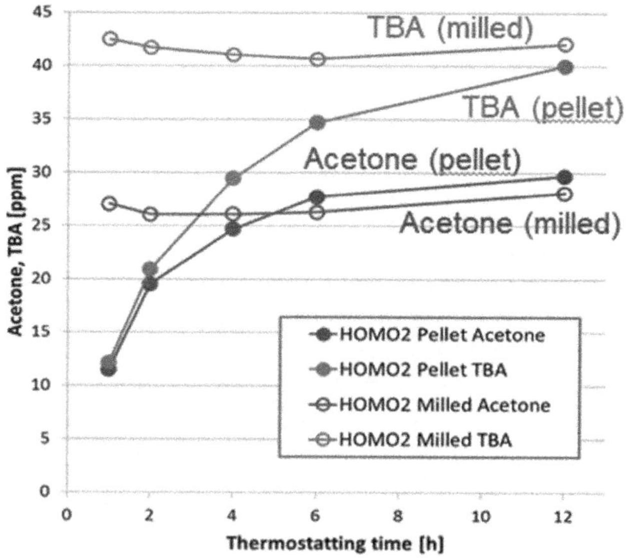

Figure 1.11 Gas-phase concentration of acetone and TBA in different matrix types, different sizes (milled and not milled pellets) at different thermostating times.

Table 1.1 *K* values for acetone and TBA in different matrices, milled or as such (pellets), at different temperatures. Acetone is co-eluted with another peroxide by-product (less than 10% of acetone area).

		K	
Temperature (°C)	Polymer	Acetone + interference	TBA
60	Base 1	2.5	3.2
	HECO 2	5.1	6.5
100	Base 1	5.0	7.0
	HECO 2	11.2	15.3

production. A peroxide is used to decrease the average MW of the polymers (viscosity reduction). It involves radical scission of high-MW chains.

Consider further that not only do acetone and TBA have different *K* values, but also *K* depends strongly on the heating temperature and on the polymeric matrix. A homopolymer (Base 1) and a heterophasic copolymer (HECO 2), both milled, were studied at different SHS temperatures (60 and 100 °C). The *K* values obtained for acetone (together with a co-eluted by-product interferent) and TBA are given in Table 1.1.

When the ESTD approach is considered, a very common error consists in preparing the reference standard calibration in an empty vial, using TVT (total vaporization technique). In this case, the entire liquid sample (a few microlitres), which contains the analytes to be quantified, is vaporized.

Only a single phase is present in the vial, hence the concentration of analyte itself in the vial is

$$C_G = \frac{W_0}{V_v} \tag{1.2}$$

where C_G = concentration of the analyte in the headspace, W_0 = amount of analyte and V_v = volume of the vial.

The described procedure does not take into account either K or matrix effects, and also the phase ratio (β) of the vial containing the unknown sample. According to good laboratory practice (GLP), heat should be provided overnight, at about 100–120 °C, with the milled matrix in a ventilated oven, to obtain a "cleaned matrix." The latter represents the same matrix as the unknown sample, but without analytes. The "cleaned matrix" can be used to prepare an external standard calibration vial, weighing the same amount of polymer of the "unknown" sample and spiking the former with a well-known number of analytes to be quantified. After the same thermostating time, at the same temperature, samples can be analysed by first injecting a gas-phase amount of the "unknown" sample and second the calibration standard. ESTD calculation can now be carried out correctly.

Considering the data in Table 1.1, if TVT is used instead of the "cleaned matrix" procedure to prepare ESTD calibration, quantitation of TBA in HECO 2 (heterophasic copolymer) at 100 °C will be underestimated by more than 50%.

1.7 Characterization: TDS (Thermal Desorption System)

TDS could be considered as a dynamic headspace technique. It is used both to identify odour and NIASs. Instead of heating a closed vial containing the sample and analysing the gas phase at equilibrium, as is done in SHS, the sample is heated at the desired temperature, an inert gas flow (usually helium) passes over it and all the stripped chemical species are cryofocused into a glass liner. After 10–20 min of extraction, analytes in the glass liner are quickly heated and injected with a PTV (programmed-temperature vaporizing injector) or MMI (multi-mode injector). A capillary GC column for separation and a mass spectrometer in scan mode will complete the analytical run.

If the same amount of different samples is thermally desorbed using the same GC conditions, a comparison can be carried out (semiquantitative analysis).

A reliable analysis is obtained if a further thermal desorption on the already analysed sample produces a chromatogram with less than 20% of the peaks obtained in the first run. In order to comply with this rule, sample heating for 20 min at 280 °C could be a good starting point. At this temperature, the polymeric matrix is molten but additives do not degrade and do not produce any by-products or "ghost peaks" (see Figure 1.12).

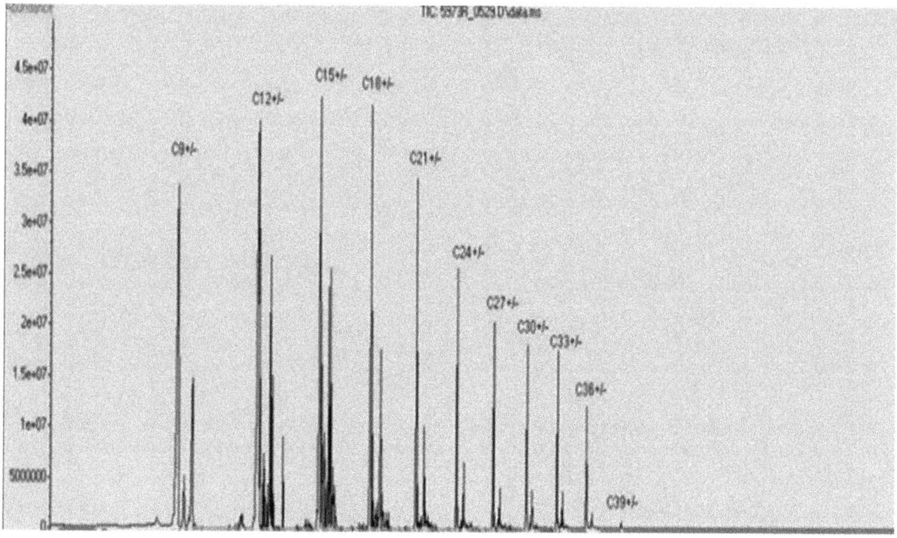

Figure 1.12 HOMO PP before extrusion. Solvent vent thermal desorption at 280 °C for 15 min, cryofocusing at −50 °C in an empty baffled liner and solvent vent injection at 300 °C into a 5% phenylmethylsilicone capillary column 0.25 mm × 60 m × 0.25 μm of film thickness.

Extracting the sample in a helium atmosphere greatly preserves it, together with the polymeric matrix. A temperature of 280 °C is sufficient to strip volatile compounds from C_3 to about C_{48}. A commonly used melt stabilizer such as Irgafos® 168 and its phosphate by-product are visible on the chromatogram when present. Their MWs are 646 and 662 Da, respectively (see Figure 1.13).

The glass liner used for cryofocusing is usually filled with a polymer-based phase (*e.g.* Tenax®). This or a similar absorbent guarantee better absorption, but full desorption is sometimes critical. If possible, baffled deactivated liners are preferred, kept at −50 °C during sample thermal desorption. If used, the chromatogram range will be reduced from C_9 to about C_{48}. Residual monomers and lighter chemicals will not be fully trapped, but no memory effects and faster desorption will be guaranteed.

The TDS technique is very effective and can give a considerable amount of information about FCMs. TDS is much more sensitive than SHS and it can analyse a broader range of compounds, not easily extracted by SHS.

In Figure 1.14, a comparison between two equal weight PP homopolymers (pellets) obtained on the same polymerization plant, but using two different catalytic systems, is shown. Zooming part of the chromatogram (see Figure 1.15), it is possible to identify different C_{12} oligomeric isomers deriving from the use of the two different catalytic systems.

The overall oligomer profiles can be obtained by summing the areas of peaks. For PP, C_9, C_{12}, C_{15} ... up to C_{48} oligomers can be desorbed and detected under suitable conditions. Eluting peaks that are attributed to isomers with the same polymerization degree are grouped. The outcome consists of a

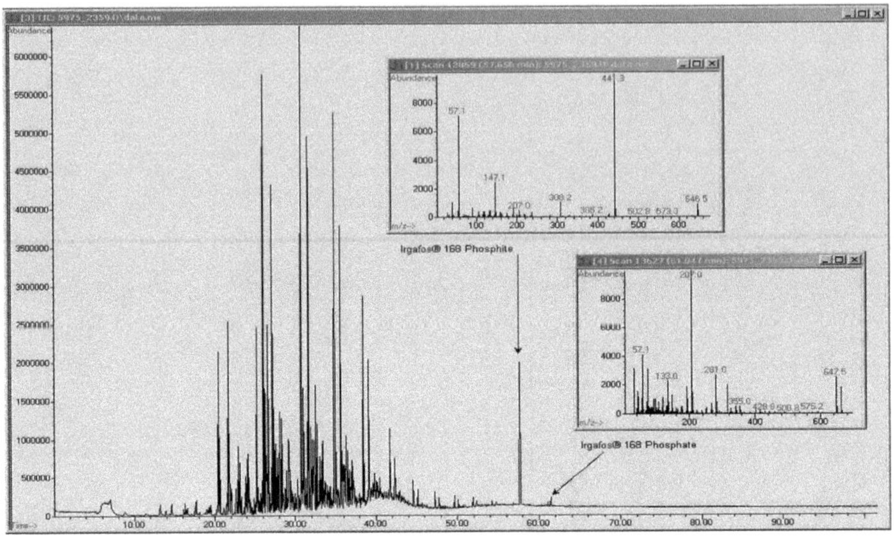

Figure 1.13 TDS GC-MS of FCM (item). It is possible to distinguish both Irgafos®168 and its oxidized by-product.

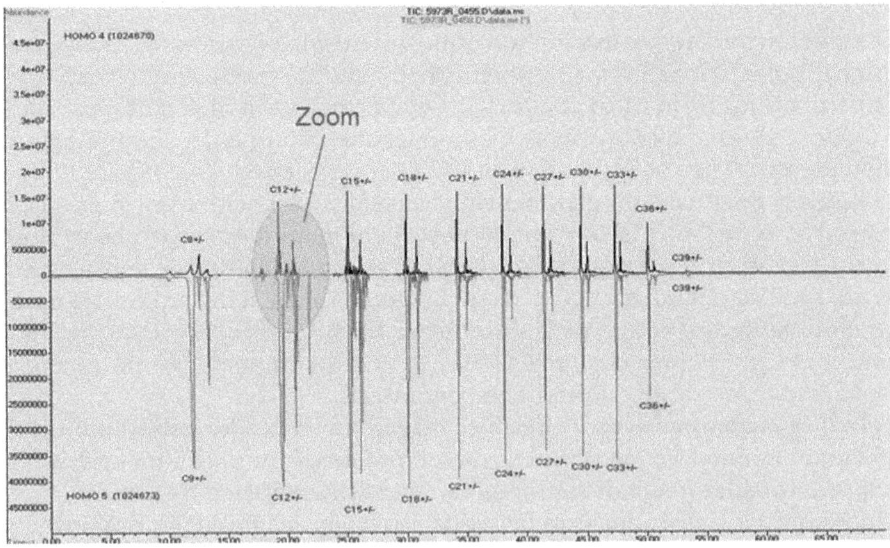

Figure 1.14 Comparison between two HOMO PP obtained using different catalytic systems. Same weight of polymer.

graph where the relative amount of the several oligomer clusters is plotted *versus* the corresponding different and increasing number of carbon atoms. To exemplify this, data from an experimental case describing three different materials are reported in Figure 1.16.

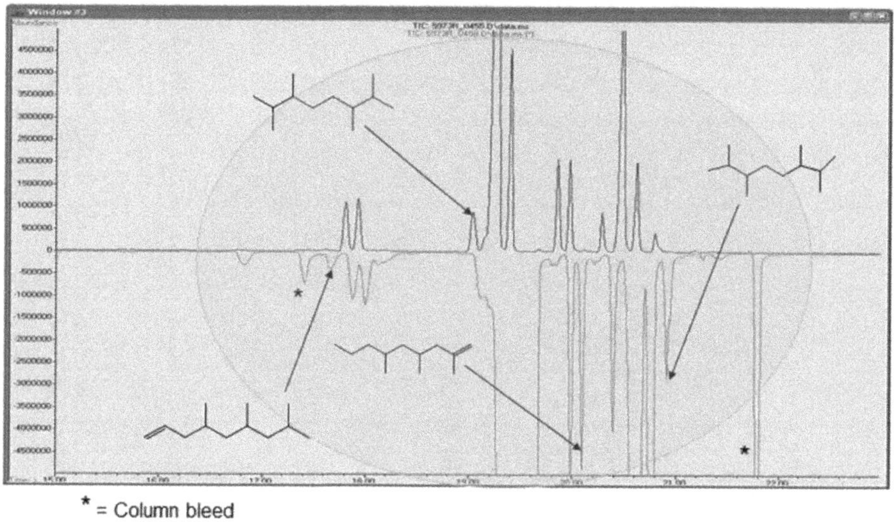

Figure 1.15 Zoom of circled part of Figure 1.14. C_{12} cluster characterization.

Using this approach, it is possible to characterize t qualitatively he oligomer distribution. Samples 2 and 1 are similar in terms of both oligomer content and oligomeric MW distribution (bottom left graph in Figure 1.16). Samples 1 and 3 have different oligomer contents and different oligomeric MW distributions (bottom centre graph). Samples 2 and 3 have different oligomer amounts but the same oligomeric MW distribution (bottom right graph).

Thermal desorption applied to polymeric samples, coupled with GC-MS, allows the detection of volatile and semivolatile species. The relevant chromatogram is generally characterized by a high number of species that hardly ever can be completely identified. Trying to characterize each single peak, starting from the very first to the last one, is not only time consuming but sometimes even useless. Frequently, it happens that even the smallest peaks cannot be ignored because they can play a crucial role in NIAS identification. A comparison between a clear monolayer film and a yellowing one, both produced using the same polymer, is shown in Figure 1.17. Notwithstanding the relatively few peaks, peak-by-peak identification appears to be a titanic task.

Working with "ionic series" in the chromatogram extraction mode could be the best approach. Quadrupole-based MS software permits the extraction of the chromatogram obtained in scan mode, considering only the ions that may be of interest in NIAS identification; *e.g.* extracting *m*/*z* 220, only peaks due to quinone-like substances, derived from some antioxidant (AO) additives, are enhanced (see Figure 1.18). The latter are NIASs and they should be quantified to comply with the risk assessment if the film is used as an FCM.

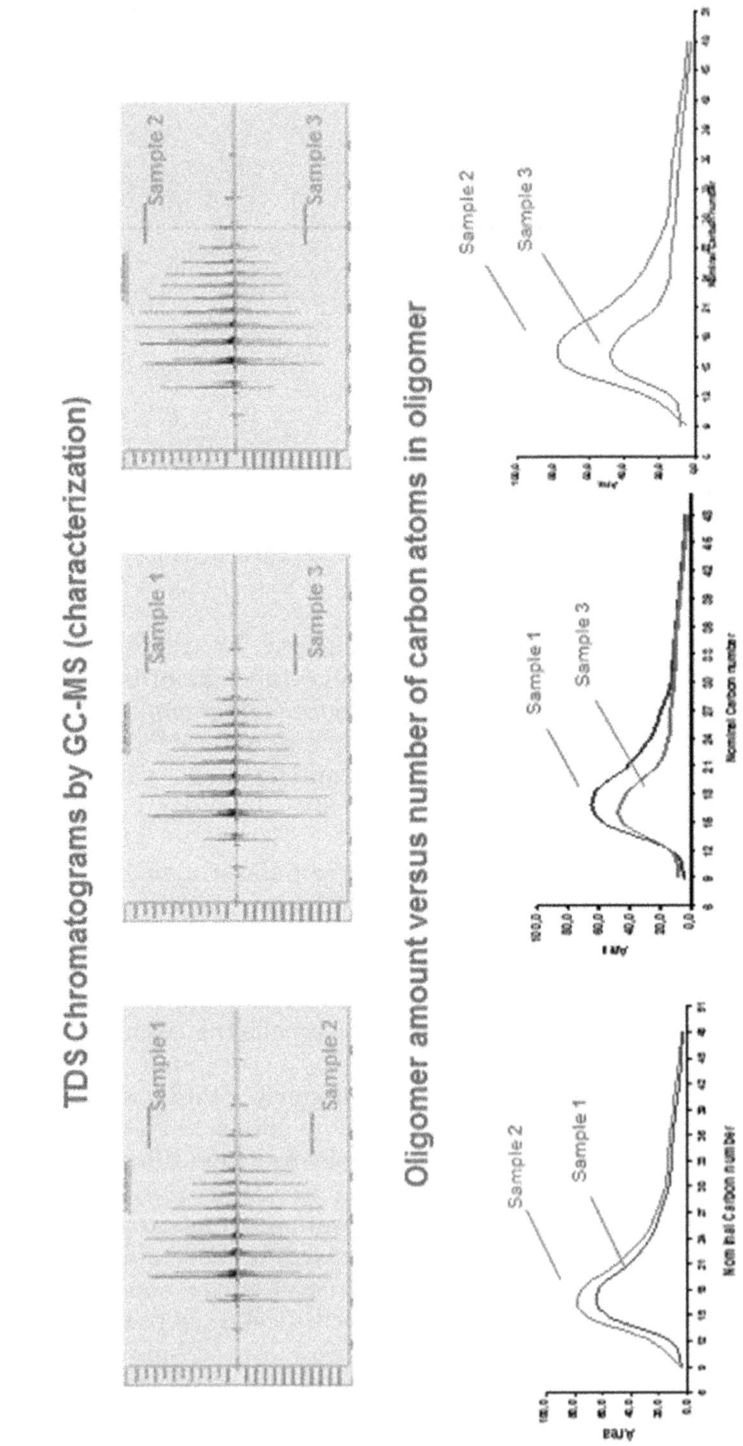

Figure 1.16 "Clusterization" of three different HOMO PP samples. Oligomer amount *versus* number of carbon atoms in oligomers.

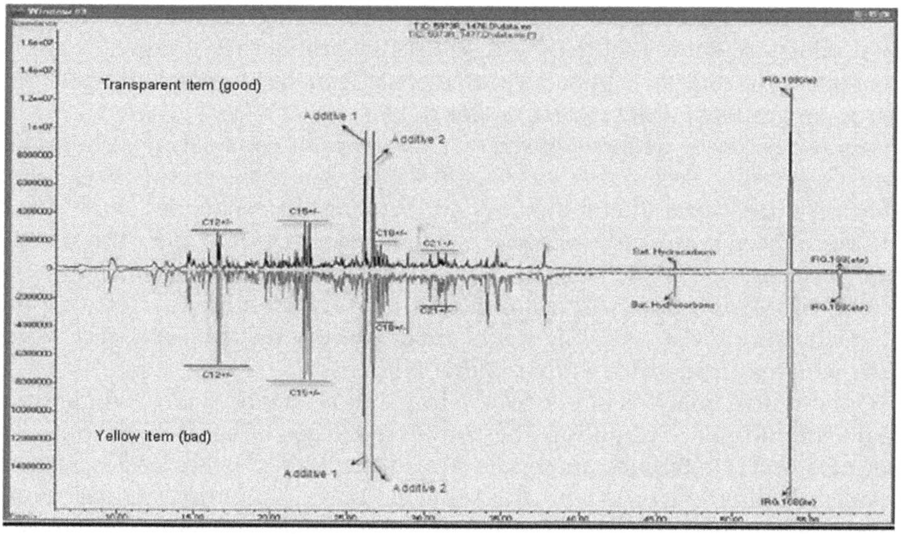

Figure 1.17 Comparison between two films produced using the same polymer lots, but with different extrusion conditions. Reference transparent sample (good) above, yellow sample (bad) below.

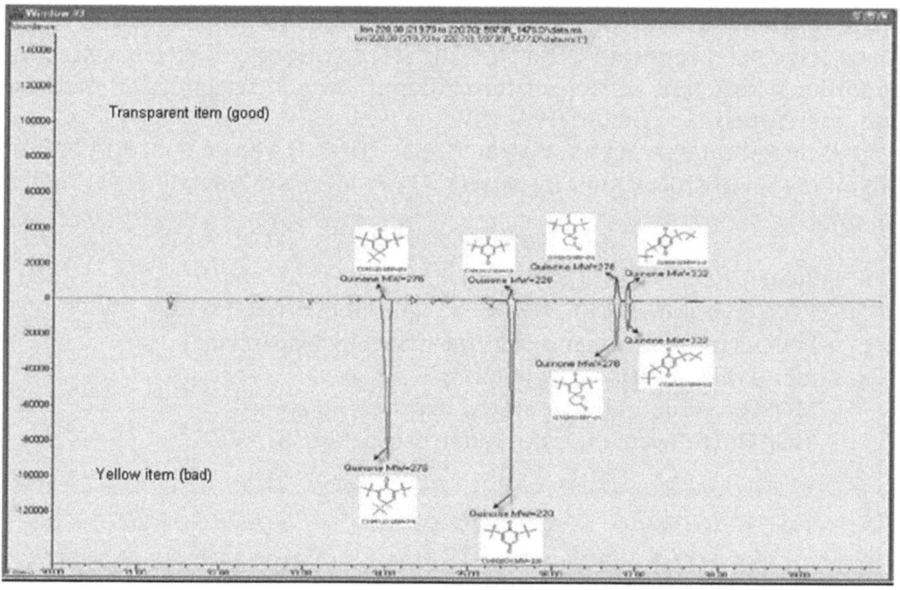

Figure 1.18 Comparison between the same samples considered in Figure 1.17, but in "extract ion mode." Coloured quinones or molecules with conjugated double bonds are much more present in the bottom (bad) sample.

If the TDS approach is used in a first screening for the presence of by-products of some additives (*e.g.* Arvin's by-product for Irganox® 1010[18]), characteristic ions of substances of interest can be used to interpret the chromatograms of different extract ions.

Ion series can be deduced based on experience or considering well-known ones: *e.g.* amine series: *m*/*z* 30, 44, 58, 72, 86, 100, 114, and it is useful to check if these types of components are present. An exhaustive book about MS could be very useful[19] for recognizing substances from their fragmentation patterns.

TDS GC-MS is a powerful technique that is very useful for performing the characterization of C_9–C_{48} oligomers, most additive by-products, some additives with low–mid MWs, white mineral oils, *etc.*

Quantitative analyses are not very simple, not only because a single thermal extraction does not allow the complete release of substances, but also because ESTD reference materials are difficult to prepare and scattered results, in different runs with the same calibration mixture, are very common. Each individual case must be optimized, considering both the analytes and matrix features.

1.8 Characterization: Solvent Extraction

Solvent extraction can be considered one of the most important techniques for quantifying NIASs and IASs with SMLs. There are no generally accepted or recognized methodologies and several different solvent extraction procedures have been reported[20] that use different solvents, different extraction conditions, different samples and different sample preparation methods and, therefore, they give different results.

First, it could be better to separate IASs with SMLs or general NIASs from oligomers or hydrocarbons in general. The latter are usually described as follows:[21]

- polyolefinic products (POHs), divided into:
  - POSHs (polyolefin oligomeric saturated hydrocarbons)
  - POHAs (polyolefins monounsaturated hydrocarbons)
- mineral oils (MOHs), divided into:
  - MOSHs (mineral oil saturated hydrocarbons)
  - MOAHs (mineral oil aromatic hydrocarbons).

1.8.1 NIASs (Other than Hydrocarbons) and IASs with SMLs

EU regulation No. 10/2011 [22] requires, when possible, analyses on base food. If this is not possible, tests have to be managed using food simulants. As an alternative, a screening approach on plastic materials should be carried out, using "worst case simulation".[23–25]

Direct analyses on foods are very complex, results are quantitative and it is almost impossible to assess the origin of oligomers and NIASs in general. This is also valid for analyses carried out on food simulants.

IASs with SMLs and mid–low MW additive by-products (NIASs) can commonly be extracted using ultrasound-assisted solvent extraction (UASE) with methylene chloride. Extraction for 4 h at room temperature of foods, milled polymers or other items could be considered suitable for this purpose. Filtration with a 0.2 µm PTFE filter is usually needed. Methylene chloride-based extraction is very common in most testing laboratories involved with NIASs.

Analytical approaches should be carried out using GC-MS in the SIM (selected ion monitoring) mode if a quadrupole is used or extraction of characteristic ions with exact mass when the time-of-flight (TOF) method is used. Triple quadrupole or TOF methods are less affected by matrix effects owing to their ion selectivity. Sometimes, characterization of NIASs is relatively easier in food matrices than in polymers. From a GC point of view, analysis of 2,4-di-*tert*-butylphenol (a very common by-product of Irgafos® 168 additive) in white bread using UASE with methylene chloride as solvent is easier than its quantitation in polymers. The white bread matrix is "cleaner" than polymers in terms of peak profile, even if it is much more complex in terms of organics content (fibre, sugars, starches, proteins, *etc.*). A typical example is shown in Figure 1.19.

As a rule of thumb, UASE with methylene chloride is acceptable for mid-MW IASs with SMLs. Heavier IASs with SMLs or heavier additive by-products require OSM or TSM. Derivatization can be used if everything else tried proved ineffective. Occam's razor is the guideline: the simpler, the better.

Sometimes, owing to a very complex polymeric matrix, it is not very easy to distinguish between oligomers and NIASs (see Figure 1.20). Figure 1.20 shows part of chromatogram of a thermally desorbed copolymer (TDS-GC-MS) (upper chromatogram). The mass spectrum of the peak at 14.9 min is not the same on the left and right parts of the peak itself, which means that at least two different substances are co-eluted. The spectra obtained are well identified as a saturated branched hydrocarbon (oligomer) and

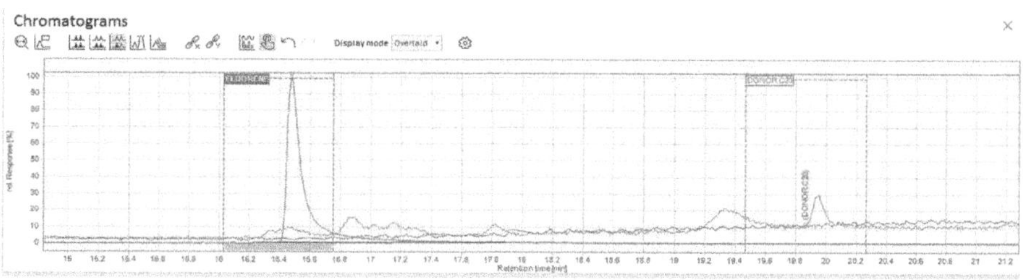

Figure 1.19 Chromatogram (on-column injection) of 0.05 mg kg^{-1} of 9,9-bis(methoxymethyl)fluorene (CAS 182121-12-6) (NIAS with SML), directly extracted from coffee (food matrix).

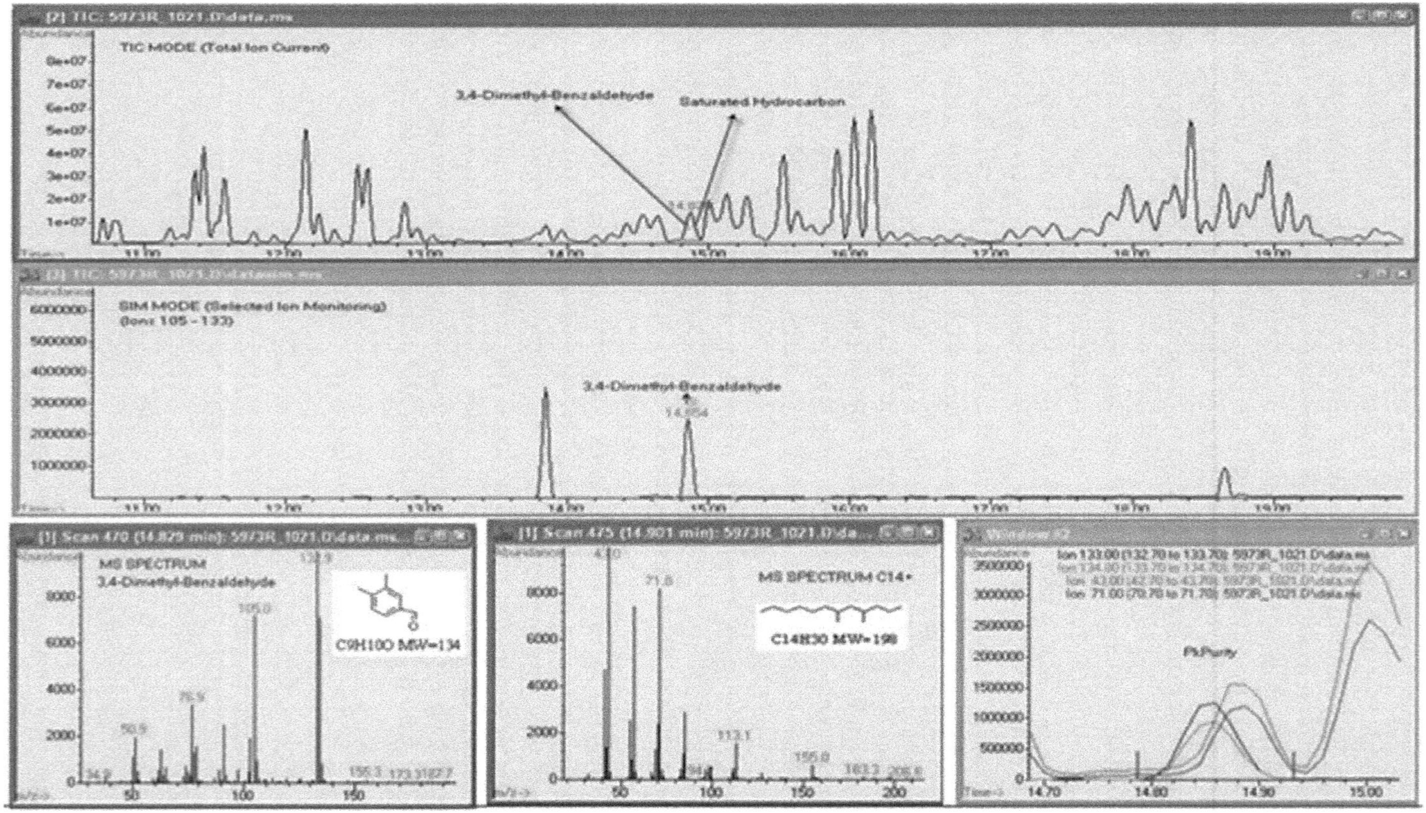

Figure 1.20 Part of TDS-GC-MS trace with two co-eluting peaks at 14.9 min: C_{14} hydrocarbon and 3,4-dimethylbenzaldehyde (additive by-product). Both are NIASs. Relative mass spectra are shown.

3,4-dimethylbenzaldehyde. The latter is a common by-product of Millad® 3988, a well-known clarifying agent for PP. Extracting the ion at 133 Da (centre chromatogram) gives a well-resolved peak. The Peak Purity software tool, bottom right, confirms the presence of two different analytes. The chromatogram was obtained using a 60 m × 0.25 mm id 5% phenylmethylsilicone capillary column with 0.25 µm film thickness. A longer capillary column, a thicker film or a stationary phase of different polarity are not the right approach to achieve better selectivity, according to tests carried out in our laboratories.

A different GC approach is needed. Multi-dimensional GC (also denoted GC×GC) could be the way to achieve the needed separation. A useful literature reference is available.[26]

When the same sample was extracted using UASE with methylene chloride and injected (on-column mode) into a GC×GC system equipped with the same capillary column as used previously, but coupled with a 1.8 m × 0.18 mm id RTX® 200 column with 0.18 µm film thickness, mounted in a secondary oven, Figure 1.21 was obtained.

Considering the 3D chromatogram obtained, it is possible to see how the situation is much easier to understand. The so-called "forest of peaks"[27,28] can now be better resolved. The *x*-axis represents time in minutes on the primary apolar column and the *y*-axis represents time in seconds (6 s of modulation) on the more polar secondary capillary column. On the right part of the chromatogram, the oligomeric pattern is clearly visible: it looks like a mountain chain, delimiting the part of the chromatogram closer to

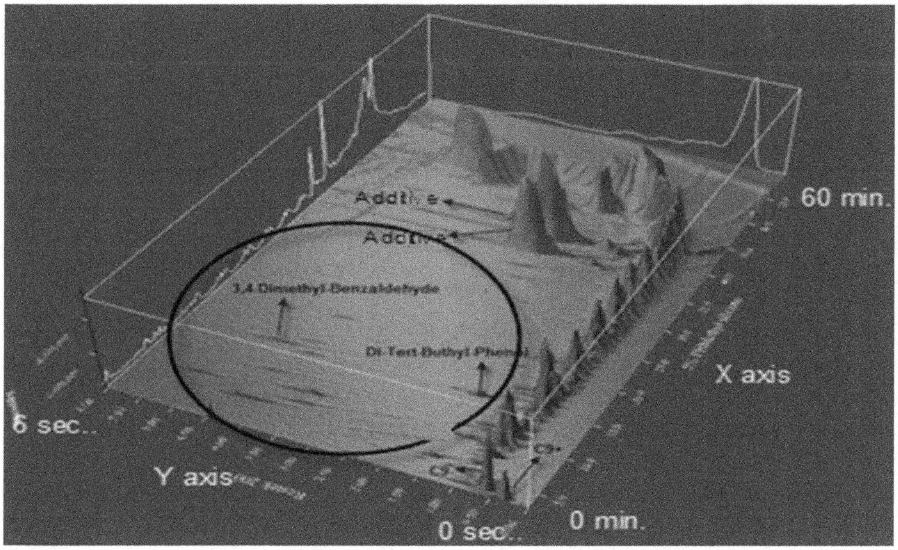

Figure 1.21 Same polymer as in Figure 1.18, but extracted by UASE with CH_2Cl_2 and injected on-column into a GC×GC instrument with TOF-MS detection.

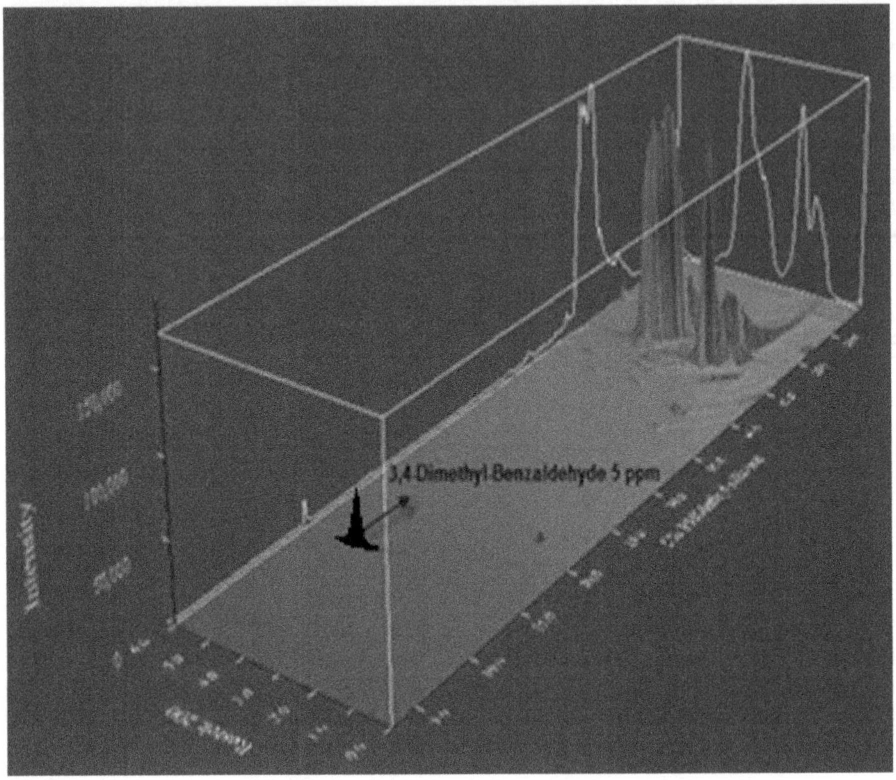

Figure 1.22 Same as in Figure 1.19, but in SIC mode (ion at 133 Da).

the *x*-axis. The more distant the substances are from the *x*-axis, the more polar they are. Close to the end of the *x*-axis run, but shifted to the *y*-axis (they are more polar than oligomers), heavier additives and some of their by-products are present. 3,4-Dimethylbenzaldehyde is now well separated along the *y*-axis. A 5 ppm (w/w) concentration based on polymer is readily visible in the scan mode. If the ion at 133 Da is extracted, the peak is well resolved in a completely empty zone of the 3D chromatogram (see Figure 1.22). The SIM mode is often denoted SIC (selected ion chromatogram) in multi-dimensional GC.

Quantitation of the aldehyde is possible by injecting an ESTD reference solution in methylene chloride containing a precisely weighed amount of the by-product. The comparison was carried out not by considering the peak area, but its relative volume. This means that, the faster the mass spectrometer, the better defined is the peak volume. A quadrupole is not fast enough, so for this purpose the TOF mode is needed. The mass spectrometer used to perform this analysis is a high-resolution TOF instrument, which can acquire spectra at a maximum of 25 Hz. Faster TOF at low resolution can reach an acquisition rate of 200–400 Hz. Usually, 200 Hz is high enough to obtain very good 3D peak shapes and correct peak volumes, together with a manageable file size.

1.8.2 POHs and MOHs

Regarding the investigation of POHs and MOHs, two different approaches can be followed:

(a) UASE in methylene chloride, followed by GC-MS, GC-FID and GC×GC-MS techniques.
(b) Static extraction in a solvent (hexane), separation of aliphatic and aromatic hydrocarbons by HPLC, followed by separate injection of the two fractions into a capillary GC-FID system.[29–33] Recently, in addition to this, GC×GC has been implemented to distinguish oils from synthetic hydrocarbons in food.[6]

When it is possible to have complete traceability of the production chain and samples of each individual step are available, approach (a) is strongly preferred. This will be considered later.

Approach (b) is usually used on food as such or on a final FCM. Unfortunately, the standard technique with HPLC-GC-FID can provide only quantitative analyses. No further specific characterization or identification of POHs and MOHs is possible.

Before starting to investigate the characterization of aliphatic and aromatic hydrocarbons in detail, it would be useful to evaluate their origin in plastic materials. "Real oligomers" (POSHs) are produced during the polymerization process and they are due to weakly activated catalytic sites or to a premature interruption of the polymeric chain, caused by hydrogen addition to limit chain transfer and polymeric growth [Ziegler–Natta (Z/N) catalysts]. If a radical reaction occurs, oligomers can be formed by termination of radical chain growth at an early stage.

Generally, pelletization and addition of polymeric particles obtained in polymerization reduce the amount of oligomers (see Table 1.2). Such components are basically saturated branched aliphatic hydrocarbons. If polyethylene (PE) is considered, saturated linear aliphatic hydrocarbons are the main components.[34] Unsaturated branched aliphatic hydrocarbons (including cyclic hydrocarbons) are usually not present or are detected at a very low concentration. MOAHs are absent, because they are not produced during olefin polymerization.

Together with Z/N catalyst, mineral oil[35] (sometimes mixed with grease) is added to prepare a catalytic paste for olefin polymerization. Added oil (and

Table 1.2 Three different PP samples obtained by the same polymerization plant, but with three different catalysts.

	Amount (ppm)		
	Sample 1 (catalyst 1)	Sample 2 (catalyst 2)	Sample 3 (catalyst 3)
Plant 1, spheres	4600	7190	3190
Plant 1, pellets	3800	6150	2450

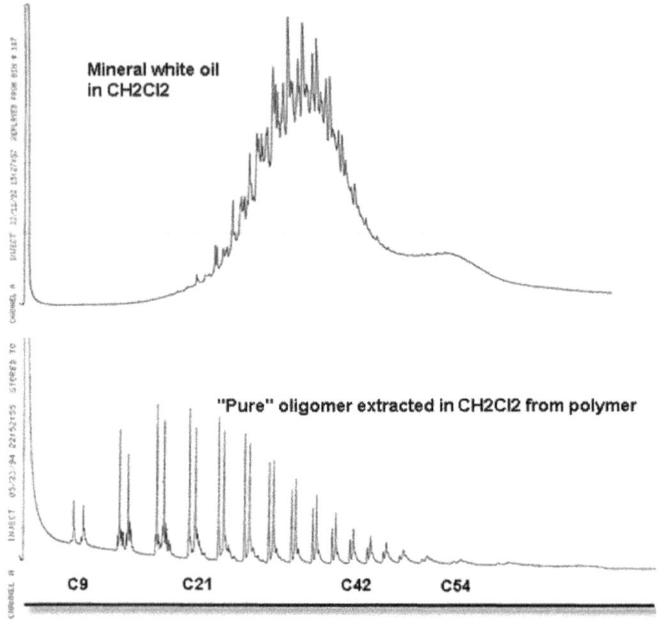

Figure 1.23 Comparison of chromatograms obtained by GC-FID between a white mineral oil (OB22AT, viscosity 2.0 cP at 100 °C) (above) and a HOMO PP obtained in an autoclave with no additives or oil. "Pure" oligomeric profile (below).

grease) can have an MW distribution very similar to those of POSHs or higher (see Figures 1.23 and 1.24), but below or very close to 1000 Da.

Oil and grease are both considered as IASs and, if they are of food grade, do not need to be risk assessed. Figure 1.25 exemplifies the situation commonly found for polyolefin pellets. The chromatogram represents a polymer obtained in an autoclave, pelletized and adding additives together with OB22AT oil (viscosity 2.0 cP at 100 °C). The relatively high amount of added oil is clearly visible. POSHs are also present and they are almost indistinguishable from the mineral oil itself. Most of the "hill" is due to mineral oil, but oligomers and one additive are also present. From the chromatographic point of view, it is almost impossible to distinguish between what are POSHs and what are MOSHs if the MW distributions are very close to each other. Additives can be separated using MS in SIM mode (their ions are very different from those of hydrocarbon series), but even if GC×GC is used, POSHs and MOSHs cannot be distinguished from each other. A mass spectrum obtained by TDS-GC-MS of a commercially available HECO is shown in Figure 1.26. The oligomeric pattern is easily recognized. The "hill" co-eluted between C_{24} and C_{33} is due to a residual mineral oil, caused by a low polymerization yield, or added as a lubricant. Both are to be considered as IASs.

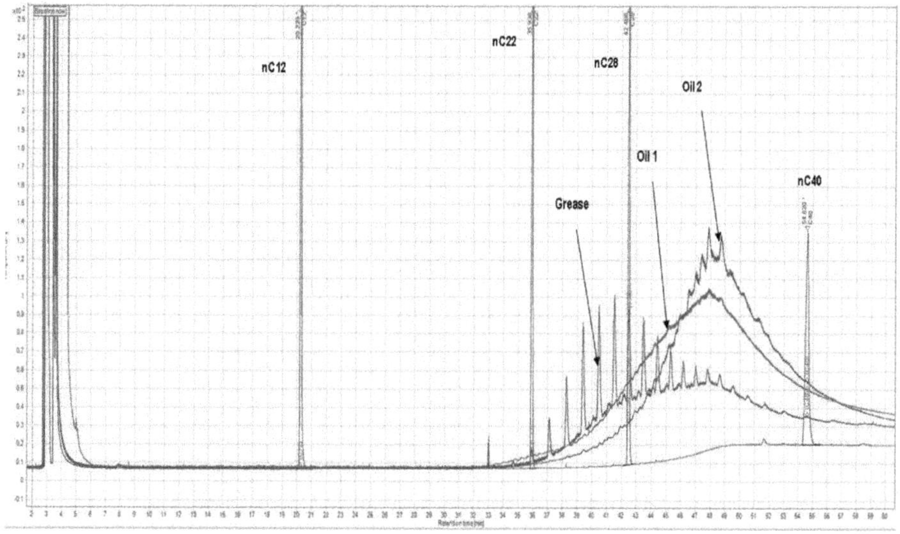

Figure 1.24 Comparison of chromatograms obtained by GC-FID among grease and two different white oils (viscosity about 8 cP at 100 °C). An ESTD calibration with linear saturated hydrocarbons (C_{12}–C_{22}–C_{28}–C_{40}) is overlapped, as a reference.

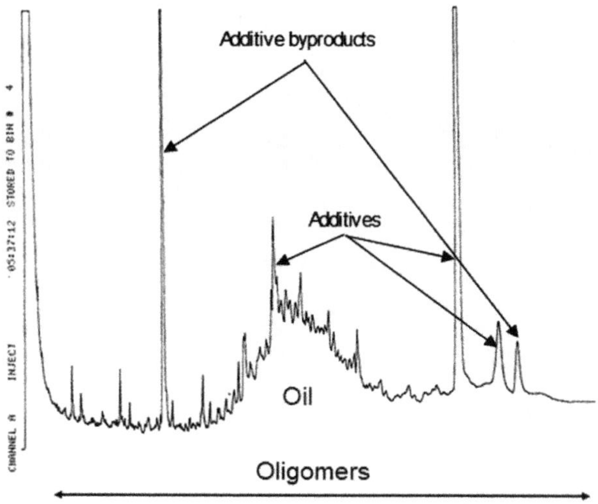

Figure 1.25 GC-FID of a polymer obtained in an autoclave and pelletized, adding additives and thousands of ppm of white oil. UASE extraction with CH_2Cl_2.

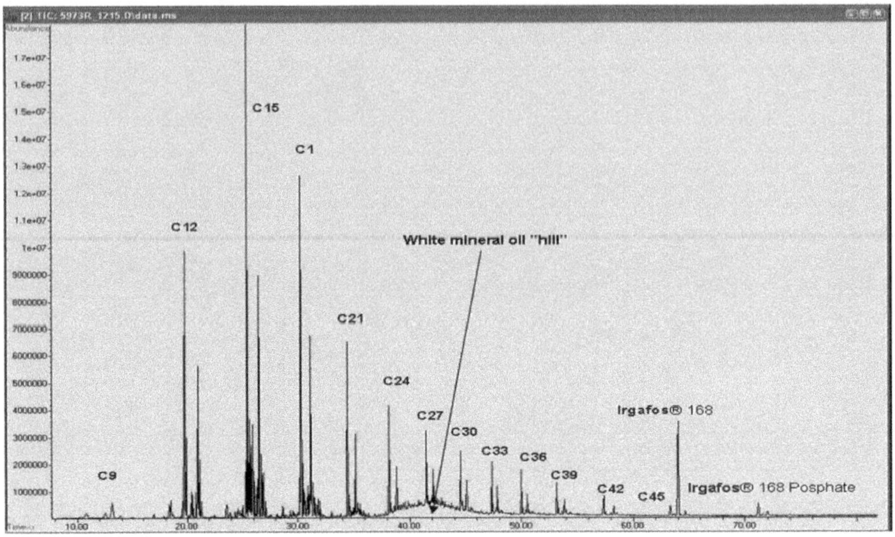

Figure 1.26 TDS-GC-MS of a commercially available HECO polymer. Mineral oil is eluted between C_{24} and C_{33} oligomers.

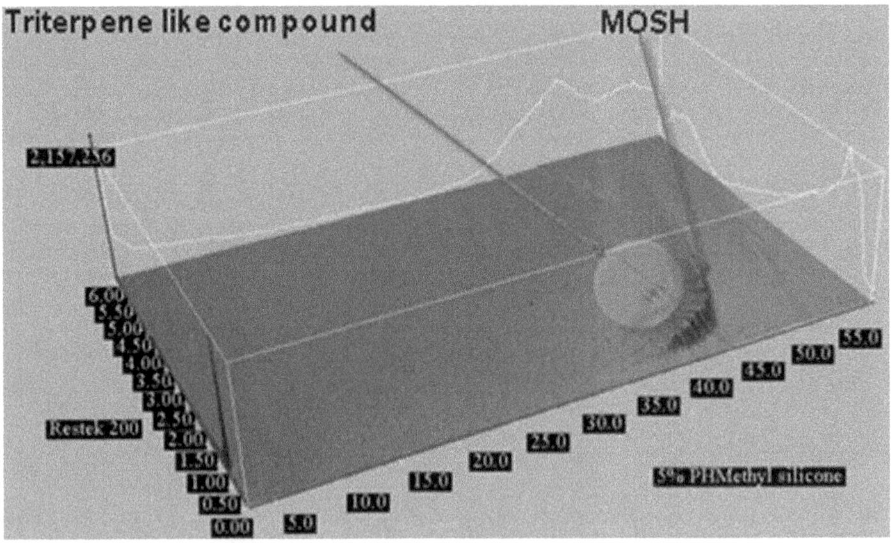

Figure 1.27 GC×GC-TOF in TIC mode of 140 ppm of Winog 70 white mineral oil in CH_2Cl_2.

When GC×GC in TIC mode was used to analyse 140 ppm of a commercially available food-grade mineral oil, Figure 1.27 was obtained. MOSHs are visible together with impurities in the oil: triterpene-like components, which are a well-known impurity in mineral oils.[6,36]

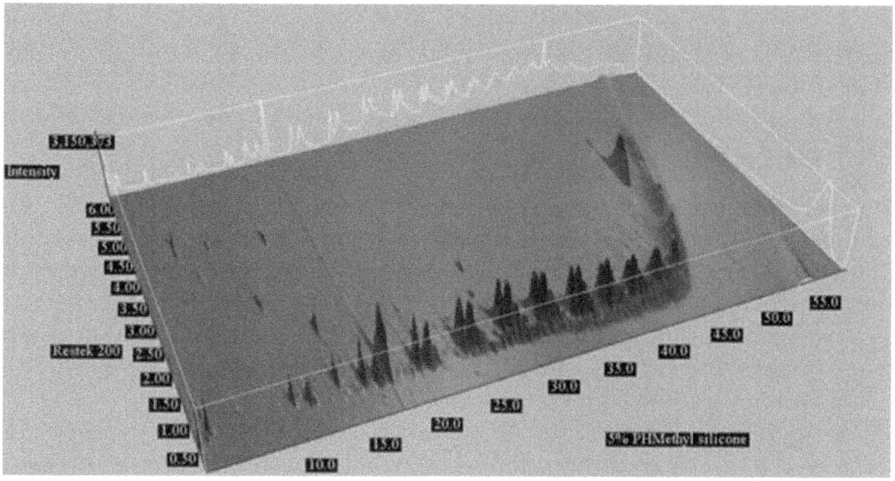

Figure 1.28 Copolymer solvent extracted and spiked with 70 ppm of Winog 70 mineral oil. Oil cannot be distinguished from oligomers.

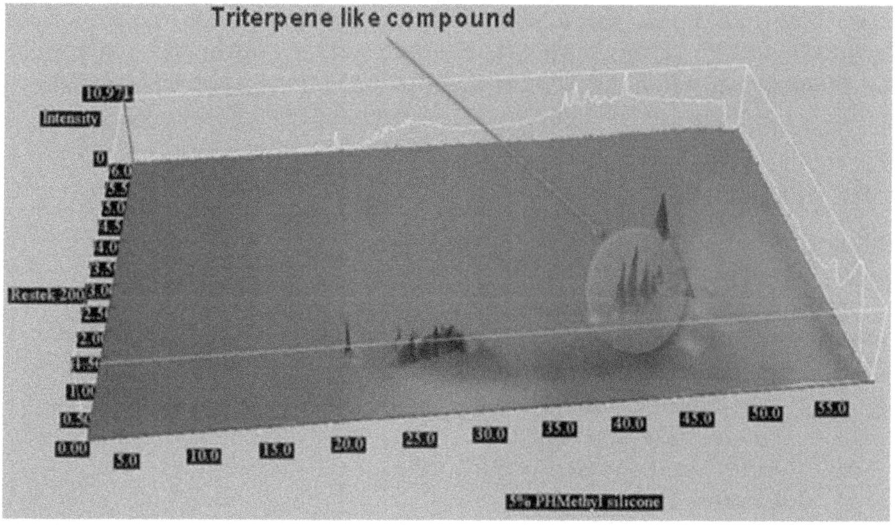

Figure 1.29 Same as Figure 1.27 but in SIC mode.

When a UASE solution of a PP copolymer was spiked in the previously analysed sample, Figure 1.28 was obtained. In the TIC mode, there is no possibility of separating MOSHs from POSHs. However, when the SIC mode was used, extracting only ions belonging to triterpene-like components, Figure 1.29 was obtained.

Monitoring white mineral oil impurities, it is now possible to quantify and separate POSHs from MOSHs, even if the separation is not efficient.

Consider also that some MOSHs and POSHs could even have the same CAS number, but they can be classified as IASs or NIASs, depending on their origin.

1.9 Conclusion

The ever-increasing need for supporting robust risk assessment of FCMs requires continual improvement of the relevant analytical characterization techniques. Generally accepted unique analytical methods are rarely available. Moreover, owing to the complexity of FCMs, even though continual improvements in sensitivity may be achieved by the analytical tools and databases, unclear traceability of all processing steps can be the cause of only partial and non-robust characterization.

A combination of analytical methods with very high sensitivity and resolution allows a thorough description of the whole species that must be monitored (IASs with SMLs and NIASs).

Regarding polyolefins, this chapter has summarized some of the knowledge achieved so far. The available results, obtained by coupling different techniques and knowledge significantly based on mass spectrometry solutions, show that MOAHs can be excluded as migrants from PP-based matrices, obtained with Z/N catalysts. MOAHs are not produced during the polymerization process and, when detected in the final packaging materials, they derive from oils – greases not food approved[36] or from external pollutants.[37]

Abbreviations

^{1}H NMR	Proton nuclear magnetic resonance
AO	Antioxidant
APCI	Atmospheric pressure chemical ionization
APPI	Atmospheric pressure photoionization
CAD	Collisionally activated dissociation
Cn	Number of carbon atoms in the possible eluted species
DAD	Diode-array detection
EFSA	European Food Safety Authority
ESI	Electrospray ionization
ESTD	External standard
FCM	Food contact material
GC-FID	Gas chromatography coupled with flame ionization detection
GC-MS	Gas chromatography coupled with mass spectrometry (quadrupole mass spectrometer)
GC×GC	Multi-dimensional (3D) gas chromatography
GLP	Good laboratory practice
GPC	Gel permeation chromatography
HPLC	High-performance liquid chromatography
IAS	Intentionally added substance
ISTD	Internal standard

LSD	Light-scattering detector
MMI	Multi-mode injector
MOH	Mineral oil
MOAH	Mineral oil aromatic hydrocarbon
MOSH	Mineral oil saturated hydrocarbon
*M*n	Polymer number-average molecular weight
*M*w/*M*n	Ratio of weight- and number-average molecular weights
*M*w	Polymer weight-average molecular weight
MW	Polymer molecular weight (Da)
*M*z/*M*w	Ratio of Z- and weight-average molecular weights
*M*z	Polymer Z-average molecular weight
NIAS	Non-intentionally added substance
OML	Overall migration limit
OSM	One-step microwave-assisted extraction
PE	Polyethylene
POH	Polyolefinic product
POHA	Polyolefin monounsaturated hydrocarbon
POSH	Polyolefin oligomeric saturated hydrocarbon
PP COPO	Polypropylene copolymer
PP HECO	Polypropylene heterophasic copolymer
PP HOMO	Polypropylene homopolymer
PPA	Polymer production aid
PTFE	Poly(1,1,2,2-tetrafluoroethylene)
PTV	Programmed-temperature vaporizing injector
SHS	Static headspace
SIC	Selected ion(s) chromatogram
SIM	Selected ion(s) monitoring
SML	Specific migration limit
TBA	*tert*-Butyl alcohol
TDS	Thermal desorption system
TIC	Total ion chromatogram
TOF	Time-of-flight
TSM	Two-step microwave-assisted extraction
TVT	Total vaporization technique
UASE	Ultrasound-assisted solvent extraction, usually with methylene chloride (CH_2Cl_2)
UV	Ultraviolet
VOC	Volatile organic compound
Z/N	Ziegler–Natta

Acknowledgements

These studies have been made possible thanks to the valuable contributions and chromatographic support of M. Cavallazzi, P. Marisaldi and G. Tani. Thanks are also due to A. Medri for discussions of regulatory aspects.

References

1. M. Al-Ali AlMa'adeed and I. Krupa, *Polyolefin Compounds and Materials Fundamentals and Industrial Applications*, Springer, 2016, pp. 181–199.
2. Official Journal of the European Union 13.11.2004, Regulation (EC) No 1935/2004 of the European Parliament and of the Council of 27 October 2004 on Materials and Articles Intended to Come into Contact with Food and Repealing Directives 80/590/EEC and 89/109/EEC.
3. EFSA Panel on Food Contact Materials, Enzymes, Flavourings and Processing Aids (CEF), Recent developments in the risk assessment of chemicals in food and their potential impact on the safety assessment of substances used in food contact materials, *EFSA J.*, January 28, 2016, **14**(1), 4357.
4. ASTM D5492-17, Standard Test Method for Determination of Xylene Solubles in Propylene Plastics.
5. International Organization for Standardization ISAO 16152:2005, Plastic: Determination of Xylene-Soluble Matter in Polypropylene.
6. M. Biedermann and K. Grob, Comprehensive two-dimensional gas chromatography for characterizing mineral oils in foods and distinguishing them from synthetic hydrocarbons, *J. Chromatogr. A*, 2015, **1375**, 146–153.
7. J. C. J. Bart, *Additives in Polymers. Industrial Analysis and Applications*, Wiley, 1997.
8. B. Marcato and M. Vianello, *J. Chromatogr. A*, 2000, **869**, 285–300.
9. W. Camacho and S. Karlsson, *Polym. Degrad. Stab.*, 2001, **71**(1), 123–134.
10. CEM Application Note E007, CEM Corporation, Matthews, NC, 1998.
11. G. Cecchin and B. Marcato, *J. Chromatogr.*, 1996, **730**, 83.
12. W. M. A. Niessen, *Liquid Chromatography–Mass Spectrometry*, CRC Taylor & Francis, Boca Raton, 3rd edn, 2006.
13. F. E. Klink, Liquid chromatography/mass spectrometry, in *Encyclopedia of Analytical Chemistry: Applications, Theory and Instrumentation*, 2010, DOI: 10.1002/9780470027318.a6013.pub2.
14. B. Kolb and L. S. Ettre, *Static Headspace-Gas Chromatography Theory and Practice*, Wiley-VHC, 1997.
15. G. Wypych, *Handbook of Odors in Plastic Material*, ChemTec Publishing, Toronto, 2013.
16. F. A. Fazzalari, *Compilation of Odor and Taste Threshold Values Data*, DS 48A ASTM, 1978.
17. G. A. Burdock, *Fenaroli's Handbook of Flavor Ingredients*, CRC Press, 6th edn, 2016.
18. D. Brocca, E. Arvin and H. Mosbæk, Identification of organic compounds migrating from polyethylene pipelines into drinking water, *Water Res.*, 2002, **36**, 3675–3680.
19. J. Throck Watson and O. David Sparkman, *Introduction to Mass Spectrometry, Instrumentation, Applications and Strategies for Data Interpretation*, Wiley, 4th edn, 2008.

20. P. Singh, A. A. Wani and H. C. Langowski, *Food Packaging Materials: Testing & Quality Assurance*, CRC Press, 2016.
21. EFSA Panel on Contaminants in the Food Chain (CONTAM), *EFSA J.*, 2012, **10**(6), 2704[185 p.].
22. Official Journal of the European Union 15.1.2011, Commission Regulation (EU) No 10/2011 of 14 January 2011 "on plastic materials and articles intended to come into contact with food, article 18 point 6.
23. FCA Guidelines on "Risk Assessment of Non-listed Substances (NLS) and Non-intentionally Added Substances (NIAS) under the Requirements of Article 3 of the Framework Regulation (EC) 1935/2004" Version 1.0 October 2016.
24. *Use of Migration Modelling for Food Contact Materials MDCTech System*, http://www.akts.ch/dwlds/sml-migration-modelling-for-food-contact-materials.
25. JRC (EC), *Practical Guidelines on the Application of Migration Modelling for the Estimation of Specific Migration*, Joint Research Center..
26. T. Taylor, A short introduction to multidimensional GC, *LCGC North Am.*, September 01, 2012, **30**(9), 870.
27. W. D. van Dongen, L. Coulier, B. Muilwijk, *et al.*, Analytical strategy to assess the safety of food contact materials, Food Contact Polymers First International Conference 21–22 February 2007 Brussels, Belgium Conference Proceedings, Smithers Rapra Ltd, 2007.
28. H. M. Hollnagel, P. van Herwijnen and P. Sura, Assessing safety of non-intentionally added substances in polymers used for food contact applications, *Toxicol. Lett.*, 2014, **229S**, S22–S39.
29. T. Populin, M. Biedermann, K. Grob, S. Moret and L. Conte, Relative hopane content confirming the mineral origin of hydrocarbons contaminating foods and human milk, *Food Addit. Contam.*, September 2004, **21**(9), 893–904.
30. S. Moret, L. Barp, G. Purcaro and L. S. Conte, Rapid and sensitive solid phase extraction-large volume injection-gas chromatography for the analysis of mineral oil saturated and aromatic hydrocarbons in cardboard and dried foods, *J. Chromatogr. A*, 2012, **1243**, 1–5.
31. K. Fiselier, F. Grundböcka, K. Schönb and O. Kappensteinb, *et al.*, Development of a manual method for the determination of mineral oil in foods and paperboard, *J. Chromatogr. A*, 2013, **1271**, 192–200.
32. S. Biedermann-Brema, N. Kasprickb, T. Simatb and K. Grob, Migration of polyolefin oligomeric saturated hydrocarbons (POSH) into food, *Food Addit. Contam.*, March 2012, **29**(3), 449–460.
33. M. Lommatzscha, M. Biedermannb, T. J. Simata and K. Grob, Argentation high performance liquid chromatography on-line coupled to gas chromatography for the analysis of monounsaturated polyolefin oligomers in packaging materials and foods, *J. Chromatogr. A*, 2015, **1402**, 94–101.
34. Polyolefin Oligomers: research project for Risk Assessment, Plasic Europe 5 July 2016.

35. D. Benetti, A. Fait, R. Flammini, G. Patroncini and L. Pedriali, Preparation of a catalytic paste for the olefin polymerization, *EP Pat.* 2054451 A1, 2009.
36. K. Grob, Mineral Oil Depicted by Gas Chromatograms. Principals for Analysis in Food, Seminar in Berlin September 23rd, 2011.
37. FEICA Factsheet, Mineral oil hydrocarbons in food: adhesives for food packaging as a source, p. 2, FEICA | FAS-EX-D08-009.

CHAPTER 2

Food Contact Paper and Paperboard: Examples of Gas and Liquid Chromatography Determinations

ANDREA STOCCHI* AND MARCO IACOPINI

Ecolstudio SpA, Via dei Bichi 293, 55100 Lucca, Italy
*E-mail: a.stocchi@ecolstudio.com

2.1 Introduction

Paper, which is mainly composed of cellulose, additives and production aids, is a material used in many articles and objects that are intended to come into contact with food. Everyone knows the numerous applications of paper and paperboard in everyday life, so it is easy to understand how important the safety of the paper products is for consumers.

Paper and board food contact materials (P&B FCMs) are regulated in Europe by the Framework Regulation (EC) No. 1935/2004 and by some national Member State legislations, such as in Italy and France. Moreover, in Germany, the BfR (Bundesinstitut für Risikobewertung – Federal Institute for Risk Assessment) Recommendations are applied to assess the compliance of food contact materials.

Food Chemistry, Function and Analysis No.10
Food Contact Materials Analysis: Mass Spectrometry Techniques
Edited by Michele Suman
© The Royal Society of Chemistry 2019
Published by the Royal Society of Chemistry, www.rsc.org

Owing to the extensive use of recycled paper, the law requires the determination of contaminants or specific substances used for different purposes in the production of P&B FCMs. Many of the standard methods used to determine these specific substances involve the use of mass spectrometry as analytical technique. Mass spectrometry techniques are also fundamental for the detection and evaluation of NIASs (not intended added substances).

This chapter is structured into two parts: in the first part, the determination of NIASs using official methods is discussed, and in the second part the detection of NIASs (and other compounds) in the absence of standard methods is reported.

Table 2.1 lists the most important chromatographic determinations, the corresponding regulations, the international standards (if present) and the detection techniques.

2.1.1 Use of Mass Spectrometry Techniques for Conformity Requirements: GC-MS and LC-MS

Considering that the legal requirement values for the compliance are usually low, gas chromatography–mass spectrometry (GC-MS) techniques should to be used in the SIM (selected ion monitoring) mode or MRM (multiple reaction monitoring) mode. The MRM or product ion scan is the working mode in liquid chromatography–mass spectrometry (LC-MS) determinations.

The first four substances listed in Table 2.1 are analyzed to evaluate the quality of recycled paper products.

2,6-Diisopropylnaphthalene (DIPN) is used as solvent in carbonless paper, and it can be found in P&B FCMs due to the use of recycled fibers. DIPN is quantified as content (mg kg^{-1}) and GC-MS analysis shows very good sensitivity (see Figure 2.1). Because of the natural presence of a series of isomers, the peak group is very characteristic. It can be found at trace levels (<1 mg kg^{-1}), and up to 30–40 mg kg^{-1}.

Phthalates are used as plasticizers; they are known to act as endocrine disruptors, interfering with the endocrine system. DEHP, DBP and DiBP can be measured as content, and the concentration is converted to a migratable quantity in order to assess the compliance (limit expressed as mg kg^{-1} food), using the grammage value (g m^{-2}) and the conventional rule of the food contact EU cube.[1]

Phthalates are generally very sensitive molecules to GC-MS determination, and they have very similar spectra. In fact, the most sensitive ion is *m/z* 149 (see Figure 2.2 for the fragmentation pattern), and other ions present differ from one compound to another.

Figure 2.3 reports a chromatogram relating to the injection of a 5 ppm analytical standard phthalate mixture. Even though it is fairly common to find phthalates in recycled paper samples, it is difficult when the concentration exceeds the limit as mg kg^{-1} food, considering the grammages of paper products (especially for tissue products).

Table 2.1 Important molecules, country and regulation, reference standard method and analytical technique for paper FCMs.

Compound: IUPAC name and CAS number	Common name or abbreviation	Country and regulation	Standard and analytical technique
2,6-Di(propan-2-yl) naphthalene, CAS 24157-81-1	Diisopropyl-naphthalene, DIPN	Germany, BfR XXXVI Recommendation	EN 14719, GC-MS
Bis(2-ethylhexyl)benzene-1,2-dicarboxylate, CAS 117-81-7	DEHP	Germany, BfR XXXVI Recommendation	EN 16453, GC-MS
Bis(2-methylpropyl) benzene-1,2-dicarboxylate, CAS 84-69-5	DIBP	Germany, BfR XXXVI Recommendation	EN 16453, GC-MS
Dibutylbenzene-1,2-dicarboxylate, CAS 84-74-2	DBP	Germany, BfR XXXVI Recommendation	EN 16453, GC-MS
3-Chloropropane-1,2-diol	MCPD	Germany, BfR XXXVI Recommendation	Amtliche Sammlung von Untersuchungsverfahren nach 64 LFGB B 80.56-2, GC-MS
1,3-Dichloropropan-2-ol	DCP	Germany, BfR XXXVI Recommendation	Amtliche Sammlung von Untersuchungsverfahren nach 64 LFGB B 80.56-2, GC-MS
—	Primary aromatic amines	Germany, BfR XXXVI Recommendation	LC-MS
Diphenylmethanone	Benzophenone	Germany, BfR XXXVI Recommendation	GC-MS
Bis[4-(dimethylamino) phenyl]methanone	4,4′-Bis(dimethylamino) benzophenone	Germany, BfR XXXVI Recommendation	GC-MS
2,3,4,5,6-Pentachlorophenol	Pentachlorophenol	France, Fiche DGCCRF	ISO 15320, GC-MS (or GC-ECD)
—	Polychlorinated biphenyls, PCBs	France, Fiche DGCCRF	ISO 15318, GC-MS (or GC-ECD)
4-[2-(4-Hydroxyphenyl)propan-2-yl] phenol, CAS 80-05-7	Bisphenol A	France, Loi No. 2012-1442 du 24 décembre 2012	LC-MS

DCP and MCPD are by-products that can be present in polyamidoamine–epichlorohydrin wet-strength resins; they are used to improve the mechanical resistance of the paper to water, and are known as possible carcinogens to humans. The method requires the determination of the two substances in a water extract. The analysis is characterized by a derivatization with 2,2,3,3,4,4,4-heptafluoro-1-imidazol-1-ylbutan-1-one (HFBI, CAS 32477-35-3) because of the chromatographic behavior of these molecules. This

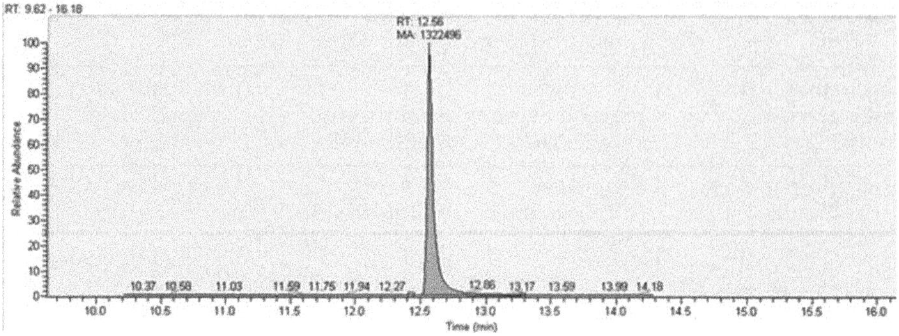

Figure 2.1 Chromatogram of a 2,6-diisopropylnaphthalene 48 μg L^{-1} standard, acquired in SIM mode.

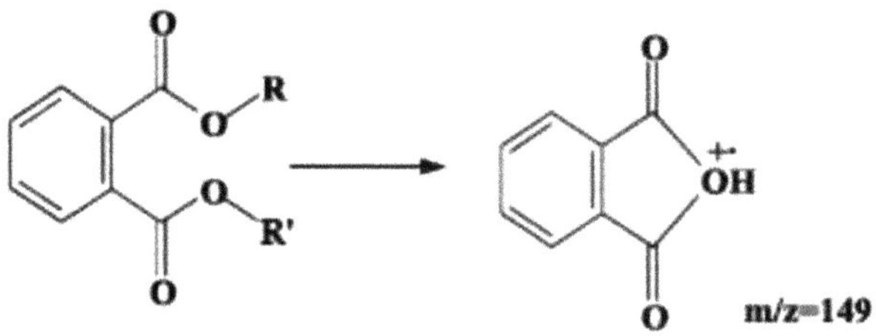

Figure 2.2 Fragmentation pattern of phthalates.

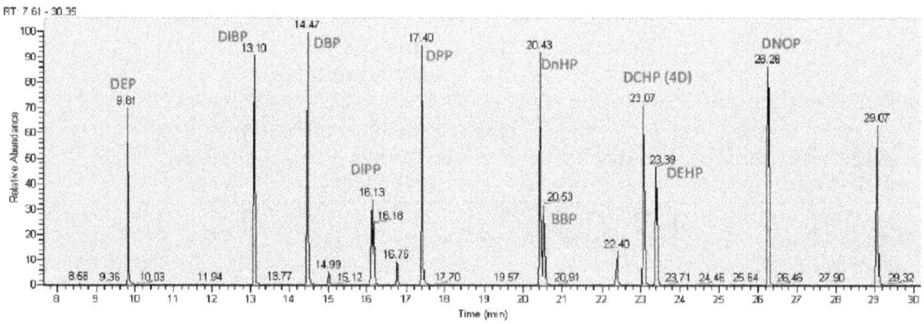

Figure 2.3 Chromatogram of a 5 ppm phthalate standard obtained by GC-MS, total ion current.

determination must be very sensitive, owing to the low limit values. Figures 2.4 and 2.5 illustrate analytical standard concentrations corresponding to 2 μg L^{-1} of the derivatized DCP and MCPD, respectively.

Benzophenone and 4,4′-bis(dimethylamino)benzophenone are widely used as photoinitiators and are searched for as contaminants in recycled paper. GC-MS in the SIM mode is the best method for this determination.

Pentachlorophenol was used as a slime-reducing agent in the paper industry. Because of its toxicological properties (it is considered as a potential

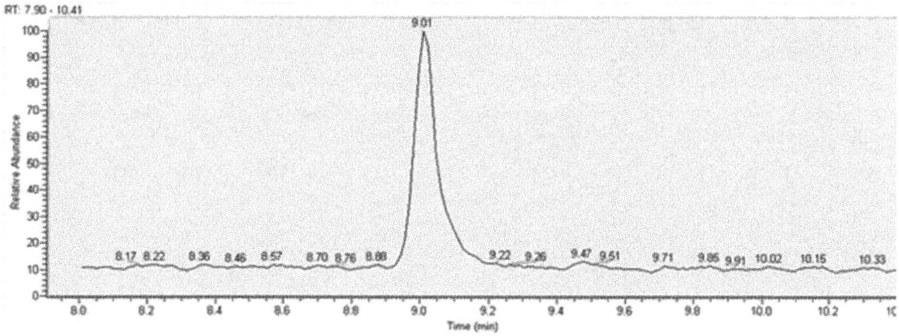

Figure 2.4 Chromatogram of a DCP standard corresponding to 2 μg L^{-1}.

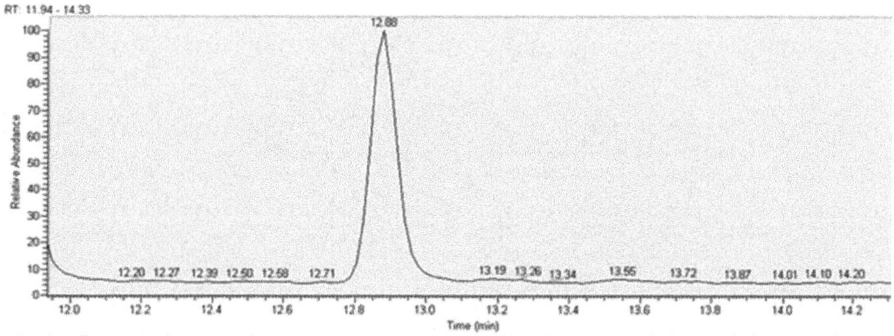

Figure 2.5 Chromatogram of an MCPD standard corresponding to 2 μg L^{-1}.

carcinogenic agent), its use is now actually forbidden, but it could be present in recycled paper. The standard method requires an aqueous extract followed by derivatization; in fact, because of its high polarity, pentachlorophenol has a non-reproducible chromatographic behavior. The analysis, executed in the SIM mode, is very sensitive (see Figure 2.6).

Polychlorinated biphenyls (PCBs) are a group of compounds characterized by a structure with two phenyl groups and chlorine as possible substituent in different positions (see Figure 2.7). These compounds, available in mixtures with different compositions, have been used for years in many sectors. They can be found in paper since they are used in ink formulations for their chemical and physical stability. Mass spectrometry is a very sensitive technique for PCBs, and they have mass spectra with many isotope clusters due to the presence of the chlorine atoms. Considering that PCBs have been used in commercial mixtures with different compositions, a positive sample will show the contemporary presence of more than one PCB isomer. The standard method (EN 15318) requires the determination of seven different PCBs.

All the aforementioned tests are performed with a gas chromatograph coupled to a mass spectrometer. Another chromatographic technique could be

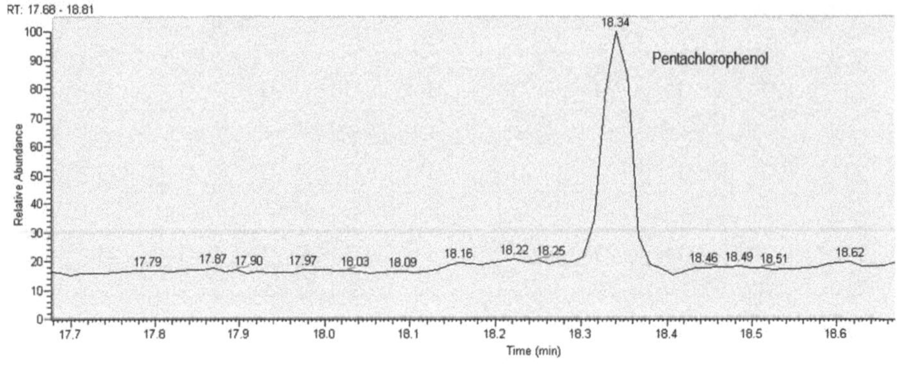

Figure 2.6 Chromatogram of a derivatized pentachlorophenol (20 ppb), SIM mode.

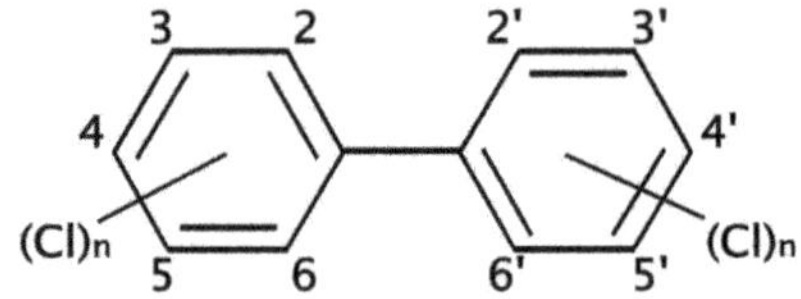

Figure 2.7 Structure of a polychlorinated biphenyl (PCB).

used for this purpose, namely liquid chromatography. Liquid chromatography can be a powerful method for the verification of paper contaminants, as detailed in Table 2.1.

Primary aromatic amines (PAAs) are known to be carcinogenic and mutagenic; they can be contaminants in color pigments used in colored products. Their presence is detected in a cold water extract or 3% acetic acid extract. BfR Recommendations specify a spectrophotometric determination to evaluate the content of PAAs, but the presence of color in the water extract of entirely colored and/or decorated paper products can cause interference and inaccurate results. Therefore, a specific test for the presence of amines such as 1A (known to have carcinogenic potential for humans) and 1B (presumed to have carcinogenic potential for humans) according to CLP (Classification, Labelling and Packaging) Regulations (Reg. CE 1272/2008) is needed. Because of the basic structure of the amines, the detection has to be performed in the positive ion mode, using electrospray ionization (ESI), and the eluent must contain an enhancer of the ionization. To achieve the reference limit, the limit of quantitation needs to be almost 2 μg L^{-1} in the extract. The best way to evaluate compliance with the BfR Recommendation for the migration of PAAs is the use of the multiple reaction monitoring (MRM) mode with a triple quadrupole (LC-QQQ) or of the product ion scan mode with Q-TOF (time of flight).[2] LC-QQQ is usually slightly more sensitive than Q-TOF, whereas the latter offers a very specific signal due to the high resolution, and the spectrum of the product ion scan can give more fragments of the ion selected in the first quadrupole. Figure 2.8 shows the LC-MS results for an example of a multicomponent standard at 10 ppb concentration.

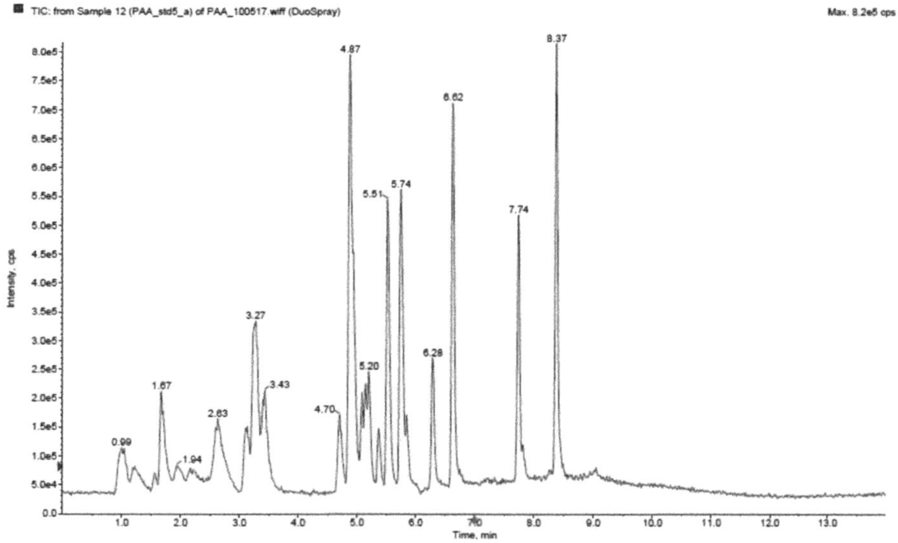

Figure 2.8 LC-MS results for a 50 ppb multicomponent standard (primary aromatic amines).

Obviously, these molecules need to be monitored by choosing an ion as quantifier, and at least two qualifiers; the qualifiers must have a ratio of intensities relative to that of the quantifier similar to the ratio verified in the standards. The criterion to be matched must be fixed and respected.

Bisphenol A (BPA) is one of the most controversial molecules in food contact determinations; it has been widely used in thermal paper for cash register receipts, and can enter the paper recycle system. Moreover, plastic materials may also contain BPA (polycarbonate in particular).[3] There has been considerable discussion of the effects of BPA on human health; it has hormone-like properties that raise concerns about potential adverse effects. The content of BPA in paper samples can be determined using a polar extraction solvent (such as methanol or acetonitrile). The ionization of BPA is facilitated at certain pH values, as reported for PAAs. Given the acidic properties of BPA, the basic pH range is the best choice for sensitive detection using ESI. Figure 2.9 shows a chromatogram of a 10 ppb standard. It is quite usual to find BPA in recycled paper samples, ranging from trace levels to 50 mg kg^{-1}.

2.2 Mass Spectrometry for Non-target Determinations

Mass spectrometry is a fundamental technique not only for target analysis, but also for non-target determinations. European Framework Regulation 1935/2004 reports in article 3 some fundamental concepts; for P&B FCMs, this can be achieved by assuring with analytical determinations the possible presence of dangerous molecules for human health or, in some cases, the

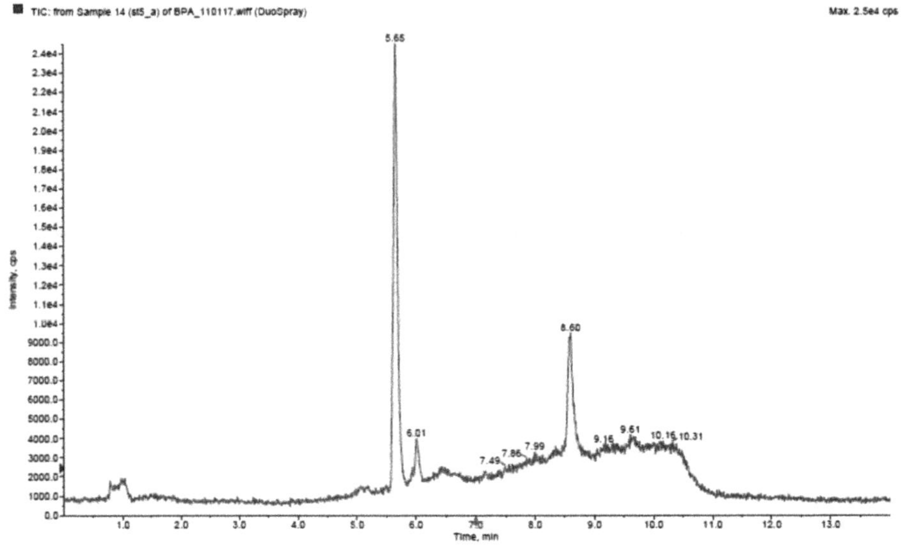

Figure 2.9 LC-MS results for a 50 ppb standard of bisphenol A (retention time 5.65 min).

possible presence of molecules responsible of the organoleptic deterioration or an alteration in the composition of a food.

For this purpose, GC-MS and LC-MS techniques are useful and complementary; LC-MS needs particular instrumentation such as Q-TOF, Orbitrap or ion trap in order to generate a tandem mass spectrum (MS/MS) that allows a possible identification of the molecule. GC-MS, equipped with libraries for compound identification, is an easier technique. Different methods of sample introduction need to be used to cover the different substance volatilities in order to carry out the final risk assessment. Headspace, SPME (solid-phase microextraction) and solvent extraction techniques are used to cover the range of volatilities in GC-MS.

Recycled paper is more subject to the presence of dangerous substances, as explained in the first part of this chapter. Compounds are identified using the available instrumental libraries, and quantified by the response factor by using an internal standard, which is added to the sample in the headspace or SPME vial, or to the extract. After this screening phase, the application of a specific and quantitative method for compound determination must be carried out.

In the following, we report some interesting examples of screening of paper samples.

Figure 2.10 shows results for an extract of a paper sample made from recycled paper. Figure 2.11 is a zoom of the same acquisition as in Figure 2.10, to provide better visualization of the peaks of interest. The compounds observed are listed in Table 2.2.

The band with a retention time in the range 26–45 min contains some hydrocarbons and could be related to the presence of mineral oil [whose components are usually known as MOAH (mineral oil aromatic hydrocarbons)

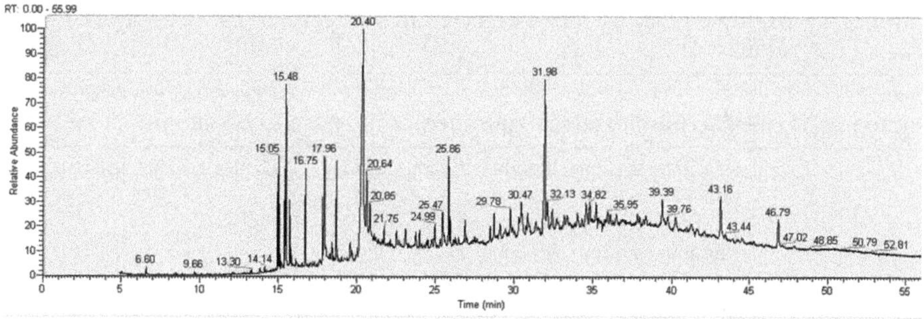

Figure 2.10 GC-MS screening acquisition of a recycled paper sample.

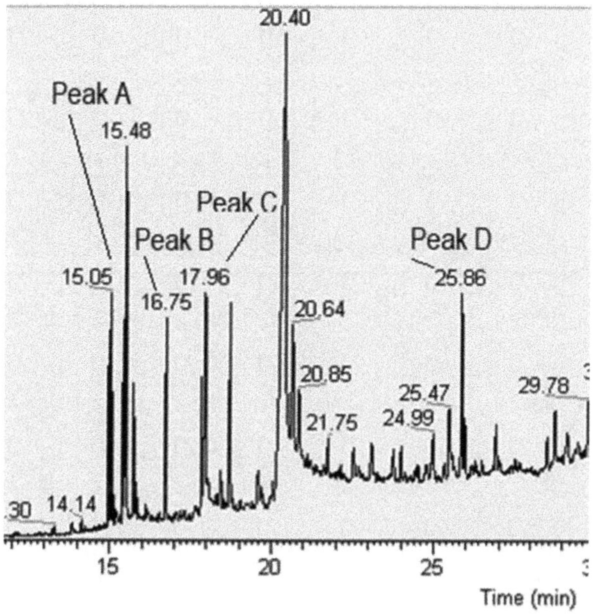

Figure 2.11 Zoom of part of the screening acquisition in Figure 2.10 with details of the most interesting peaks.

and MOSH (mineral oil saturated hydrocarbons)], as confirmed by the ion current of *m/z* 71 and 85 (see Figure 2.12).

Figure 2.13 shows another extraction screening of a recycled paper product. Some of the most important identified peaks are as follows:

- Peak A: DIPN.
- Peaks B, C, D, E: phthalates – DiBP, DBP, DEHP, dioctyl phthalate.
- Peak F: eicosanol (IUPAC name icosan-1-ol, CAS number 629-96-9), used as antifoaming agent in the paper industry. This compound is known to have an LD_{50} for rats of 10 000 mg kg^{-1}, but no data are available about its carcinogenic or mutagenic activity.

Table 2.2 Example of compounds detected in a screening test of a recycled paper sample.

Peak (Figure 2.11)	IUPAC name and CAS number	Common name or abbreviation
A	2,6-Di(propan-2-yl)naphthalene, CAS 24157-81-1	Diisopropylnaphthalene
B	Bis(2-methylpropyl)benzene-1,2-dicarboxylate, CAS 84-69-5	DiBP
C	Dibutylbenzene-1,2-dicarboxylate, CAS 84-74-2	DBP
D	Bis(2-ethylhexyl)benzene-1,2-dicarboxylate, CAS 117-81-7	DEHP

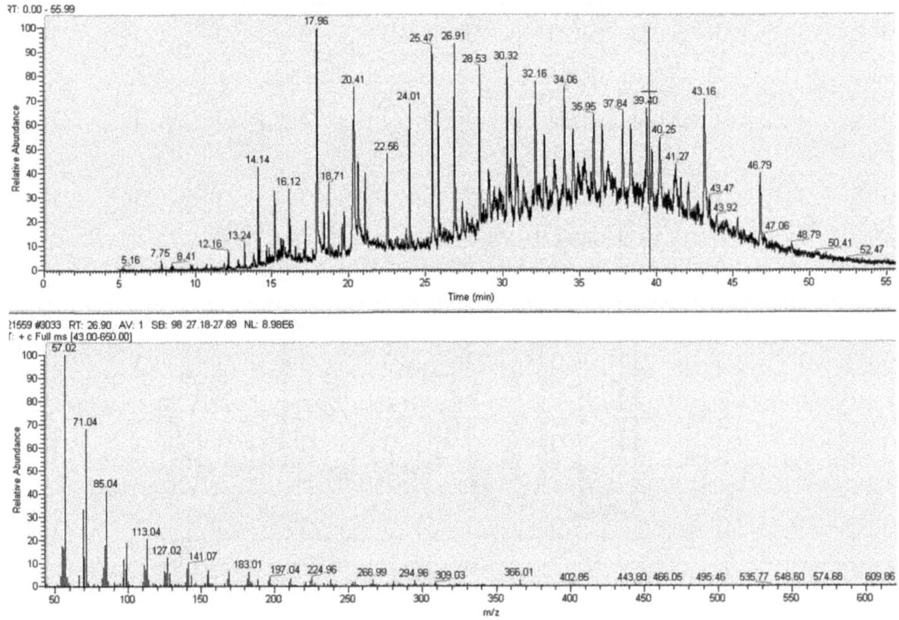

Figure 2.12 *m/z* 75 and 81 ion current in a screening acquisition of a recycled paper sample (presence of hydrocarbons) with (bottom) mass spectrum of a detected hydrocarbon.

- Peak G: fatty acids such as oleic acid, stearic acid, *etc.*, deriving from lubricants or tall oil.
- Peaks H, I, L: ketones formed after alkaline hydrolysis of an alkylketene dimer (AKD). These ketones have CAS numbers 502-73-8, 15740-35-9 and 504-53-0, respectively. AKD is a sizing agent used to provide water resistance to paper. The German BfR reports a limit of 5 mg kg^{-1} of food contamination by these three dimers. It is known that AKD and the ketones produced by hydrolysis do not have adverse effects on human health.

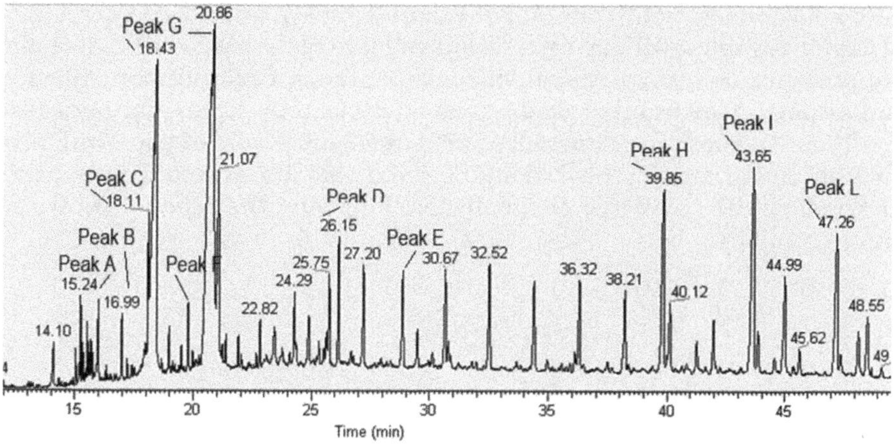

Figure 2.13 Chromatogram of a recycled paper sample.

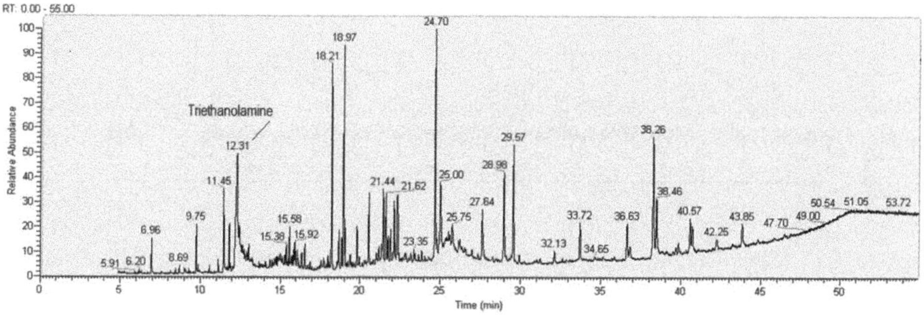

Figure 2.14 Chromatogram of a printed pure cellulose sample containing triethanolamine.

In some cases, substances that are dangerous with respect to their structure and/or concentration may be found related to the presence of inks or glues. For example, the extract for the screening determination of a sample of printed pure cellulose revealed the presence of triethanolamine (IUPAC name 2-[bis(2-hydroxyethyl)amino]ethanol, CAS number 102-71-6). This compound is regulated under Reg. 10/2011 for plastic materials with a specific migration limit of 0.05 mg kg^{-1} in food. Figure 2.14 shows a chromatogram with the presence of triethanolamine at a retention time of 12.31 min. The estimated concentration, considering the grammage of the material, indicates a migratable quantity of 0.09 mg kg^{-1} in food, suggesting a more precise estimation of the content with a specific measurement method because of the semiquantitative estimation of the concentration.

Having regard to organoleptic controls, mass spectrometry may also be useful for evaluating which molecules could create odor problems. The most commonly present compounds that cause unpleasant odors in paper have a low olfactory threshold. For example, SPME and GC-MS screening of napkin samples, with an unpleasant smell in the sensorial test, detected the presence of two volatile aldehydes (pentanal, CAS number 110-62-3, and hexanal, CAS number 66-25-1) characterized by a pungent odor (see Figure 2.15). These two substances are typical off-flavors of papermill production; some studies have demonstrated that unsaturated fatty acids in wood can be converted to aldehydes, following the scheme shown in Figure 2.16.

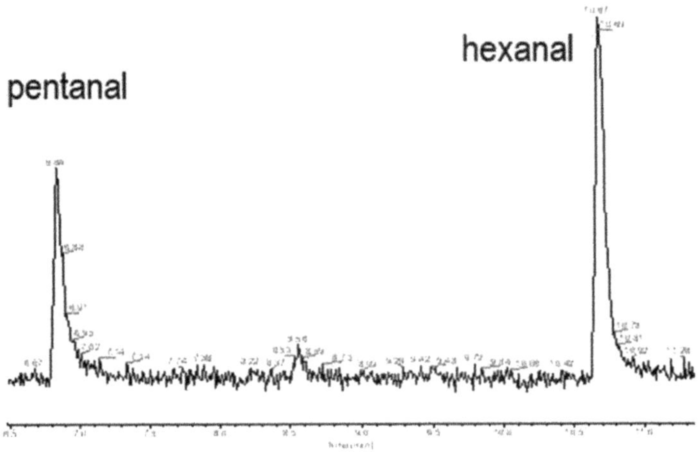

Figure 2.15 Peaks related to the presence of pentanal and hexanal in a napkin sample.

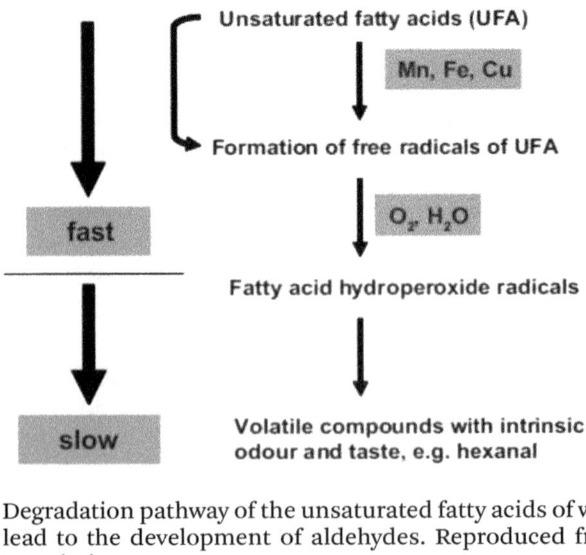

Figure 2.16 Degradation pathway of the unsaturated fatty acids of wood lipids that lead to the development of aldehydes. Reproduced from ref. 4 with permission.

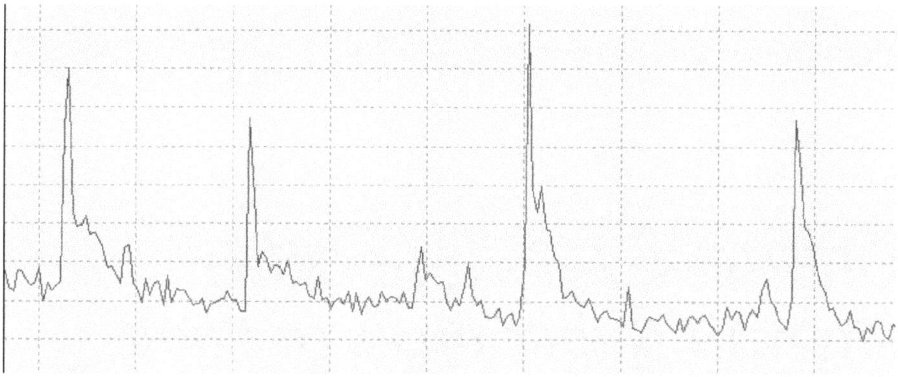

Figure 2.17 Chromatogram of traces of low molecular weight fatty acids.

Aldehydes are not the only group of molecules that can cause odors in paper: other substances may also be present, such as low molecular weight fatty acids (usually produced by anaerobic microorganisms in water and pulp systems). These molecules can be formed in the water and pulp systems under anaerobic conditions, and have a very low odor threshold. For this reason, the determination must be very sensitive (see Figure 2.17, for example). Usually, modifications to the production system can solve the problem.

References

1. FCA Guidelines on Risk Assessment of non-listed substances (NLS) and non-intentionally added substances (NIAS) under the requirements of Article 3 of the Framework Regulation (EC) 1935/2004, version 1.0, October 2016.
2. O. Yavuz, S. Valzacchi, E. Hoekstra and C. Simoneau, Determination of primary aromatic amines in cold water extract of coloured paper napkin samples by liquid chromatography-tandem mass spectrometry, *Food Addit. Contam., Part A*, 2016, **33**(6), 1072–1079.
3. T. Mendum, E. Stoler, H. VanBenschoten and J. C. Warner, Concentration of bisphenol A in thermal paper, *Green Chem. Lett. Rev.*, 2011, **4**(1), 81–86.
4. H. Jung and J. Kappen, Odor control in papermaking, *PaperAge*, July/August 2010, pp. 18–21.

CHAPTER 3

Additives, Inks and Other Migrant Substances in Food Contact Materials

RAFAEL PASEIRO-CERRATO*[a,b], LOWRI DEJAGER[a] AND TIMOTHY H. BEGLEY[a]

[a]US FDA, Center for Food Safety and Applied Nutrition, 5001 Campus Drive, College Park, MD 20740, USA; [b]Joint Institute for Safety and Applied Nutrition (JIFSAN), University of Maryland, College Park, MD 20742, USA
*E-mail: rafael.cerrato@fda.hhs.gov

3.1 Introduction to Food Contact Materials

Advances in food contact materials (FCMs), particularly the materials used for food packaging, have revolutionized the field of food safety. Food packaging protects food from external agents, such as oxidation, light, humidity and microorganisms. This protection also helps maintain desirable organoleptic properties of food (flavor, color, aromas) and freshness. As these features also increase the shelf life of packaged products, higher quality foods become available to more households and food waste is reduced. However, food packaging has roles beyond protecting the food until it reaches the consumer. Packaging needs to attract the consumer's attention and provide information about the product. Therefore, the color and shape of packages matter, and labels on these packages must convey information about brand,

Food Chemistry, Function and Analysis No.10
Food Contact Materials Analysis: Mass Spectrometry Techniques
Edited by Michele Suman
© The Royal Society of Chemistry 2019
Published by the Royal Society of Chemistry, www.rsc.org

product composition and expiration date in addition to required nutritional information, instructions for food preparation and proper conservation after opening. Meeting all these goals may require an FCM to have special features that go beyond those necessary for food protection.

Although several types of food packaging (*i.e.* glass, ceramic, paper and wood) are used, polymeric materials (PMs) are probably the most commonly used in the manufacture of FCMs, including food packaging. These PMs are synthesized using starting substances called monomers, which are reacted to form polymeric chains and subsequently the PM. During this process, substances that impart certain properties to the PM, such as elasticity, stability, resistance to heat and opacity, are added, hence the term "additives," which we will use in the remainder of this chapter.

There are two main groups of PMs: thermoplastics and thermosets. Thermoplastics are formed by polymers with molecular weight (MW) of thousands of daltons that are easily moldable at a certain temperature, but become solid after cooling. These PMs can be reformed if heated to the proper temperature again (Table 3.1). In contrast, the chemical bonds in thermosets become irreversible during the polymerization process. Thermosets used in food contact materials are mainly coatings (*e.g.* epoxy resins, acrylic resins) and adhesives (*e.g.* polyurethane). Both groups are used in the manufacture of FCMs.

Meeting both the food protection and consumer goals for food packaging can require building a wide variety of specialized characteristics into PMs. This chapter examine the additives used in the manufacture of FCMs to achieve these goals. The reader will become familiarized with types of additives used in FCMs and the different properties they have and learn about analytical techniques for analyzing FCMs using mass spectrometry.

3.2 Additives

Food additives are defined partially by exclusion in the US Code of Federal Regulations (CFR) as "all substances not exempted by section 201(s) of the Act, the intended use of which results or may reasonably be expected to result, directly or indirectly, either in their becoming a component of food

Table 3.1 List of main thermoplastics used in FCMs.

Polyethylene (PE)	Poly(4-methylpentene-1) (P4MP1)
Polypropylene (PP)	Ionomers
Polystyrene (PS)	Poly(vinylidene chloride) (PVDC)
Poly(vinyl chloride) (PVC)	Poly(methyl methacrylate) (PMMA)
Poly(ethylene terephthalate) (PET)	Polysulfone (PSU)
Poly(butylene terephthalate) (PBT)	Poly(ether sulfonate)
Polycarbonate (PC)	Polyamide (PA)
Polybutene-1 (PB-1)	Poly(vinyl ether)
Polyisobutene	Tritan copolyester
	Polytetrafluoroethylene

or otherwise affecting the characteristics of food."[1] According to European Union (EU) regulations, an additive is "a substance which is intentionally added to plastics to achieve a physical or chemical effect during processing of the plastic or in the final material or article; it is intended to be present in the final material or article".[2] Additives may have a synthetic origin, such as bis(2-ethylhexyl) phthalate (DEHP), or a natural origin, such as epoxidized soybean oil (ESBO) and α-tocopherol. Although these additives belong to a wide variety of chemical groups (*e.g.* amines, carboxylic acids, esters, alcohols), each brings specific properties (different polarity, molecular weight, *etc.*) and confers certain desired properties to a completed FCM. Therefore, we classify additives according to the type of property that each imparts to the final polymer.

3.2.1 Types of Additives

3.2.1.1 Antioxidants

One of the most important categories of additives is antioxidants. Oxygen can have a detrimental effect on polymers, since the process of oxidation can degrade polymer structures. Therefore, antioxidants are used to decrease that oxygen degradation. Antioxidants can be classified into two types, depending on their mechanism of action: (1) antioxidants that capture free radicals, which are mainly phenols and secondary aryl amines, and (2) decomposers of peroxides, namely organophosphates and thioesters. Both types of antioxidants can be used together to provide a synergistic effect.

3.2.1.2 Plasticizers

Plasticizers are one of the most frequently used additives, as they enhance the flexibility and processability of PMs. For example, plasticizers decrease the glass transition temperature (T_g), the melt viscosity and the stiffness of the polymers without changing the chemical structure of the PM. In certain polymers, such as poly(vinyl chloride) (PVC), plasticizers may constitute up to 40% of the weight of a PM.[3] Although some studies have suggested that certain plasticizers may have adverse effects on the human body, such toxicological effects are disputed.[4]

3.2.1.3 Nucleating Agents/Fillers

Nucleating agents are used during polymer synthesis to influence the growth of the polymeric structure and enhance the crystallinity of the PM. These agents are highly important, as the initial polymer structure helps determine the structural properties of the finished PM. In addition, some nucleating agents may be used as fillers to enhance the physical and mechanical properties of PMs. Talc and silicon dioxide are examples of nucleating agents that are also fillers.

3.2.1.4 UV Absorbers

Ultraviolet (UV) light can generate the formation of free radicals that degrade polymer properties, resulting in a loss of color and flexibility of the PM. The presence of free radicals can also cause polymers to become unstable, allowing unwanted oxygen, humidity or contaminants into food packaging. UV absorbers can reduce these kinds of damage by preventing UV radiation from interacting with the polymers. UV absorbers can also be added to protect the food from light. Examples of substances used as UV absorbers are Tinuvins (*e.g.* Tinuvin P) and Chimassorbs (*e.g.* Chimassorb 81).

3.2.1.5 Lubricants

These additives are mainly used to reduce the friction of the polymers, which improves the synthesis and conversion to the final package. Fatty acid amides, such as erucamide and oleamide, are among the most commonly employed lubricants in FCM manufacture.

3.2.1.6 Antimicrobials

It is very important to prevent the growth of bacteria on the surface of FCMs, as established bacterial colonies could be transmitted onto food and consequently to consumers. An additional concern is that microorganisms could damage the structural integrity of the polymers, leading to performance failures in the FCM. Antimicrobials protect FCMs by preventing the growth and/or by killing microorganisms.

3.2.1.7 Optical Brighteners

Consumers are very strict when selecting their products and flaws in product appearance will affect those choices. No one would buy a bottle of water if the container had a yellow discoloration due to aging or light exposure; such features may raise questions about overall product quality. Therefore, additives such as 2,2′-(2,5-thienediyl)bis[5-(2-methyl-2-propanyl)-1,3-benzoxazole] (commonly known as Uvitex OB) serve as optical brighteners to protect polymers against discoloration, yellowing and opacity caused by external agents (*e.g.* UV light). By allowing the finished PM to reflect more visible light from the desired areas of the spectrum, optical brighteners also improve the overall appearance of plastics.

3.2.1.8 Other Types of Additives Commonly Used in FCMs

Many additional types of additives are now available to impart other unique properties to polymers. For example, nanoparticles are becoming widely used in FCMs. Nanoparticles can be tuned in size and shape to provide functionality as antimicrobials, nucleating agents, fillers, *etc.* Because nanoparticles also have high MW, they do not tend to migrate from their original place of deposition. Other additives commonly used in the manufacture of FCMs are surfactants, accelerators of vulcanization, heat stabilizers and anti-acids.

3.2.1.9 *Printing Inks: Photoinitiators*

Packaging manufacturers know that it is more appealing for consumers to buy a precooked food wrapped in a shiny or otherwise attractive shape, featuring a photograph of the food it contains. In addition, that packaging must be properly labeled with nutrition information. Both of these goals require effective printing inks. Photoinitiators such as benzophenone are widely used as curing substances in printing inks; these substances react to UV light, thereby activating the polymerization and curing process of those inks. The use of photoinitiators implies a low cost and are considered environmentally friendly because of the absence of organic solvents in their formulation. There are two types of photoinitiators:

Type 1. These compounds break down as a result of exposure to UV light according to the following scheme:

$$AB \xrightarrow{\text{UV light}} A* + B*$$

Type 2. These compounds absorb energy from UV light and thereby form an excited molecule according to the following scheme:

$$AB \xrightarrow{\text{UV light}} AB*$$

These excited molecules will interact with the monomers to start the curing process and polymerization of the PM.

Photoinitiators have attracted the attention of FCM scientists because of their potential migration into food *via* the gas phase, direct contact, and set-off (meaning ink contamination from the printed surface to the internal layer when stored), and because of food recalls due to possible contamination.[5,6,70] Photoinitiator compounds have not yet been specifically regulated in most countries, including the USA and EU.

3.3 Non-intentionally Added Substances (NIASs)

Understanding non-intentionally added substances (NIASs) is important to the field of FCMs. When we talk about monomers and additives in the manufacture of PMs, we are basically talking about chemical reactions that occur during manufacturing processes. While the forming polymers are subjected to controlled pressures and temperatures to achieve specific effects, chemical side reactions may occur that form unknown substances; to perform appropriate risk assessments, these substances need to be identified properly. NIASs may come from the following:

1. Incomplete reactions producing polymer chains, which may lead to very short-chain substances called oligomers.

2. Complex substances that may have degraded into smaller ones. For example, some oligomers may suffer hydrolysis, thereby yielding multiple smaller oligomers in the final FCM that were not present at the beginning of the process and sometimes not expected in the end product.
3. Impurities that may be present in the starting material. For example, most of the starting substances are not 100% pure. Therefore, if the reagent used is 99% pure, the remaining 1% consists of impurities that may be present in the final PM.
4. Substances coming from other materials. For example, recycled materials potentially contain extraneous substances remaining from their previous uses.
5. Substances coming from the machinery, storage containers or exposure to other contact materials in the manufacturing process.
6. Substances generated from side reactions. For example, an unreacted oligomer could react with an impurity introduced from a starting substance to form another compound of unknown nature.
7. Compounds resulting from the degradation of monomers or additives as a consequence of their desired actions. For example, in the process of setting printing inks, new substances maybe formed by Type 1 photoinitiators after UV interaction.
8. Compounds formed by unreacted substances that have migrated and reacted with food compounds, thereby forming new unknown substances that may be difficult to detect and identify.

In conclusion, NIASs may be present in the final FCM and food and therefore may represent a risk to consumers. It is therefore necessary to identify these substances and perform risk assessments as part of a safety evaluation.

3.4 Migration from FCMs into Food or Simulants

Whenever two surfaces are placed in direct contact, a process of mass transfer always occurs. This mass transfer is governed by Fick's second law of diffusion. This is especially important when evaluating FCMs, as all the substances used in the manufacture of FCMs, and also the products of their reaction, impurities, contaminants, *etc.*, could potentially migrate into food and therefore, potentially represent a food safety issue. To determine the extent of this migration, food safety researchers are permitted by several legislations to use food simulants, such as water, alcoholic mixtures and vegetable oils. This is allowed because food analysis is often a complex and expensive task, and wasting food to assess migration patterns is inappropriate. Therefore, to calculate migration concentrations, FCMs are usually exposed to simulants for different periods of time and at different temperatures. Generally, the most common migration test conditions are 10 days of exposure at 40 °C.

There is another type of migration that must also be considered, which occurs with no direct contact. Some substances, depending on their physical and chemical characteristics, such as the boiling point and vapor pressure, can migrate through the vapor phase. This is a phenomenon that is particularly relevant for components of printing inks and that should be considered when carrying out risk assessment in foods.

3.5 Legislation on FCMs

Many countries have legislation regarding FCMs. European and US forms of legislation are frequently used as reference points. The EU describes the basic requirements for FCMs in EU 1935/2004.[7] Within this legislation frame, probably, the most important is Regulation 10/2011 on plastic materials,[2] which provides a positive list of authorized compounds that can be used for the manufacture of plastic materials. All these compounds must comply with specific migration limits (SMLs) and the appropriate migration tests, simulants, storage and use conditions are also described. Other FCMs are also governed by this legislation frame.[8–11] Although there are no specific European regulations for coatings, these must comply with the good manufacturing practice (GMP) established in Regulation 2023/2006[12] and with the frame Regulation 1935/2004.[4]

In the USA, all FCMs or food contact substances (FCSs) must be approved by the Food and Drug Administration (FDA) before being placed on the market. Two systems govern the approval process in the USA. The first is the 21 Code of Federal Regulations (CFR) (Part 170–199),[1] which provides standard definitions for food additives, thresholds of regulation and so on. The CFR specifies which types of materials, such as coatings and paper, may be used in the manufacture of the FCM, along with the testing requirements and use limitations (if any) for each. Provided that materials comply with these specifications and tests, the FCM is considered regulated for use. The second is through the Food Contact Notification Process (FCN). The FCN is a proprietary process that is specific for the notifier only. Guidance documentation provides details, including migration testing and analytical methods, that can be used by manufacturers in order to comply with FDA chemical requirements.[13] A petitioner interested in having an FCM or FCS approved for use must first supply chemical, environmental and toxicological information, along with the appropriate administrative documentation, to the FDA for review. Once the material has been reviewed and approved, it can be placed on the market.

3.6 Analysis of Potential Migrants

3.6.1 Selection of Model Substances

As mentioned previously, additives added to PMs serve important purposes. However, it is possible that these additives may migrate into foods. The extent to which this occurs is likely dependent upon the storage and conditions of use (mainly time of exposure and temperature). Therefore, product testing to determine which substances are migrants and the extent of their migration is required to ensure food safety. Such analyses can be performed on either the material or the food, although only migration concentrations in food and food simulants can be used for risk assessments or to estimate consumer exposure. Nonetheless, analyzing FCMs may be useful for analytical chemists, as using these for identification will avoid the interference of matrix effects from foods. Mathematical models can be used to apply the Fick's second law and Crank equations to estimate migration into food. However, these estimates must be supported by experimental data.

Obviously, there are hundreds of additives, impurities, contaminants and NIASs that could be present in FCMs and might potentially migrate into foods. These may belong to a variety of chemical groups (*e.g.* amines, alcohols, acids) and may have differing chemical properties (*e.g.* MW, polar surface). In addition, there is a wide range of FCM matrices, including plastics (PET, PVC), paper, ceramics and glass; these substances may be in contact with foods with a range of chemical compositions (*e.g.* fatty, aqueous, high or low acidity). From an analytical point of view, it would require many chapters to detail the complete range of analyses for additives in FCMs and foods. However, in this chapter we will just describe a research-based methodology that may be used as a guideline to the analysis of both additives and NIASs. For this purpose, we selected several model compounds that could migrate from FCMs into foods. These model compounds exhibit different chemical groups, MW, polarity and chemical properties. The compounds and their chemical properties, classified by MW, are presented in Table 3.2. We will describe these compounds and why we selected them as models, then discuss sample preparation and proceed to analysis using chromatography and mass spectrometric (MS) detection. We will also discuss the application of MS techniques for the identification and determination of NIASs, and present information about the methods developed by other researchers and the results they obtained.

Titanium dioxide (TiO_2) can be used in a vast range of products including food and food packaging. It also serves a wide range of purposes, including as a nucleating agent, an antimicrobial and an absorber of UV radiation. Because of these multiple uses, and because it also has extensive uses in its nanoparticle forms, the production and use of TiO_2 has been increasing. It is listed as a color additive exempt from certification in the CFR[14] and it is also an approved food additive (E171) in the EU.[15] Here, we will use TiO_2 as the model for migrants that have *lower MW* and that can be used as nanoparticle. The detection of this potential migrant will be mostly performed using inductively coupled plasma mass spectrometry (ICP-MS) (Figure 3.1).

2-Methyl-4-isothiazolin-3-one (MI) (Figure 3.2) was selected to represent *antimicrobial* additives. It is used in a variety of cosmetics and also in some FCMs because it inhibits the growth of fungi and bacteria, thus preventing product contamination. It is one of the two compounds that we selected that contains sulfur in its chemical structure.

Table 3.2 Selected model compounds and chemical properties.[a,c]

Compound	Type of additive	SML (mg kg^{-1})	CAS No.	MW	Bp (°C)	Mp (°C)	Vp (mm Hg)	Log *P*
Titanium dioxide (TiO_2)	Nucleating agent	60	13463-67-7	79.9	2900	1840	0 (approximate)	2.23^b
2-Methyl-3-isothiazolinone (MI)	Antimicrobial agent	0.5	2682-20-4	115.2	182.8 ± 23.0 at 760 mm Hg^b	43	0.8 ± 0.3 at 25 °C^b	8.24×10^{-2b}
Benzophenone	UV absorber/photoinitiator	0.6	119-61-9	182.2	305	49	1.93×10^3 at 25 °C^b	3.18
Butylated hydroxytoluene (BHT)	Antioxidant	3	128-37-0	220.4	264	71	5.16×10^{-3}	5.10
Erucamide	Lubricant	60	112-84-5	337.6	474.2 ± 14.0 at 760 mm Hg^b	79	7.13×10^{-8b}	8.08^b
Bis(2-ethylhexyl) phthalate (DEHP)	Plasticizer	1.5	117-81-7	390.6	459.846 at 760 mm Hg	−55	1.42×10^{-7} at 25 °C	7.6
Uvitex OB	Optical brightener	0.6	7128-64-5	430.6	531.2^b	201	4.26×10^{-10} at 25 °C^b	8.98
Irganox 1076	Antioxidant	6	2082-79-3	530.9	568.1 ± 45.0 at 760 mm Hg^b	53	1.11×10^{-8} at 25 °C^b	11.9^b
Perfluorododecanoic acid (PFDoA)	Oil and moisture repellents	Not listed	307-55-1	614.1	246.292 at 760 mm Hg	108	1.14×10^{-2b}	8.3^b
Epoxidized soybean oil (ESBO)	Plasticizer	60/30	8013-07-8	975.4	885.6 ± 30.0 at 760 mm Hg^b	-	0.0 ± 0.3 at 25 °C^b	16.43^b

[a]SML, Specific migration limit;[2] MW, molecular weight; Bp, boiling point; Mp, melting point; Vp, vapor pressure.
[b]Predicted.
[c]Sources: Pubchem, Chemspider and EPA.

Figure 3.1 Structure of titanium dioxide.

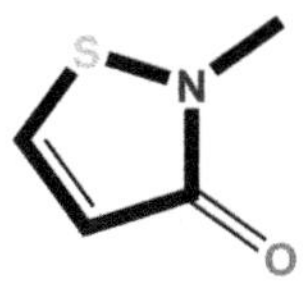

Figure 3.2 Structure of 2-methyl-4-isothiazolin-3-one.

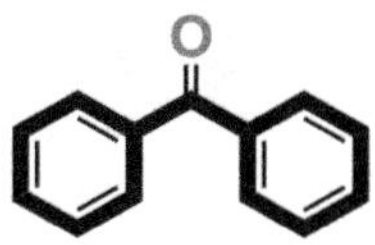

Figure 3.3 Structure of benzophenone.

Benzophenone (Figure 3.3) is a common additive used in food packaging. It is included in the CFR and in the positive list of the EU. It can have multiple uses (*e.g.* as a synthetic flavoring, plasticizer in rubbers, UV stabilizer) and it can also be used as a *Type 2 photoinitiator*, meaning that after interaction with UV light, benzophenone returns to its initial state. Several publications have demonstrated that benzophenone can migrate through the gas phase, which has been linked to its vapor pressure.[16] The analysis of benzophenone is very interesting from a migration point of view.

Butylated hydroxytoluene (BHT) (Figure 3.4) is a primary *antioxidant* approved for use in FCMs and is also approved for direct addition to foods. It acts as a scavenger of free radicals and thereby prevents PM oxidation. It has been analyzed in several types of polymers (*e.g.* polypropylene, polyethylene) and is widely used both as a model migrant and as an internal standard for analyses using gas chromatographic (GC) techniques.

The category of *plasticizers* has attracted the attention of researchers and food safety professionals because some plasticizers have demonstrated high rates of migration into foods.[17] To demonstrate the range of additives that serve as plasticizers, we have selected two plasticizers that differ in both functional groups and MW. The first is *bis(2-ethylhexyl) phthalate (DEHP)*, a common phthalate (Figure 3.5). The second is *epoxidized soybean oil (ESBO)*, a plasticizer derived from natural products (Figure 3.6). From an analytical point of view it is an interesting compound because of its high MW (975.4) and its low solubility in aqueous matrices ($\log P = 16.43$).

The chosen representative of *slip agents* is *erucamide*, a fatty acid amide (Figure 3.7). It is one of the most common slip agents used in plastics. Despite its MW (337.6), it has a $\log P$ of 8.08, most likely because it has a long aliphatic chain with an unsaturation at C_9–C_{10}. This combination of properties makes erucamide an interesting subject for analyses.

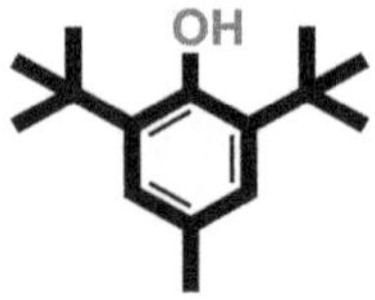

Figure 3.4 Structure of BHT.

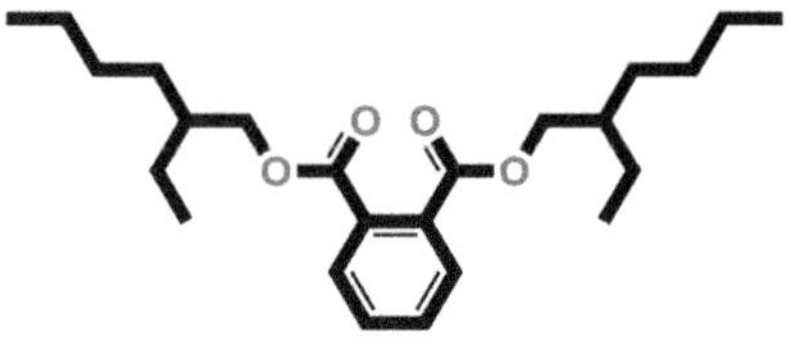

Figure 3.5 Structure of DEHP.

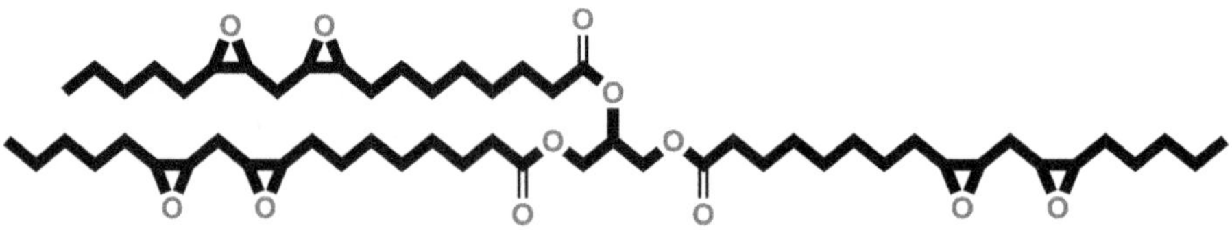

Figure 3.6 Structure of ESBO.

Figure 3.7 Structure of erucamide.

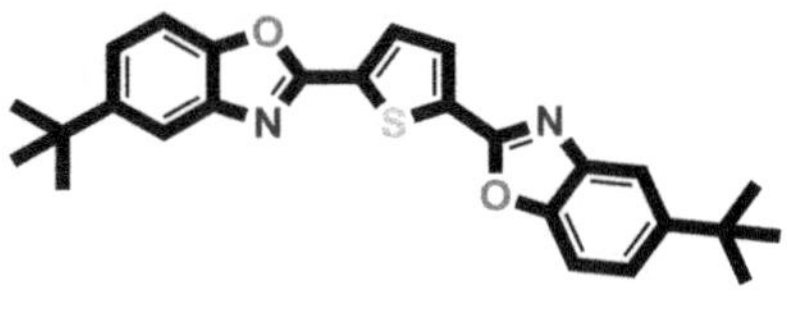

Figure 3.8 Structure of Uvitex OB.

Uvitex OB is an *optical brightener* having a low polarity (Figure 3.8). It has been widely determined in FCMs[18–20] and also as a surrogate for modeling migration. It has an SML of 0.6 mg kg^{-1} in the EU.[2]

Octadecyl 3-(3,5-di-tert-butyl-4-hydroxyphenyl)propanoate (commonly known as Irganox 1076) is an antioxidant commonly used in FCMs to prevent polymer degradation due to oxygen exposure (Figure 3.9).

Fluorochemicals are substances that may be used as *oil and moisture repellents* in a wide array of industrial products, including some FCMs. Because

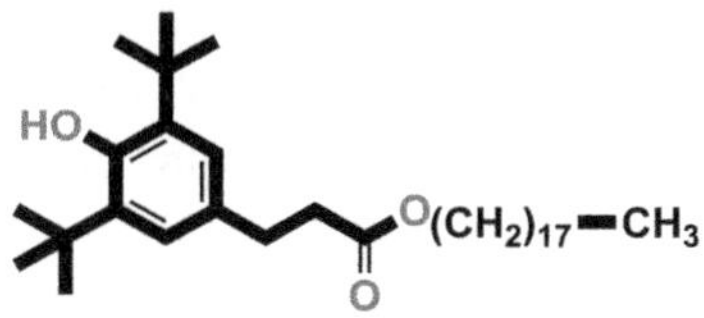

Figure 3.9 Structure of Irganox 1076.

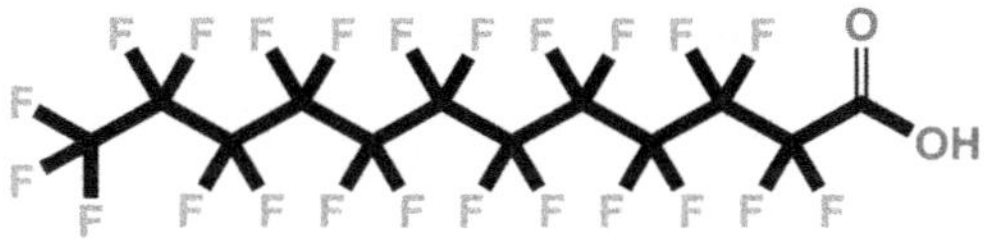

Figure 3.10 Structure of perfluorododecanoic acid.

they are usually bio-persistent has raised concerns in the scientific community.[21] We selected *perfluorododecanoic acid (PFDoA)* as a model because of its chemical structure (it is an acid containing fluorine) and its relatively high MW (Figure 3.10). Perfluorododecanoic acid is a contaminant in FCM production and is not included on the positive list of accepted additives in EU Regulation 10/2011 and is also not listing by the FDA.

3.6.2 Sample Preparation

3.6.2.1 Extraction: Selecting Appropriate Solvents

Extraction is the first step in separating target compounds from the matrix and determining the concentrations present in FCMs or food. This process requires identifying and using the appropriate solvent for this purpose. Each target and each matrix may have different requirements and there are many solvents available (acetonitrile, hexane, tetrahydrofuran, toluene, *etc.*). To choose the ideal solvent for specific needs, the following should be considered:

- Know the composition and properties of the FCM concerned. An expert in food packaging might easily recognize the type of FCM. If properly labeled, the number of the plastic provided on the surface of the FCM (1 for PET, 2 for HDPE, *etc.*) can also identify the type of polymer used. Nonetheless, as scientists we must confirm these observations to ensure that we are applying the appropriate analytical techniques. Infrared spectroscopy techniques are usually the most suitable for this task and Fourier transform infrared (FTIR) spectroscopy is most often the method used to identify polymers. Once these data have been acquired, identification of the material is a relatively simple task of comparing the results obtained with libraries containing spectra of comparator polymers, coatings, rubbers and other known FCMs. Then the appropriate solvent for that FCM can be selected.

- The compound of interest (the potentially migrating additive) also must be soluble in the solvent selected. As a rule of thumb, polar solvents dissolve polar compounds and non-polar solvents dissolve molecules with low polarity.
- The food or food simulant must also be soluble or at least partially soluble in the solvent selected. The choices here may be more intuitive: if a food is liquid or solid, if it is fatty or aqueous, these features can guide the choice of solvent. For example, a fatty food (such as butter or oil) would be soluble in a non-polar solvent such as hexane, rather than a polar solvent such as water.

Once the solvent that meets these criteria have been identified, one must then make sure that the analytical technique of choice will work with that solvent. For example, hexane is an appropriate solvent for GC, but is not useful for reversed-phase liquid chromatography (LC). Additional parameters may be adjusted to improve the efficiency of the extraction process, including time, temperature, mixing method and the use of ultrasound. The next three subsections provide examples of extraction procedures applied to FCMs, food simulants and foods.

3.6.2.1.1 Extraction from FCMs. Several authors have used acetonitrile or methanol to extract a wide variety of chemicals from different FCM matrices. These solvents have good extraction properties and they are compatible with both LC and GC. Lin *et al.* used methanol to extract MI from paper.[22] Methanol has also been used to extract erucamide and Irganox 1076 from polyethylene (PE) films/multilayer FCMs.[23] Lago *et al.* used acetonitrile to extract photoinitiators, including benzophenone, from can linings, paper, plastic and cardboard.[24] On the other hand, a standard test method for the determination of antioxidants and erucamide slip additives in PE used isopropanol or cyclohexane as extraction solvent.[25]

Organic solvents, such as dichloromethane (DCM) and hexane, are appropriate for extracting a wide variety of chemicals, providing high volatility and appropriate expansion volumes for optimal GC analyses. For example, BHT, benzophenone, Uvitex OB and Irganox 1076 have all been successfully extracted from low-density PE (LDPE) using DCM,[20] demonstrating how DCM can extract additives that have different chemical structures, MW and polarity.

As mentioned earlier, ultrasound can be used to assist extraction procedures. Soxhlet extraction uses ultrasound in combination with pressurized solvent extraction with isopropanol and chloroform; Lv *et al.* used this technique to determine fatty acid amides and erucamide in PE packaging film.[26] There are cases in which extraction of the desired target additive will require the additional step of digesting the FCM matrix. This is the case for the extraction of TiO_2 nanoparticles. Lin *et al.*[27] used HNO_3, H_2O_2 and hydrofluoric acid for microwave-assisted digestion of PE films. Then the samples were diluted with water for analysis by ICP-MS.

3.6.2.1.2 Extraction from Food Simulants. Usually food simulants are relatively easy to analyze, since some of them can be injected directly into the chromatographic system without sample preparation or after a simple liquid–liquid extraction. Vaclavikova *et al.*[28] directly analyzed aqueous food simulants (*e.g.* water) to determine migrant additives, using ultra-high-performance liquid chromatography–high-resolution mass spectrometry (UHPLC-HRMS). Beldi *et al.*[29] also analyzed Irganox 1076 in isooctane (a fatty food simulant) without any sample preparation by GC-MS. However, for other food simulants, such as vegetable oils (fatty food simulants) or poly(2,6-diphenylphenylene oxide) or Tenax® (an EU-recommended dry food simulant), an additional extraction process is usually required. For example, to extract MI from Tenax, Lin *et al.* used methanol and ethanol and obtained good recoveries (up to 95%).[30] For fatty food simulants, Monteiro *et al.* used a Bio-Beads S-X3 column and cyclohexane–dichloromethane (70 : 30 v/v) for BHT determination.[31]

3.6.2.1.3 Extraction from Food. Analysis of food is usually tedious, complicated and expensive, as there are multiple complex matrices that must be managed to perform effective extractions. However, sometimes it is the only way to determine the concentrations of migrating additives in foods. Here we describe several methods that have been found to be effective for extracting our model targets.

Microwave digestion allowed satisfactory recoveries (81–87%) for the analysis of TiO_2 in several types of food.[32] Experiments to analyze the potential migration of printing ink additives into foods employed acetonitrile, DCM and hexane to extract benzophenone from several samples, obtaining excellent recoveries (97%).[33] Extraction of BHT from vegetable oil and bread spread using *n*-hexane, acetonitrile and 2-propanol has also been successful.[34] Nakashima *et al.*[35] used methanol and water for the determination of BHT, erucamide and DEHP in processed food by GC-MS. For the determination of Uvitex OB in chocolate, diethyl ether and acetonitrile were used for analysis by UHPLC-HRMS.[36] Aqueous and organic solvents, in combination with transesterification, enabled Suman *et al.* to extract ESBO, typically used in jar lid gaskets, from pesto and tomato food sauces.[37]

3.6.2.2 QuEChERS

Since 2003, the "quick, easy, cheap, effective, rugged and safe," or QuEChERS, method has been used to facilitate chemical analyses.[38] Although initially applied to pesticide analysis, QuEChERS is also useful for food analyses and can be applied to the determination of additives in different types of matrices. For example, Vavrous *et al.* used QuEChERS to extract benzophenone to analyze it for multiple additives from paper, obtaining recoveries from 99 to 120%.[39] Xu *et al.* used a similar protocol for determining DEHP in several food samples, obtaining good recoveries (81–107%).[40]

3.6.2.3 Cleanup and Concentration of the Sample

The process of extracting target compounds can also extract other compounds that may have similar MW as the targets(s), causing difficulties in the analysis. Cleanup procedures are necessary to remove these extraneous materials to prevent interference with the remainder of the analysis. Cleanup procedures also help prevent dirty samples from disrupting the sensitivity of a given method and/or damaging expensive instrumentation. Certain cleaning methods, such as solid-phase extraction (SPE), may be also used for removing or reducing matrix effects. Suman *et al.* used a C_{18} SPE for cleanup for ESBO determination in food samples.[37] However, there are cases in which target analytes are present not only in the samples, but also already in laboratory materials and/or analytical instruments. To avoid sample contamination caused by such compounds, guard columns may also be used for this purpose. For example, Hayasaka used a hold-back column to separate phthalates from different contamination sources.[41]

Just as it is important to eliminate non-target compounds, it is also sometimes necessary to make a phase change in the solvent (sometimes the extraction solvent may not be directly compatible with the chromatographic system needed for analysis) or to concentrate the sample to obtain the appropriate level of sensitivity. This can be performed by evaporating the solvent under a stream of nitrogen and then dissolving the compound(s) of interest in a more appropriate solvent. It is also possible to use SPE to retain the compound of interest and then elute it with a more appropriate solvent and also more concentrated to obtain the appropriate sensitivity.

3.6.2.4 Filtration

Samples are typically filtered prior to being injected into a chromatographic system, often using PTFE membranes, to improve the analysis and prevent contaminants from entering the system. All filtration devices should be checked for possible sources of analyte contamination in order to prevent false-positive results.

3.6.3 Mass Spectrometry for Determination and Identification of Migrants

First developed in 1899 by Thompson, MS has become one of the most important tools in the field of analytical chemistry.[42] The basic procedure generates multiple ions from a given sample, which are then separated according to their particular mass-to-charge ratio (m/z) and their relative abundance is recorded by the instrument. To improve method performance and to minimize errors, internal standards often used. Refinements to MS instrumentation and techniques have revolutionized this analysis and the detection and identification of substances. The most common separation techniques used with MS are GC (GC-MS) and LC (LC-MS). In these techniques, the interface between the chromatograph and the detector acts as the ion generator. However, other ion sources can also be applied to analyses of additives in FCMs, including electrospray ionization (ESI), atmospheric pressure chemical ionization (APCI), matrix-assisted laser desorption/ionization (MALDI), chemical ionization (CI), electron ionization (EI) and inductively coupled plasma source (ICP). More novel techniques, such as atmospheric pressure secondary ion MS, direct analysis in real time (DART) and desorption electrospray ionization (DESI), also have their place in the analysis of FCMs and potential migrants.

Several mass analyzers are especially appropriate for determining migrants from FCMs; the most frequently used are quadrupoles, ion traps, Orbitraps and time-of-flight analyzers. These mass detectors are generally sensitive, reliable, robust and selective. Tandem mass spectrometry (which enhances the detection of m/z ratios and improves the ability to obtain products from the parent product) and high-resolution MS (which can achieve up to four decimal places of mass accuracy) are other valuable separation techniques. It is important to understand that there is no single "best" technique for determining additives in FCMs. Depending on the specific chemical properties of a substance, the appropriate separation technique will be different. What is appropriate for volatile substances (GC-MS) is different from what is appropriate for non-volatile substances (LC-MS). As one might predict, semivolatile substances may be analyzed by either GC-MS or LC-MS. In the following sections, we will explore MS methods in more detail.

3.6.3.1 GC-MS

Detection of volatile and semivolatile compounds is best performed using GC-MS. As shown in Table 3.3, helium is the most frequently used carrier gas, at flow rates typically ranging from 1 to 1.3 mL min^{-1}. Alternatively, Fujii *et al.* used methane as a reagent gas at a rate of 2 mL L^{-1} for the analysis of PFDoA, using electron-capture negative ionization (ECNI).[43] Splitless or slip mode were both used and the injector temperatures were up to 280 °C. The oven temperature ranged from 40 to 350 °C. The most commonly used columns for GC-MS are DB-5MS and HP-5MS (30 m × 0.25 mm id, 1 μm film thickness). Notable examples of GC-MS analysis include the use of Rtx-1 (60 m × 0.25 mm id, 0.25 mm film thickness) for analyzing benzophenone[44] and DB-17 HT to determine Irganox 1076.[29] Motas *et al.* used a DB-35-MS column to determine the presence of PFDoA and other perfluorinated carboxylic acids (PFCAs).[45]

Typical settings used in GC-MS analyses are presented in Table 3.4. Current MS devices usually have the appropriate sensitivity to detect target migrants in the matrices of interest, allowing one to assess whether the presence of those migrants is within acceptable limits. The first decision to make with these settings is whether to perform a full spectrum scan (scan mode) or to use the selected ion monitoring (SIM) mode. The SIM mode is more

Table 3.3 GC settings employed for the analysis of the model compounds in the reviewed literature.

Compound	Mobile phase	Flow rate (mL min^{-1})	Column	Oven temperature (°C)	Injector[a]	Reference
2-Methyl-3-iso-thiazolinone	Helium	1.2	DB-5MS (30 m × 0.25 mm id, 1 μm)	50–280	IT: 270	Rafoth *et al.*, 2007[63]
Benzophenone	Helium	—	HP-5MS (30 m × 0.25 mm id, 0.25 μm)	100–300	Inlet: 280 °C Splitless	Lloyd *et al.*, 2012[33]
	Helium	1.2	Rtx-1 (60 m × 0.25 mm id, 0.25 μm)	50–280	IT: 250 Splitless	Anderson and Castle, 2003[44]
	Helium	1.3	DB-5MS (30 m × 0.25 mm id, 0.1 μm)	100–250	IT: 250 Split 1:40	Koivikko *et al.*, 2010[64]
Butylated hydroxytoluene	Helium	110 kPa	DB 1701 (60 m × 0.25 mm id, 0.25 μm)	100–280	IT: 280	Monteiro *et al.*, 1999[31]
	Helium	1.3	HP-5MS (30 m × 0.25 mm id, 0.25 μm)	70–325	IT: 280 Splitless	Veyrand *et al.*, 2017[65]
Erucamide	Helium	1	HP-5MS (30 m × 0.25 mm id, 0.25 μm)	200–250	—	Lv *et al.*, 2009[26]
DEHP	Helium	1	DB-5MS (30 m × 0.25 mm id, 0.25 μm)	100–300	—	Chatonnet *et al.*, 2014[66]
Irganox 1076	Helium	1.2	DB-17 HT (30 m × 0.25 mm id, 0.15 μm)	70–350	Splitless	Beldi *et al.*, 2012[29]
PFDoA	Helium	1	DB-35 MS (30 m × 0.25 mm id, 0.25 μm)	40–170	IT: 250 Splitless	Motas Guzmàn *et al.*, 2016[45]
	Helium–methane	Methane: 2 mL min^{-1}	DB-5MS (30 m × 0.25 mm id, 0.1 μm)	70–280	IT: 220 Splitless	Fujii *et al.*, 2012[43]
ESBO	Helium	1	DB-5MS (30 m × 0.25 mm id, 0.25 μm)	200–300	Inlet: 280 Split 20:0	Choi *et al.*, 2018[46]

[a]IT, injector temperature (°C).

Table 3.4 GC-MS settings for the model compounds and levels determined in samples.[a]

Compound	Detector settings	Spectral data (*m*/*z*)	LOD	Concentrations	Reference
2-Methyl-3-isothiazolinone	TL: 280 ST: 230	61, 87, 115	—	Not applicable	Rafoth *et al.*, 2007[63]
Benzophenone	TL: 280 ST: 280	MRM: 182–105/152/77	—	56 μg kg^{-1} in fish sample	Lloyd *et al.*, 2012[33]
	—	192, 110 182, 105	0.01 mg kg^{-1}	<0.05–3.3 mg dm^{-2} <0.01–7.3 mg kg^{-1}	Anderson and Castle, 2003[44]
	—	—	—	ND to 3.99 mg dm^{-2} in paper board	Koivikko *et al.*, 2010[64]
Butylated hydroxytoluene	ST: 280	SIM	—	—	Monteiro *et al.*, 1999[31]
	TL: 280 ST: 230	Full scan: 30–800	—	—	Veyrand *et al.*, 2017[65]
Erucamide	—	Full scan: 50–400 SIM: 57, 97, 124	105 ng g^{-1}	201–337 μg g^{-1}	Lv *et al.*, 2009[26]
DEHP	TL: 300 ST: 250	SIM	0.004 mg L^{-1}	15% of samples >0.01 mg kg^{-1}	Chatonnet *et al.*, 2014[66]
Irganox 1076	—	SIM: 530, 515	—	Migration higher in simulant	Beldi *et al.*, 2012[29]
PFDoA	TL: 300	SIM: 69, 131, 169, 181	0.6 ng μL^{-1}	16–26 ng L^{-1}	Motas Guzman *et al.*, 2016[45]
	ST: 230 (EI), 150 (ECNI)	Scan: 613, 704	0.4 (EI), 0.005 (ECNI) pg	ND to 17 ng per day	Fujii *et al.*, 2012[43]
ESBO	—	277, 309	1 μg L^{-1}	ND to 3 8 μg mL^{-1} PVC/PVDC: 0.2–7% w/w	Choi *et al.*, 2018[46]

[a]LOD, limit of detection; TL, transfer line, ST, source temperature; ND, not detected.

sensitive and can be used for compound confirmation. Matrix effects within the samples or the nature of the substances being analyzed for may affect the instrument sensitivity. Sometimes a derivatization process is required to achieve adequate sensitivity. For example, among the model substances, those having high MW (ESBO and PFDoA) usually require derivatization prior to GC-MS analysis.[43,45,46]

Once the analysis by GC-MS has been conducted, the concentrations migrants can be calculated and discussed. In general, concentrations in the FCMs are higher than in food. Anderson and Castle[44] determined the concentrations of the additive benzophenone ranging from <0.05 to 3.3 mg dm^{-2} in FCMs and from <0.01 to 7.3 mg kg^{-1} in food samples. It is important to note that to make comparisons between data for FCMs and food, it is commonly assumed that in the EU and the USA that 6 and 6.4 dm^2, respectively are in contact with 1 kg of food. Lv *et al.*[26] determined erucamide concentrations in PE from 201 to 337 μg g^{-1}. Beldi *et al.*[29] found that Irganox 1076 migration into simulants was higher than in real food. This was expected since simulants are supposed to represent the worst-case migration scenario. On the other hand, Carlos *et al.*[47] found that DEHP and ESBO concentrations in PVC ranged from 15 to 46%. These plasticizer concentrations are consistent with the literature.[3,4] ESBO does not migrate into polar food simulants such as water.[46] Similar results were observed in another migration study.[28] This is expected because ESBO has low polarity (predicted log $P = 16.43$) and high MW (975.4). The concentrations of additives found in the samples are usually low and most comply with specific legislations.

3.6.3.2 LC-MS

LC has become a routine technique for analyzing migrants in FCMs and also in foods and food simulants. It is also a powerful tool for compound identification. LC-MS analyses supply information about the mass of the target substance and about collision products from the parent ion of a molecule; these are very useful confirmatory techniques that can increase method sensitivity. Analyte determination is typically performed by reversed-phase LC-MS, using a non-polar C_{18} stationary phase column and a polar mobile phase such as water, acetonitrile or methanol, usually containing a volatile buffer (*e.g.* 0.1% formic acid) to control pH and enhance ionization and MS detection. Columns used for LC-MS can have different dimensions; as shown in Table 3.5, lengths of 50–250 mm and internal diameters of 2–4.6 mm are typical, with flow rates of 0.25–1.0 mL L^{-1}. The particle size ranges from 1.7 to 5 μm. These systems are usually temperature controlled between 20 and 40 °C to ensure reproducible retention times. Settings should be selected with care, as these can affect the extent of separation and the sensitivity of the analyses. For example, when the particle size is smaller, there is more interaction between the sample and the stationary phase, and also an increase in column pressure. Consequently, an increase in sensitivity caused

Table 3.5 LC settings employed for the analysis of the model compounds in the reviewed literature.

Compound	Column	Temperature (°C)	Mobile phase	Flow rate (mL min^{-1})	Interphase/mode	Settings[a]	Reference
2-Methyl-3-isothiazolinone	Aqua C_{18} 125 Å (250 × 4.6 mm id, 5 μm)	1	0.1% acetic acid in water–methanol	1	APCI+	CT-275	Speksnijder *et al.*, 2010[67]
	UPLC HSS T3 (50 × 2.1 mm id, 1.8 μm)	35	Methanol–water	0.3	ESI+	3.5 kV, N_2 = 350 °C	Lin *et al.*, 2011[30]
	UPLC HSS T3 (50 mm × 2.1 mm id, 1.8 μm)	35	Methanol–water	0.3	ESI+	3.5 kV, N_2 = 350 °C	Lin *et al.*, 2010[22]
Benzophenone	Kromasil C_{18} (150 × 3.2 mm id, 5 μm)	30	0.1% formic acid (methanol–water)	0.5	ESI+	3 kV, CT-340-350	Lago *et al.*, 2014[24]
Benzophenone Uvitex OB	Lune C_{18} (50 × 2 mm id, 3 μm)	40	5 mM ammonium acetate–methanol	0.4	ESI+/–	1.5 kV	Wu *et al.*, 2016[53]
BHT	RP-MS (100 × 2.1 mm id, 2.6 μm)	30	10 mM ammonium acetate (water–acetonitrile)	0.25	ESI–	4.8 kV, CT-300	Kim and Lee, 2016[34]
BHTErucamide-Irganox 1076	Pak C_{18} (150 × 3.9 mm id, 4 μm)		Methanol–water	1	APCI+/–	–	Block *et al.*, 2006[52]
DEHP	Poroshell 120 EC-C_{18} (100 × 4.6 mm id, 2.7 μm)	35	Water with 0.1% formic acid–methanol	0.7	ESI+	5.5 kV, AT = 550	Xu *et al.*, 2014[40]
	Synergi Hydro-RP HPLC (75 × 2 mm id, 4 μm) 80 Å	25	10 mM ammonium acetate–methanol	0.3	ESI+	4 kV, TG = 500	Hayasaka, 2014[41]
Uvitex OB	BEH C_{18} (50 × 2.1 mm id, 1.7 μm)	30	0.1% aqueous formic acid–methanol	0.3	ESI+	1 kV, N_2 = 400 °C	Wu *et al.*, 2018[48]
PFDoA	Varian PursuitXRs C_{18} (150 × 2 mm id, 3 μm)	35	2 mM ammonium acetate–methanol	0.3	ESI–	1 kV, N_2 = 300 °C	Young *et al.*, 2012[68]
	Poroshell EC C_{18} (150 × 3 mm id, 2.7 μm)	40	2 mM ammonium acetate (methanol–water)	0.4	ESI–	3 kV, SG = 325, DG = 200	Vavrous *et al.*, 2016[39]
	Zorbax Extend C_{18} (150 × 2.1 mm id, 5 μm)	35	2 mM ammonium acetate– methanol	0.3	ESI–	750 V	Xu *et al.*, 2013[49]
ESBO	Acquity UPLCTM BEH C_{18} (50 × 2.1 mm id, 1.7 μm)	30	10 mM ammonium formate–methanol	0.3	ESI+	3.5 kV, DT = 350	Suman *et al.*, 2010[37]

[a]DT is desolvation temperature, AT is atomization temperature, TG is turbo gas temperature, SG is sheath gas temperature

by the chromatographic conditions will occur at the expense of an increased column pressure.

In LC-MS, the most common ionization methods are ESI and APCI, which can be set in positive and/or negative ion modes. ESI has a wider range of applications and is therefore more frequently used. However, of these two ion sources, APCI is more useful for analyzing compounds that are not easily ionized, such as acrylics, which may be present in FCMs.

For these ionization techniques, parameters such as cone voltage, capillary temperature, capillary voltage and gas temperature can be modified to improve the ionization of the analyte and thereby improve the MS output. These settings are displayed, along with notable uses, in Table 3.6. Collision energy (CE) and cone voltage are useful for obtaining the products of additives. In the literature cited in Table 3.6, the cone voltage was set between 20 and 40 V. For our model compounds, the CE was set from 8 to 45 V. The capillary voltage is set between 3.5 and 5 kV, nitrogen is the desolvation gas and the source temperature is between 300 and 500 °C. MS may be used in different modes, from the full-scan mode to MS/MS.

The detection limits presented in Table 3.6 range from 0.04 to 9.8 μg L^{-1}. This is one of the main advantages of MS detection in LC, which show that it is a good technique for detecting trace concentrations of migrants. Regarding t the concentration of the analytes in the samples presented in Table 3.6, very low concentrations were determined for MI. Benzophenone was found at a concentration of 560 μg kg^{-1} in a fish sample. If the benzophenone were to have migrated from an FCM, the concentration would be compliant with the SML established by the EU.[2] Similarly, concentrations of BHT in vegetable oil and bread spread were found by Kim *et al.*[34] to range from undetectable (ND) to 62 mg kg^{-1}. Since BHT is also approved for use as a food additive, when detected in food it is difficult to prove that the concentration obtained is from migration from an FCM or from direct addition to food. Besides, Vaclavikova *et al.*[28] tested for possible migration of erucamide from vinyl can coatings into aqueous food simulants, under specific combinations of exposure time and temperature conditions, no migrant was found. These results are likely due to the low solubility of erucamide, a fatty acid amide, in the polar simulants used (log P = 8.08). Xu *et al.*[40] found 0.02–2685 mg kg^{-1} of the plasticizer DEHP in food samples. On the other hand, OB was determined in flour and the concentrations found were between 31 and 60 μg kg^{-1},[48] which is well below the SML for that substance in the EU.[2] For PFDoA, concentrations from 698 to 824 μg kg^{-1} were quantified in paper. Migration into food simulants was lower than 20%.[49] For ESBO, low concentrations were found in pesto and tomato sauce.[37]

3.6.3.3 Multi-method Analyses Using Both LC-MS and GC-MS

MS techniques are also a valuable tool for developing multi-analytical methods for the analysis of several additives. Many "multi-method" analyses based on either GC and/or LC can now be found in the literature. Of special note is the work of Seiler *et al.*, who used two GC methods for multi-analyte

Table 3.6 LC-MS settings of the model compounds and levels determined in samples.[a]

Compound	Sample	Concentration in sample	Analytical technique	LOD	Spectral data (m/z)	Settings	Reference
2-Methyl-3-isothiazolinone	Drinking water	<0.04 μg L^{-1}	HPLC-MS/MS. MRM	0.04 μg L^{-1}	116–101, 116–71	CE = 20–30	Speksnijder *et al.*, 2010[67]
	Tenax, paper	0.21–3.4%	UPLC-MS/MS. MRM	0.002 mg kg^{-1}	116–101, 116–71	CV = 40 CE = 14–18	Lin *et al.*, 2011[30]
	Paper	ND	UHPLC-MS/MS. MRM	0.002 mg kg^{-1}	116–101, 116–99	CV = 40 CE = 14–18	Lin *et al.*, 2010[22]
Benzophenone	Cans, paper, plastic, cardboard	ND to 15.2 μg dm^{-2}	HPLC-MS/MS. MRM	—	183–105, 183–77		Lago *et al.*, 2014[24]
Benzophenone, also Uvitex OB	Paper	ND	UPLC-MS/MS. MRM	0.5 μg L^{-1}	183–105, 183–77	CV = 20 CE = 15–25	Wu *et al.*, 2016[53]
BHT	Vegetable oil/bread spread	ND to 62 mg kg^{-1} in one sample	LC-MS/MS. SRM	—	219–203/163	CE = 27–32	Kim& Lee, 2016[34]
BHT, erucamide and Irganox 1076	LDPE	—	LC-MS	—	Scan: 220–1200	CV = 20, 40 CE = 20, 25, 30	Block *et al.*, 2006[52]
DEHP	Several food samples	0.02–2685 mg kg^{-1}	LC-MS/MS	<1 μg kg^{-1}	391–149, 391–113	Cluster: 27 0/250 V CE = 14–25	Xu *et al.*, 2014[40]
	White and red wine	1.7–4 μg L^{-1}	LC-MS/MS. SRM	6.1–9.8 μg L^{-1}	391–167, 391–279	CE = 11–35	Hayasaka, 2014[41]
Uvitex OB	Flour	31–60 μg kg^{-1}	UPLC-MS/MS. MRM	0.2 μg L^{-1}	431–415, 431–401	CE = 40–45	Wu *et al.*, 2018[48]
	Paper	—	HPLC-MS/MS. MRM	0.0032 mg kg^{-1} LOQ	613–269, 613–569	CE = 17–35	Vavrous *et al.*, 2016[39]
PFDoA	Paper/food simulants	Paper: 698–824 μg kg^{-1} Migration into simulants <20%	HPLC-MS/MS.	0.02 μg L^{-1} LOQ	613–169, 613–569		Xu *et al.*, 2013[49]
ESBO	Pesto and tomato food sauces	Low levels	UPLC-MS/MS. Scan/MRM	0.04 μg L^{-1}	Precursor ions: 363, 399. Products: 277, 291, 309, 345	CE = 10–17	Suman *et al.*, 2010[37]

[a]LOD, limit of detection; LOQ, limit of quantification; CE, collision energy; CV, cone voltage; MDL, method detection limit; ND, not detected.

determinations of migrants (including the model migrants benzophenone, Irganox 1076, BHT and Uvitex OB) in LDPE to establish a correlation between foodstuffs and food simulants.[20] Mutsuga *et al.*[50] tested several non-polar capillary columns (length 15.1–16.5 m) for the analysis of 96 additives using GC-MS in the scan mode (*m*/*z* 40–800) and obtained satisfactory results. Ezerskis *et al.*[51] used GC-MS in the SIM and full scan mode to analyze PVC gaskets and food samples for additives (including plasticizers). Vavrous *et al.* used GC and LC-MS techniques to analyze for PFDoA, benzophenone and several other substances of interest.[39] In 2006, Block *et al.*[52] successfully used LC-MS for the analysis of polymer additives (including Irganox 1076, erucamide and BHT) using an APCI ionization source in the positive and negative ion modes and a scan mode of *m*/*z* 220–1200. An ultra-performance LC (UPLC)-MS/MS method was used by Wu *et al.* for the analysis of whitening agents (such as Uvitex OB) and photoinitiators (*i.e.* benzophenone) in paper.[53] A Luna C_{18} stationary phase column (50 × 2 mm id, 3 μm film thickness) was used with 5 mM ammonium acetate–methanol as the mobile phase. An ESI source was used in the negative and positive ion modes and multiple reaction monitoring (MRM) was used for determination of the additives. Benzophenone was not detected in paper and Uvitex OB (referred as FWA 184) was detected in only one paper sample at a concentration of 151 μg kg^{-1}.

Overall, these studies provide ample evidence that chromatographic techniques coupled with MS are highly effective for multi-analyte analyses. Not only do these multi-analyses reduce the time and effort necessary to identify migrating substances, they also reduce costs, waste production and environmental impacts.

3.6.3.4 Inductively Coupled Plasma Mass Spectrometry (ICP-MS)

For the analysis of metal-containing nanoparticles, such as TiO_2, the appropriate detection technique is ICP-MS. Table 3.7 lists a selection of relevant studies that used this technique to analyze for titanium dioxide. Candas-Zapico *et al.*[54] analyzed the migration of TiO_2 in chewing gum, achieving a limit of detection (LOD) of 0.0019 μg L^{-1} at *m*/*z* 114, 0.0042 μg L^{-1} at *m*/*z* 131 and 0.006 μg L^{-1} at *m*/*z* 148. An analysis by Lin *et al.* of PE films to determine particle migration into aqueous food simulants achieved an LOD of 0.04 μg L^{-1} and identified migration of 249–255 mg kg^{-1}.[27] Maximum Ti migration amounts for the food simulant 3% acetic acid were 1.4, 6.3 and 12 μg kg^{-1} and for 50% ethanol 0.5, 0.6 and 0.1 μg kg^{-1} at 25, 70 and 100 °C, respectively. Weir *et al.*[32] detected a concentration of TiO_2 in coconut curd of 3.59 μg mg^{-1} and in the rest of the analyzed food samples concentrations from 0.00077 to 210 μg mg^{-1}.

3.6.3.5 High-resolution Mass Spectrometry (HRMS)

Advances in both instrumentation and computing power have changed our expectations of what "high resolution" can mean. Increasingly precise high-resolution mass spectrometers now provide accurate information on

Table 3.7 Analysis of titanium dioxide by ICP-MS.

Sample	Levels in sample	Extraction solvent	Recovery (%)	Analytical technique	Detector sensitivity	Spectral data (*m*/*z*)	Reference
Chewing gum	No. of particles: 6.75×10^{11}–7.28×10^{11}	Water	82	ICP-MS	Size detection limit: 26 nm 0.006 μg–0.0042 μg L^{-1}	114, 131, 148	Candas-Zapico *et al.*, 2018[54]
Water	0.602 ng mL^{-1}	Microwave-assisted digestion	—	Single-particle ICP-MS	3.69×10^{-3} ng mL^{-1}; 0.058×10^{-3} ng mL^{-1}	—	Vidmar *et al.*, 2017[69]
PE films/food simulants	Films: 249–255 mg kg^{-1-1}Simulants: 0.1–12 μg kg^{-1}	HNO_3, H_2O_2, HF, microwave	89–116	ICP-MS	0.04 μg L^{-1}	—	Lin *et al.*, 2014[27]
Several food types	0.00077–359 μg mg^{-1}	HNO_3, H_2O_2, HF, microwave	81–87	ICP-MS	0.0001 μg mg^{-1}	—	Weir *et al.*, 2012[32]

the exact mass of chemical compounds of interest. Importantly, for results to be reliable, the mass error must not be higher than 5 ppm. Among the most frequently used detectors for these applications are the Orbitrap and the quadrupole time-of-flight (Q-TOF) mass analyzer. Because of their sensitivity and discrimination power, these detectors are very useful for the analysis of migrants and specially for the identification of unknowns such as NIASs.

3.6.3.5.1 Application of HRMS to the Analysis of Target Substances. As with the original MS systems, HRMS can be paired with LC for target compound analysis. Moreta *et al.* used UHPLC coupled to an Q-TOF instrument to analyze plastic additives in PE film/multi-layer packaging. They used a Cortecs UHPLC C_{18} column (100 × 2.1 mm id, 1.6 µm film thickness) at 50 °C and 1 mM ammonium formate with 0.5% formic acid in methanol and water as eluent at a flow rate of 0.4 mL min^{-1}; ESI was used in the positive ion mode and the detector was set in full-scan mode (*m*/*z* 50–1200). This resulted in a resolving power of ≥17 500 FWHM. For our model additives, Moreta *et al.* found the accurate mass *m*/*z* 675.6784 for erucamide, corresponding to $(2M + H)^+$, and *m*/*z* 548.5051 for Irganox 1076, corresponding to the adduct NH_4^+. The measured concentrations of the migrants erucamide and Irganox 1076 in the matrix were 13–183 and 38–353 µg g^{-1}, respectively.[23]

A similar method was employed by Rosero-Moreano *et al.* to analyze for the presence of 2-methyl-3-isothiazolinone in acrylic packaging; for this purpose, the cone voltage was 30 V and the collision energy was 20–40 eV, the LOD was 0.004 µg g^{-1} and the measured accurate mass for this compound was *m*/*z* 115.0173, which allowed the detection of concentrations from below the limit of quantification to 55 µg g^{-1}.[55] These techniques are especially valuable for analyses of migrants in complex food matrices, such as chocolate: the optical brightener Uvitex OB was analyzed in chocolate by Bignardi *et al.* using HRMS.[36] Setting the detector to full MS/MS – *m*/*z* 80–500 – and tSIM-MS/MS (targeted SIM tandem MS), this team achieved an LOD from 0.2 to 5 ng g^{-1} and a resolving power between 17 500 and 70 000 FWHM, which allowed them to measure concentrations of Uvitex from below the LOD to 0.06 ng g^{-1}.

3.6.3.6 Application of MS to the Analysis and Identification of Unknowns

Having the capability to detect specific substances of interest does not eliminate the problem of the reaction and degradation products of monomers and additives, contaminants and other unknowns. Nonetheless, these substances can represent a risk to public health; therefore, methods to identify and assess the concentrations of these substances are necessary. The first step is understanding the nature of the FCM to be analyzed, as knowing the specific composition of each packaging type can narrow the range of likely additives and migrants, thereby simplifying the selection of appropriate analytical techniques. For example, polyester coatings are likely to contain monomers used in the manufacture of polyesters, such as phthalic acid and alcohols. Generally, it is also likely that the manufacturer of the FCM will have used approved substances for the synthesis of the packaging. Techniques such as Fourier transform infrared (FTIR) spectroscopy can be used for polymer identification or confirmation and with that knowledge it is possible to hypothesize which types of starting substances were used and which side reactions might occur. For example, an unfinished polymeric reaction could yield an unreacted oligomer that is likely to migrate into foods. Another example is a degradation product from a Type 1 photoinitiator. As these NIASs and other undesired materials can be volatile, semivolatile and or non-volatile substances, we will discuss next the analytical options for each type.

3.6.3.6.1 Identification of Volatile and Semivolatile Compounds. As described previously, GC-MS a suitable analytical technique for volatile and semivolatile substances. When attempting to detect and identify NIASs, GC methods that can capture a wide variety of diverse substances should be chosen to maximize the likelihood of successful identification. This is illustrated by the work of Paseiro-Cerrato *et al.*, who developed a GC method that could cover a large mass range, using an HP-5MS column (30 m × 0.32 mm id, 0.25 µm film thickness) at a low initial oven temperature (30 °C), which was ramped to high temperature (300 °C). As a further optimization, an inlet temperature gradient from 70 to 350 °C was also employed. The detector mass range was from m/z 50 to 500.[56] The goal was to create conditions that could elute and capture compounds having a high polarity and low MW and also those compounds having high MW and lower polarity. Because analytical standards are not available for unknowns, Paseiro-Cerrato *et al.* validated this separation method using a mixture D-(−)-2,3-butanediol, decane, dicyclohexylamine, 2,6-dimethylaniline, 2,6-dimethylphenol, 2-ethylcaproic acid, methyl decanoate, methyl laurate, methyl undecanoate, 1-octanol and dodecane in hexane and methylene chloride (known collectively as "Test Mixture 2 for apolar capillary columns according to Grob"), which represents a variety of chemical groups, polarities and MWs. This mixture was also used to verify that the detector settings were indeed appropriate for capturing a wide variety of chemical entities, and therefore would be suitable for identifying unknowns.[56]

One notable advantage of GC methods is the number of libraries available that provide reliable information about the nature of detected analytes. The data in these libraries can be supplemented with in-house GC-MS databases. For example, Nakashima *et al.* constructed a database to support the identification of BHT, erucamide, DEHP and other potential migrants in processed foods. The column used for this project was an HP-5MS (30 m × 0.25 mm id, 0.25 µm film thickness) and the oven temperature ranged from 70 to 320 °C. GC-MS was performed in the scan mode (m/z 35–550). The database was successfully used for screening analyses of additives.[35]

3.6.3.6.2 Identification of Non-volatile and Semivolatile Compounds. Identifying non-volatile and semivolatile substances is appropriately performed using LC-HRMS. Specific analytical methods should be selected to cover a wide range of diverse chemical entities. Unlike GC-MS, libraries are not available; therefore, in-house libraries, databases of compounds and specialized software must be used to identify detected compounds correctly. Two series of experiments by Paseiro-Cerrato *et al.* using UHPLC-HRMS demonstrate how these techniques can tentatively identify unknown migrants from polyester, acrylic–phenolic and epoxy can coatings.[56,57] This group used a C_{18} column with a long gradient (starting with a high water content) to capture most of the potential migrants (from low to high polarity). Because it is important to elute compounds with good resolution and peak shape, samples were injected into the system with a high water content but also with a high concentration of the organic solvent acetonitrile. The mobile phase contained a volatile buffer to achieve both good separation and good ionization of the chemical entities. Using the full-scan mode (positive and negative ion modes), species up to m/z 1000 appeared, capable of capturing most of the potential migrants. The ionization source selected was ESI, because this method produces a softer ionization and consequently reduces fragmentation of the eluted substances, which facilitates identification. To narrow the candidate identities for these migrants, an in-house library was constructed, consisting of the set of approved monomers and additives in both EU and US legislations, the products of potential reactions among these entities and previously identified migrants reported in the literature.[56,57] Vaclavikova *et al.* successfully applied a similar method to identify plasticizers commonly used in the manufacture of PVC, along with other potential migrants, using databases to identify potential migrants in vinyl can coatings.[28] Other kinds of detectors, *e.g.* a diode-array detector (DAD), may be used to provide complementary supporting data to enhance confidence in the results of these analyses. The final step in any identification process is to obtain analytical standard(s) for the tentatively identified compounds to confirm their presence in the sample using NMR spectroscopy. This is not always easy, since most of the unknowns are not common chemicals and therefore must be custom synthesized for this purpose.

3.6.3.7 Direct Analysis in Real Time (DART-HRMS)

One of the main obstacles to the efficient identification of migrant substances is sample preparation. Some substances can be lost during the extraction process, the procedures can be time consuming and the demands of staffing and equipment necessary for these procedures can make these important food safety investigations expensive. However, in 2005, Cody *et al.* pioneered a new ambient ionization technique that does not require sample preparation: direct analysis in real time (DART), a technique that uses an ion source and an HRMS detector.[58] The source used for DART analyses is a tube in which a heated plasma is generated and then interacts with the sample (food, FCM, drugs, *etc.*) by thermally desorbing the molecules of the sample for introduction into the HRMS system. This method provides initial results in minutes rather than hours or days, making it a highly attractive alternative to more traditional techniques for analyzing migrants and NIASs. An extensive review of how DART methods can be applied to analyze food matrices was presented by Hajslova *et al.* in 2011.[59] That review also discusses relevant parameters such as gas temperature, the use of dopants, matrix effects and other valuable information about how to obtain data most effectively.

Other workers have also used this technique to analyze potential migrants in FCMs. Ackerman *et al.* analyzed 13 additives that may be used in FCMs.[60] All the additives employed could be identified using DART. More recently, Lago *et al.* used DART to identify print-related molecules.[61] Paseiro-Cerrato *et al.* used DART to identify rapidly potential migrants in can coatings[56,57] and as a rapid screening method to identify primary aromatic amines in kitchen utensils.[62]

3.7 Conclusion

Assessing FCMs and foods/food simulants for the presence of migrating additives is necessary to ensure that their concentrations remain within internationally accepted limits. Novel applications of MS have revolutionized our capability to identify and measure the presence of these substances in food matrices and FCMs. The literature reviewed in this chapter shows the wide range of available tools and techniques, along with parameters that have been found especially useful for the analysis of known or unknown migrating substances. These techniques are under continual development and are already very powerful tools for identifying compounds that are difficult to analyze. Proper use of these approaches greatly facilitates the analysis of these compounds in either FCMs, foods and/or food simulants. The analysis of migrating substances that travel from FCMs into foods is just one of the ways in which MS technology can help protect consumer health. Identifying which undesired substances are present and how these migrants occur may also assist with improvements in FCM design and production.

Acknowledgements

The authors thank Lili Fox Vélez, Office of Regulatory Science, CFSAN, for scientific writing support.

References

1. Code of Federal Regulations, 2017, *Parts 170–199, Title 21*, U.S. Government Printing Office, Washington (DC).

2. [EC] European Commission, Commission Regulation (EU) No. 10/2011 of 14 January 2011 on plastic materials and articles intended to come into contact with food, *Off. J. Eur. Union*, 2011, **L. 12**, 1–89.
3. H. C. Erythropel, M. Maric, J. A. Nicell, R. L. Leask and V. Yargeau, Leaching of the plasticizer di(2-ethylhexyl)phthalate (DEHP) from plastic containers and the question of human exposure, *Appl. Microbiol. Biotechnol.*, 2014, **98**(24), 9967–9981.
4. D. Y. Bang, M. Kyung, M. J. Kim, B. Y. Jung, M. C. Cho, S. M. Choi, Y. W. Kim, S. K. Lim, D. S. Lim, A. J. Won, S. J. Kwack, Y. Lee, H. S. Kim and B. M. Lee, Human risk assessment of endocrine-disrupting chemicals derived from plastic food containers, *Compr. Rev. Food Sci. Food Saf.*, 2012, **11**(5), 453–470.
5. G. Morlock and W. Schwack, Determination of isopropylthioxanthone (ITX) in milk, yoghurt and fat by HPTLC-FLD, HPTLC-ESI/MS and HPTLC-DART/MS, *Anal. Bioanal. Chem.*, 2006, **385**(3), 586–595.
6. A. Sanches-Silva, S. Pastorelli, J. M. Cruz, C. Simoneau, I. Castanheira and P. Paseiro-Losada, Development of an analytical method for the determination of photoinitiators used for food packaging materials with potential to migrate into milk, *J. Dairy Sci.*, 2008, **91**(3), 900–909.
7. Regulation (EC) No. 1935/2004 of the European Parliament and of the Council of 27 October 2004 on materials and articles intended to come into contact with food and repealing Directives 80/590/EEC and 89/109/EEC.
8. Commission Regulation EC 450/2009 on Active and Intelligent Materials and Articles Intended to Come into Contact with Food.
9. Directive 2005/31/EC on the Declaration of Compliance and Performance Criteria of the Analytical Method for Ceramic Articles Intended to Come into Contact with Foodstuffs.
10. Directive 2007/42/EC Relating to Materials and Articles Made of Regenerated Cellulose Film Intended to Come into Contact with Foodstuffs.
11. Regulation 282/2008 on Recycled Plastics Intended to Come into Contact with Food.
12. Commission Regulation (EC) No 2023/2006 of 22 December 2006 on Good Manufacturing Practice for Materials and Articles Intended to Come into Contact with Food.
13. http://www.fda.gov/Food/GuidanceRegulation/GuidanceDocuments-RegulatoryInformation/ucm081818.htm5/9/2018.
14. Code of Federal Regulations, *Parts 73, Title 21*, U.S. Government Printing Office, Washington (DC), 2017.
15. Regulation (Ec) No 1333/2008 of the European Parliament and of the Council of 16 December 2008 on Food Additives.
16. A. Rodriguez-Bernaldo de Quiros, R. Paseiro-Cerrato, S. Pastorelli, R. Koivikko, C. Simoneau and P. Paseiro-Losada, Migration of photoinitiators by gas phase into dry foods, *J. Agric. Food Chem.*, 2009, **57**(21), 10211–10215.

17. G. McCombie, A. Harling-Vollmer, M. Morandini, G. Schmaeschke, S. Pechstein, W. Altkofer, M. Biedermann, S. Biedermann-Brem, M. Zurfluh, G. Sutter, M. Landis and K. Grob, Migration of plasticizers from the gaskets of lids into oily food in glass jars: a European enforcement campaign, *Eur. Food Res. Technol.*, 2012, **235**(1), 129–137.
18. P. Otero-Pazos, D. A. Pereira de Abreu, R. Sendon, A. R. Bernaldo de Quiros, I. Angulo, J. M. Cruz and P. Paseiro-Losada, Determination of partition coefficients of selected model migrants between polyethylene and polypropylene and nanocomposite polypropylene, *J. Chem.*, 2016, 3952631/1–3952631/10.
19. R. Paseiro-Cerrato, A. Rodriguez-Bernaldo de Quiros, P. Otero-Pazos, R. Sendon and P. Paseiro-Losada, Determination of diffusion and partition coefficients of model migrants by direct contact and vapour phase transfer from low-density polyethylene films into cake, *Food Addit. Contam., Part A*, 2018, **35**(3), 555–562.
20. A. Seiler, A. Bach, M. Driffield, P. Paseiro Losada, P. Mercea, V. Tosa and R. Franz, Correlation of foodstuffs with ethanol–water mixtures with regard to the solubility of migrants from food contact materials, *Food Addit. Contam., Part A*, 2014, **31**(3), 498–511.
21. T. H. Begley, W. Hsu, G. Noonan and G. Diachenko, Migration of fluorochemical paper additives from food-contact paper into foods and food simulants, *Food Addit. Contam., Part A*, 2008, **25**(3), 384–390.
22. Q. B. Lin, T. J. Wang, H. Song and B. Li, Analysis of isothiazolinone biocides in paper for food packaging by ultra-high-performance liquid chromatography-tandem mass spectrometry, *Food Addit. Contam., Part A*, 2010, **27**(12), 1775–1781.
23. C. Moreta and M.-T. Tena, Determination of plastic additives in packaging by liquid chromatography coupled to high resolution mass spectrometry, *J. Chromatogr. A*, 2015, **1414**, 77–87.
24. M. A. Lago, A. Rodriguez-Bernaldo de Quiros, R. Sendon, J. Bustos, M. I. Santillana and P. Paseiro, Simultaneous chromatographic analysis of photoinitiators and amine synergists in food contact materials, *Anal. Bioanal. Chem.*, 2014, **406**(17), 4251–4259.
25. ASTM 6953, Standard Test Method for Determination of Antioxidants and Erucamide Slip Additives in Polyethylene Using Liquid Chromatography (LC).
26. G. Lv, L. Wang, J. Liu and S. Li, Method for determination of fatty acid amides in polyethylene packaging materials—Gas chromatography/mass spectrometry, *J. Chromatogr. A*, 2009, **1216**(48), 8545–8548.
27. Q.-B. Lin, H. Li, H.-N. Zhong, Q. Zhao, D.-H. Xiao and Z.-W. Wang, Migration of Ti from nano-TiO_2–polyethylene composite packaging into food simulants, *Food Addit. Contam., Part A*, 2014, **31**(7), 1284–1290.
28. M. Vaclavikova, R. Paseiro-Cerrato, L. Vaclavik, G. O. Noonan, J. DeVries and T. H. Begley, Target and non-target analysis of migrants from PVC-coated cans using UHPLC-Q-Orbitrap MS: evaluation of long-term migration testing, *Food Addit. Contam., Part A*, 2016, **33**(2), 352–363.

29. G. Beldi, S. Pastorelli, F. Franchini and C. Simoneau, Time- and temperature-dependent migration studies of Irganox 1076 from plastics into foods and food simulants, *Food Addit. Contam., Part A*, 2012, **29**(5), 836–845.
30. Q. B. Lin, T. J. Wang, H. Song and R. Z. Wang, Kinetic migration of isothiazolinone biocides from paper packaging to Tenax and Porapak, *Food Addit. Contam., Part A*, 2011, **28**(9), 1294–1301.
31. M. Monteiro, C. Nerin and F. G. R. Reyes, Migration of Tinuvin P, a UV stabilizer, from PET bottles into fatty-food simulants, *Packag. Technol. Sci.*, 1999, **12**(5), 241–248.
32. A. Weir, P. Westerhoff, L. Fabricius, K. Hristovski and N. von Goetz, Titanium dioxide nanoparticles in food and personal care products, *Environ. Sci. Technol.*, 2012, **46**(4), 2242–2250.
33. A. Lloyd, C.-A. Salanson, H. Semizer, J. Leak, M. Driffield, E. Bradley, *Analysis of Printing Ink Components from Food Packaging Materials by GC/MS/MS*, Agilent Technologies UK Ltd, 2012.
34. J.-M. Kim, S.-H. Choi, G.-H. Shin, J.-H. Lee, S.-R. Kang, K.-Y. Lee, H.-S. Lim, T. S. Kang and O.-H. Lee, Method validation and measurement uncertainty for the simultaneous determination of synthetic phenolic antioxidants in edible oils commonly consumed in Korea, *Food Chem.*, 2016, **213**, 19–25.
35. S. Nakashima, H. Ji, T. Yamagami, K. Asai, K. Kadokami, M. Mutsuga, Y. Kawamura, R. Shinohara and K. Arizono, Development of a novel GC/MS database for the determination of additives for food packaging into the processed foods, *Nippon Shokuhin Kagaku Gakkaishi*, 2013, **20**(1), 42–51.
36. C. Bignardi, A. Cavazza, C. Lagana, P. Salvadeo and C. Corradini, Optimization of mass spectrometry acquisition parameters for determination of polycarbonate additives, degradation products, and colorants migrating from food contact materials to chocolate, *J. Mass Spectrom.*, 2018, **53**(1), 83–90.
37. M. Suman, E. De Dominicis and I. Commissati, Trace detection of the chlorohydrins of epoxidized soybean oil in foodstuffs by UPLC-ESI-MS/MS, *J. Mass Spectrom.*, 2010, **45**(9), 996–1002.
38. M. Anastassiades, S. J. Lehotay, D. Stajnbaher and F. J. Schenck, Fast and easy multiresidue method employing acetonitrile extraction/partitioning and "dispersive solid-phase extraction" for the determination of pesticide residues in produce, *J. AOAC Int.*, 2003, **86**(2), 412–431.
39. A. Vavrous, L. Vapenka, J. Sosnovcova, K. Kejlova, K. Vrbik and D. Jirova, Method for analysis of 68 organic contaminants in food contact paper using gas and liquid chromatography coupled with tandem mass spectrometry, *Food Control*, 2016, **60**, 221–229.
40. D. Xu, X. Deng, E. Fang, X. Zheng, Y. Zhou, L. Lin, L. Chen, M. Wu and Z. Huang, Determination of 23 phthalic acid esters in food by liquid chromatography tandem mass spectrometry, *J. Chromatogr. A*, 2014, **1324**, 49–56.

41. Y. Hayasaka, Analysis of phthalates in wine using liquid chromatography tandem mass spectrometry combined with a hold-back column: chromatographic strategy to avoid the influence of pre-existing phthalate contamination in a liquid chromatography system, *J. Chromatogr. A*, 2014, **1372**, 120–127.
42. S. Borman, H. Russell and G. Siuzdak, A mass spectrometry timeline, *J. l: Today's Chem. Work*, 2003, 47–49.
43. Y. Fujii, K. H. Harada and A. Koizumi, Analysis of perfluoroalkyl carboxylic acids in composite dietary samples by gas chromatography/mass spectrometry with electron capture negative ionization, *Environ. Sci. Technol.*, 2012, **46**(20), 11235–11242.
44. W. A. C. Anderson and L. Castle, Benzophenone in cartonboard packaging materials and the factors that influence its migration into food, *Food Addit. Contam.*, 2003, **20**(6), 607–618.
45. M. Motas Guzmàn, C. Clementini, M. D. Pérez-Cárceles, S. Jiménez Rejón, A. Cascone, T. Martellini, C. Guerranti and A. Cincinelli, Perfluorinated carboxylic acids in human breast milk from Spain and estimation of infant's daily intake, *Sci. Total Environ.*, 2016, **544**, 595–600.
46. M. S. Choi, S. U. Rehman, H. Kim, S. B. Han, J. Lee, J. Hong and H. H. Yoo, Migration of epoxidized soybean oil from polyvinyl chloride/polyvinylidene chloride food packaging wraps into food simulants, *Environ. Sci. Pollut. Res.*, 2018, **25**(5), 5033–5039.
47. K. S. Carlos, L. S. de Jager and T. H. Begley, Investigation of the primary plasticisers present in polyvinyl chloride (PVC) products currently authorised as food contact materials, *Food Addit. Contam., Part A*, 2018, **35**(6), 1214–1222.
48. Y. Wu, Y. Xian, X. Guo, L. Chen, X. Zhao, B. Wang and L. Wang, Development and validation of a screening and quantification method for simultaneous determination of seven fluorescent whitening agents in commercial flour using UPLC-MS/MS, *Food Chem.*, 2018, **243**, 162–167.
49. Y. Xu, G. O. Noonan and T. H. Begley, Migration of perfluoroalkyl acids from food packaging to food simulants, *Food Addit. Contam., Part A Chem. Anal. Control Exposure Risk Assess.*, 2013, **30**(5), 899–908.
50. M. Mutsuga, M. Yamaguchi, Y. Abe and H. Akiyama, Evaluation of the equality of non-polar capillary columns in GC/MS analysis of food contact plastics, *Am. J. Anal. Chem.*, 2013, **4**(9), 476–487.
51. Z. Ezerskis, V. Morkunas, M. Suman and C. Simoneau, Analytical screening of polyadipates and other plasticisers in poly(vinyl chloride) gasket seals and in fatty food by gas chromatography-mass spectrometry, *Anal. Chim. Acta*, 2007, **604**(1), 29–38.
52. C. Block, L. Wynants, M. Kelchtermans, R. De Boer and F. Compernolle, Identification of polymer additives by liquid chromatography-mass spectrometry, *Polym. Degrad. Stab.*, 2006, **91**(12), 3163–3173.
53. Z. Wu, Y. Xu, M. Li, X. Guo, Y. Xian and H. Dong, Simultaneous determination of fluorescent whitening agents (FWAs) and photoinitiators (PIs)

in food packaging coated paper products by the UPLC-MS/MS method using ESI positive and negative switching modes, *Anal. Methods*, 2016, **8**(5), 1052–1059.
54. S. Candas-Zapico, D. J. Kutscher, M. Montes-Bayon and J. Bettmer, Single particle analysis of TiO_2 in candy products using triple quadrupole ICP-MS, *Talanta*, 2018, **180**, 309–315.
55. M. Rosero-Moreano, E. Canellas and C. Nerin, Three-phase hollow-fiber liquid-phase microextraction combined with HPLC-UV for the determination of isothiazolinone biocides in adhesives used for food packaging materials, *J. Sep. Sci.*, 2014, **37**(3), 272–280.
56. R. Paseiro-Cerrato, S. MacMahon, C. D. Ridge, G. O. Noonan and T. H. Begley, Identification of unknown compounds from polyester cans coatings that may potentially migrate into food or food simulants, *J. Chromatogr. A*, 2016, **1444**, 106–113.
57. R. Paseiro-Cerrato, J. DeVries and T. H. Begley, Evaluation of short-term and long-term migration testing from can coatings into food simulants: epoxy and acrylic–phenolic coatings, *J. Agric. Food Chem.*, 2017, **65**(12), 2594–2602.
58. R. B. Cody, J. A. Laramee and H. D. Durst, Versatile new ion source for the analysis of materials in open air under ambient conditions, *Anal. Chem.*, 2005, 77(8), 2297–2302.
59. J. Hajslova, T. Cajka and L. Vaclavik, Challenging applications offered by direct analysis in real time (DART) in food-quality and safety analysis, *TrAC, Trends Anal. Chem.*, 2011, **30**(2), 204–218.
60. L. K. Ackerman, G. O. Noonan and T. H. Begley, Assessing direct analysis in real-time-mass spectrometry (DART-MS) for the rapid identification of additives in food packaging, *Food Addit. Contam., Part A*, 2009, **26**(12), 1611–1618.
61. M. A. Lago and L. K. Ackerman, Identification of print-related contaminants in food packaging, *Food Addit. Contam., Part A*, 2016, **33**(3), 518–529.
62. R. Paseiro-Cerrato, G. O. Noonan and T. H. Begley, Development of a rapid screening method to determine primary aromatic amines in kitchen utensils using direct analysis in real time mass spectrometry (DART-MS), *Food Addit. Contam., Part A*, 2014, **31**(3), 537–545.
63. A. Rafoth, S. Gabriel, F. Sacher and H.-J. Brauch, Analysis of isothiazolinones in environmental waters by gas chromatography-mass spectrometry, *J. Chromatogr. A*, 2007, **1164**(1–2), 74–81.
64. R. Koivikko, S. Pastorelli, A. Rodriguez-Bernaldo de Quiros, R. Paseiro-Cerrato, P. Paseiro-Losada and C. Simoneau, Rapid multi-analyte quantification of benzophenone, 4-methylbenzophenone and related derivatives from paperboard food packaging, *Food Addit. Contam., Part A*, 2010, **27**(10), 1478–1486.
65. J. Veyrand, M. Marin-Kuan, C. Bezencon, N. Frank, V. Guerin, S. Koster, H. Latado, J. Mollergues, A. Patin, D. Piguet, P. Serrant, J. Varela and

B. Schilter, Integrating bioassays and analytical chemistry as an improved approach to support safety assessment of food contact materials, *Food Addit. Contam., Part A*, 2017, **34**(10), 1807–1816.

66. P. Chatonnet, S. Boutou and A. Plana, Contamination of wines and spirits by phthalates: types of contaminants present, contamination sources and means of prevention, *Food Addit. Contam., Part A*, 2014, **31**(9), 1605–1615.
67. P. Speksnijder, J. van Ravestijn and P. de Voogt, Trace analysis of isothiazolinones in water samples by large-volume direct injection liquid chromatography tandem mass spectrometry, *J. Chromatogr. A*, 2010, **1217**(32), 5184–5189.
68. W. M. Young, P. South, T. H. Begley, G. W. Diachenko and G. O. Noonan, Determination of perfluorochemicals in cow's milk using liquid chromatography-tandem mass spectrometry, *J. Agric. Food Chem.*, 2012, **60**(7), 1652–1658.
69. J. Vidmar, R. Milacic and J. Scancar, Sizing and simultaneous quantification of nanoscale titanium dioxide and a dissolved titanium form by single particle inductively coupled plasma mass spectrometry, *Microchem. J.*, 2017, **132**, 391–400.
70. M. A. Lago, A. Rodriguez-Bernaldo de Quiros, R. Sendon, J. Bustos, M. T. Nieto and P. Paseiro, Photoinitiators: a food safety review, *Food Addit. Contam., Part A*, 2015, **32**(5), 779–798.

CHAPTER 4

Adhesives in Food Packaging

MARÍA TERESA TENA*

Department of Chemistry, University of La Rioja, c/ Madre de Dios 51, 26006 Logroño (La Rioja), Spain
*E-mail: maria-teresa.tena@unirioja.es

4.1 Introduction

Adhesives can be used to glue different substrates to form multilayer materials with specific properties, known as laminates. In other cases, the adhesive is used only in part of the packaging material to build up the final food packaging, for instance a box, a bag or a sachet. Adhesives can be also used to stick labels on packaging or directly on the food. Depending on the final use of a packaging, different materials can be used as substrates to form the multilayer packaging. Among them are paper, cardboard, plastic polymers such as polyethylene (PE), polypropylene (PP), polyesters (PES) and polyamides (PA) and barrier materials such as aluminium (Al), poly(ethylene terephthalate) (PET) and ethylene–vinyl alcohol (EVOH).

Food contact materials (FCMs) must comply with the European Regulation 1935/2004, which requires that adhesives used in food packaging must not affect the food, make the food harmful and change the nature or quality of the food.[1] However, unlike plastics that have a positive list of authorised ingredients,[2] adhesives do not yet have a specific regulation.

The study of adhesives is a difficult task, mainly because there are many different formulations with many different compounds involved and, moreover,

Food Chemistry, Function and Analysis No.10
Food Contact Materials Analysis: Mass Spectrometry Techniques
Edited by Michele Suman
© The Royal Society of Chemistry 2019
Published by the Royal Society of Chemistry, www.rsc.org

no information is provided about them by the manufacturers. In addition, their migration behaviour is usually unknown and has seldom been tested. To solve these problems, in the frame of the European Union Project MIGRESIVES (http://www.migresives.eu), new analytical methods for different types of adhesives have been developed and applied to migration studies from a large number of food contact packaging materials.[3]

There are many different types of adhesives available for any specific application. They may be classified according to different criteria: origin, chemical composition, intended function, method of curing and physical form. According to their origin, they can be natural or synthetic adhesives. Natural adhesives are manufactured from animal or agricultural products, for instance, proteins, starch or natural rubber. Synthetic adhesives include a broad range of chemicals and can be sub-classified according to their method of curing: chemical reaction, cooling from a melt or loss of solvent (an organic solvent or water). For instance, polyurethanes (PU) and epoxy resins are adhesives that harden by a chemical reaction.

PU adhesives are produced by the reaction of the NCO group of isocyanates and any organic compound containing a reactive hydroxyl group such as polyether and polyester polyols in the presence of a catalyst. They represent around 90% of the adhesives used for manufacturing flexible laminates for FCMs. The isocyanate components are usually methylene bis(phenyl isocyanate), toluene diisocyanate and isophorone diisocyanate.

Hotmelt (HM) adhesives harden by cooling from the melt. They are solvent-free polymer-based thermoplastics, which means they are a solid at room temperature but liquefy upon heating. Therefore, they should be applied at higher temperatures (160–180 °C) than other adhesives.

Adhesives that harden by water or solvent loss can be further classified according to their chemical composition, such as acrylic adhesives (AC) or vinylic adhesives (VI). In this case, the adhesive is an acrylate or a vinylic polymer, respectively.

To summarize, adhesives are complex mixtures of substances with specific functions, and can include polymers, monomers, tackifiers, inhibitors, catalysts, solvents, antioxidants, plasticizers, fillers, antifungals, surfactants, *etc.* Moreover, the packaging manufacturers do not know the adhesive composition and even for adhesive producers it is difficult to know exactly the final composition of their products owing to impurities from raw materials, degradation products and by-products formed by reactions between ingredients when the adhesive is applied or during the curing process.

Mass spectrometry (MS) is a powerful analytical technique that is very useful for the analysis of adhesives, particularly in the field of food packaging. On the one hand, there are many examples of its use to determine potential adhesive migrants from packaging to food. On the other hand, it has also been used to study the causes of delamination phenomena caused by the migration of some chemicals into the packaging.

4.2 Mass Spectrometry Techniques Used for the Analysis of Food Packaging Adhesives

Mass spectrometers with single, triple or ion-trap quadrupole analysers were used in initial studies. However, they do not provide adequate sensitivity in full-scan mode when applied to complex samples. Their sensitivity can be improved in selected ion monitoring (SIM), precursor scan, neutral loss scan and multiple reaction monitoring (MRM) modes but with these acquisition modes the selectivity and identification capability are lower. Moreover, these instruments do not provide sufficient mass accuracy and mass resolution for screening unknown and non-targeted compounds. The resolving power [full width at half-maximum (FWHM)] of these instruments is around 1000, hence they provide a resolution of 1 Da for low molecular mass ($M \leq 1000$ Da) monocharged ions.

Conversely, high-resolution mass spectrometers with a time-of-flight analyser (TOF-MS) and the combination of quadrupole and time-of-flight analysers (Q-TOF-MS), with FWHM >10 000, provide the sensitivity and selectivity required for non-targeted screening in complex samples. The later hybrid tandem mass spectrometry (MS/MS) instruments allow one to acquire full-scan mass spectra with high sensitivity and mass accuracy for any ionizable components in the sample. This combination provides accurate masses for both parent and fragment ions. Accurate mass information enables chemical formulae for precursor and fragment ions to be proposed to enhance the identification of unknown compounds.[4,5]

Structural information on unknown compounds can be elucidated from the low-energy exact mass precursor ion and elevated-energy exact mass fragment ion spectra. First, the software provides the elemental composition of the precursor ion based on the isotopic ratios of the most common elements (namely C, H, O, N, Cl and S, which can be selected by the analyst) and considering the possibility of Na adduct formation. Then, molecular structures for the most probable formula provided by the software can be searched for in chemical databases such as Chemspider and Scifinder. This information, together with the exact mass fragment ion data, can be used to confirm the identity of the compound. If the accurate masses of the fragments of the molecular structure proposed for the known compound coincide with the masses measured in the high-energy fragment ion spectra, the identification is almost confirmed. The final step for identification is to compare the retention time and the mass spectrum of the unknown compound with those obtained for pure standards of the proposed candidates.

Another interesting approach for the identification of non-targeted adhesive migrants from food packaging is the coupling of ion mobility spectrometry (IMS) with TOF-MS for high-resolution mass spectrometry (HRMS).[6] In this coupled system, the first generation of fragments are separated by ion mobility, which gives a rapid separation of gas-phase ions on the milliseconds time scale. The ions are separated because of their different mobilities when they are accelerated through a drift tube by a weak electric field gradient and in the presence of a neutral gas such as nitrogen or helium. Fragmentations of precursor ions can take place in the trap and transfer regions. First, the selected precursor ions are fragmented in the trap region, these fragments are separated by IMS, then a second-generation of fragments are produced in the transfer region. Finally, the first- and second-generation ions are time aligned.

4.3 Migration Studies of Adhesive Components

The migration of adhesive compounds from a laminate into food depends on the chemical and physical properties of the compound and the food and also on the materials used to manufacture the laminate. Although mathematical models can be used to estimate the migration levels from laminates into foods, it is more likely for migration studies to be performed for each specific laminate using different food simulants.

In these studies, MS is always involved. For instance, in the case of volatile or semivolatile potential migrants, the chosen technique is gas chromatography coupled with mass spectrometry (GC-MS), usually following headspace solid-phase microextraction (HS-SPME)[7,8] because this technique provides a high preconcentration factor of the analytes prior to the GC-MS analysis.

In other studies, intended to establish the migration levels of non-volatile polar adhesive compounds (*e.g.* primary aromatic amines), liquid chromatography coupled with mass spectrometry (LC-MS)[4,9,10] is the technique selected.

4.3.1 Volatile and Semivolatile Compounds from Adhesives

The odorous compounds from different types of adhesives [hotmelt, vinyl acetate ethylene, starch, poly(vinyl acetate) and acrylic] have been identified and quantified by gas chromatography–olfactometry–mass spectrometry (GC-O-MS).[11] To obtain a substantial number of compounds, two extraction methodologies, headspace solid-phase extraction (HS-SPE) and headspace solid-phase microextraction (HS-SPME), were compared. The best SPME fibre was polydimethylsiloxane (PDMS) for the hotmelt adhesive and divinylbenzene (DVB)–carboxen (CAR)–PDMS for the other adhesives. The two methodologies were complementary, as several compounds were trapped either in the SPE cartridge or in the SPME fibre, but not in both. Thirty-three compounds with characteristic odours were identified; some of them, such as butyric acid, acetic acid, methyl butyrate, 1-butanol and nonanal, were present in most of the adhesives, producing cheesy, rancid, sour, medicinal and green aromas, respectively.

Migration of these off-flavours from the adhesives was tested in a subsequent study[12] to evaluate the possible human risk and to determine if this migration could affect the organoleptic properties of packed food. Thus, different laminates used for packaging dry food (tomatoes, cakes, cookies, breadcrumbs, flour or salt) or fresh food (pizza and pastry) manufactured with different adhesives were analysed. A total of 25 different compounds from adhesives were detected in these laminates by GC-O-MS and 76% of them migrated into dry food simulant used (Tenax®). Acetic acid, butyric acid and cyclohexanol with vinegar, cheese and camphor odours were the most abundant compounds.[12]

Aznar *et al.*[7] carried out a screening of different adhesives types from market samples by HS-SPME-GC-MS to obtain a list of the possible migrant compounds in laminates containing adhesives. They studied 45 three-layer laminate glued samples with a structure [substrate 1–adhesive–substrate 2], which comprised multilayer films, folding boxes and corrugated board samples; and which were prepared using 29 different adhesives (including PU, AC, HM, VI and starch) and a wide variety of substrates such as cardboard, PP-coated cardboard, corrugated paper, PP, PE, PA, PET, metallized PET, PES, Al and EVOH. The migration data obtained were used to test the applicability of a mathematical model designed for predicting migration to food from laminates containing adhesives.

In a first step, laminates were analysed by HS-SPME-GC-MS and compared with similar sized cut-outs of the substrates to identify the potential migrant compounds from the adhesive. The pure adhesive was also analysed by HS-SPME-GC-MS. Analytes were concentrated in a PDMS fibre before injection into the GC column and the detection was performed with a quadrupole MS detector in scan mode (m/z 50–350). For the migration experiments, the surface of the laminate was fully covered with Tenax® and kept at 40 °C for 10 days. After this period, the Tenax® was extracted with acetone and the extract analysed by GC-MS, using the SIM mode for quantification. For two laminates with PU adhesives, isooctane was also used as food simulant.

A total of 55 different compounds were detected in the adhesives analysed and 57% of them were found that had migrated into the dry food simulant through the food contact layer. Thirty-three compounds were identified and four of them (2,6-toluene diisocyanate, 3,3′-dimethyl-4,4′-biphenylene diisocyanate, decamethylcyclopentasiloxane and retene) were classified as toxic compounds (see Table 4.1). Diacetin and triacetin, which are commonly used as plasticizers or humectants in the adhesives, were also found in vinyl adhesives.[13]

4.3.1.1 Volatile and Semivolatile Compounds from Polyurethane Adhesives

HS-SPME-GC-MS has been used for the identification of NIASs coming from PU adhesives in multilayer packaging materials and their migration into food simulants (Tenax® and isooctane).[8] The MS instrument used was an ion trap operated in scan acquisition mode (m/z 40–400) with electron ionization (EI) and the identification was based on mass spectral databases

Table 4.1 Compounds identified in different types of adhesives by HS-SPME-GC-MS. Vinyl acetate–ethylene (VAE)-, poly(vinyl acetate) (PVAc)- and poly(vinyl alcohol) (PVOH)-based adhesives. Toxicity class III compounds (according to Cramer rules) are shown in bold. Data taken from ref. 7.

Adhesive type	Compound (quantification ion, *m/z*)
PU	Dodecane (168.2), glyceryl triacetate (103.1), **2,6-toluene diisocyanate** (174.0), butylated hydroxytoluene (205.2), adipate – structure unknown 1 (129.0), adipate – structure unknown 2 (129.0), isocyanate – structure unknown 3 (99.0), adipate – structure unknown 4 (82.1), phthalate – structure known 5 (149.0), phthalate – structure known 6 (149.0), **3,3′-dimethyl-4,4′-biphenylene diisocyanate** (250.1)
VAE	**Decamethylcyclopentasiloxane** (355.1), glyceryl triacetate (103.1), biphenyl (154.1), isobutyl nonyl phthalate (149.1), dibutyl phthalate (149.0), docosane (57.1), tricosane (57.1), tetracosane (57.1), bis(2-ethylhexyl) adipate (147.1), pentacosane (57.1), diethylene glycol dibenzoate (105.1), 2-ethylhexyl phthalate (149.0)
PVAc	**Decamethylcyclopentasiloxane** (355.1), glyceryl triacetate (103.1), diethylene glycol butyl ether (57.1), 2-(2-butoxyethoxy)ethyl acetate (87.1), biphenyl (154.1), 2,2,4-trimethyl-1,3-pentanediol diisobutyrate (111.2), isobutyl nonyl phthalate (149.1), dibutyl phthalate (149.0), docosane (57.1), tricosane (57.1), tetracosane (57.1), pentacosane (57.1), diethylene glycol dibenzoate (105.1), 2-ethylhexyl phthalate (149.0)
PVOH	Isobutyl nonyl phthalate (149.1), dibutyl phthalate (149.0), bis(2-ethylhexyl) adipate (147.1), 2-ethylhexyl phthalate (149.0)
Starch	Dibutyl phthalate (149.0), 2-ethylhexyl phthalate (149.0)
AC	Decahydronaphthalene (257.2), isobutyl nonyl phthalate (149.1)
VI + starch	Biphenyl (154.1)
HM	2,5-Di-*tert*-butyl-1,4-benzoquinone (220.1), butylated hydroxytoluene (205.2), hexadecane (57.1), octadecane (57.1), eicosane (57.1), 4b,8-dimethyl-2-isopropylphenanthrene (241.2), dehydroabietin (227.2), docosane (57.1), 1-methyl-10,18-bisnorabieta-8,11,13-triene (241.2), **retene** (219.0), dehydroabietal (269.2), dehydroabietic acid methyl ester (239.2), tetracosane (57.1)

and Kováts retention indices. HS-SPME using a PDMS–DVB (polydimethylsiloxane–divinylbenzene) fibre was used as sample treatment prior to GC-MS for the screening qualitative analysis, but liquid injection was carried out for quantitative analysis in the migration studies. A typical chromatogram obtained by HS-SPME-GC-MS is shown in Figure 4.1. The use of HS-SPME has the advantage of direct analysis of plastic materials without an aggressive sample treatment that could form artefacts, which might be attributed to the original material.

In this study,[8] different PU adhesives and plastic films (PET, OPA, cast PP, PE and PE–EVOH) were qualitatively analysed to identify volatile and semivolatile compounds. More than 63 volatile and semivolatile compounds were detected in the pure PU adhesives. Some of them were silane compounds, such as trimethoxy[3-(oxiranylmethoxy)propyl]silane, trimethylsilanol

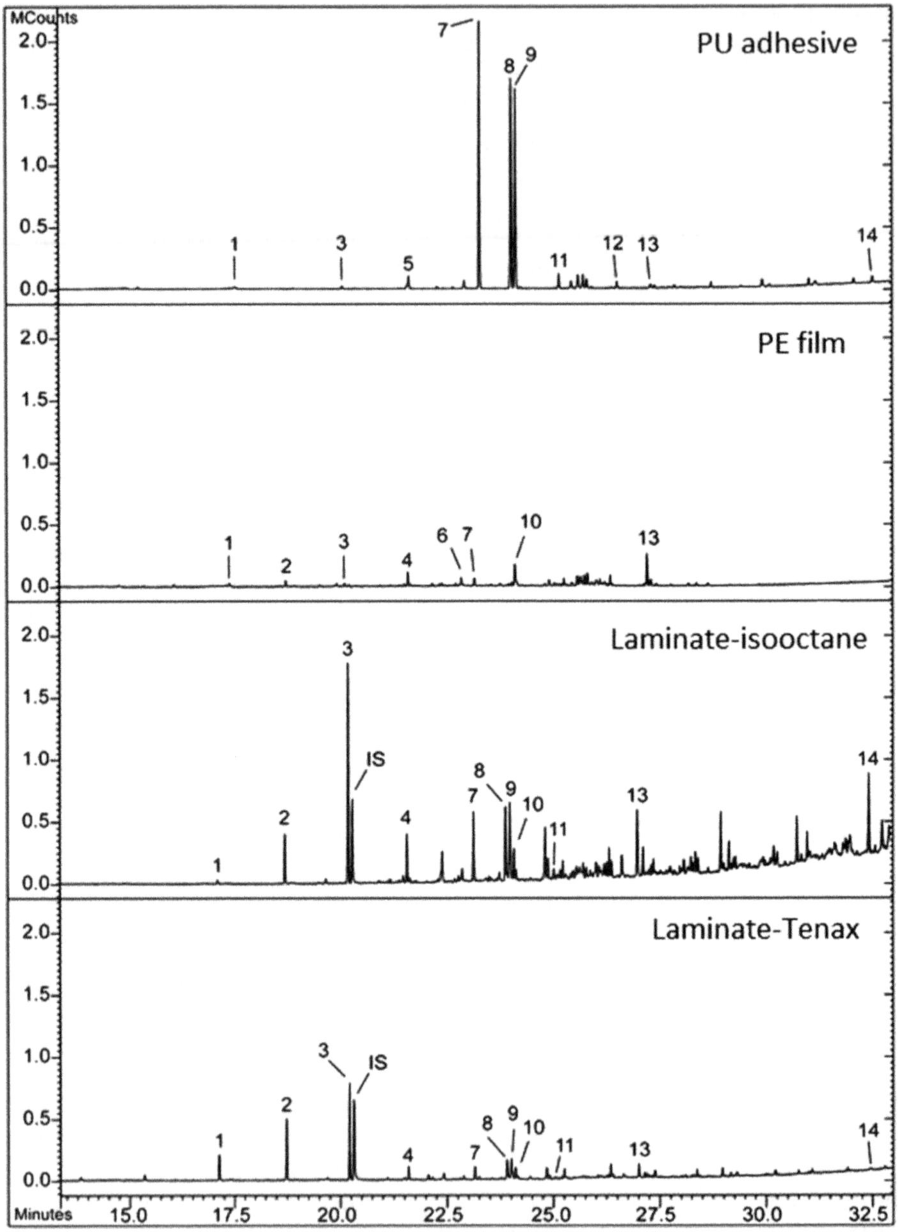

Figure 4.1 Typical chromatograms obtained by HS-SPME-GC-MS from a PU adhesive and a PE film and of the migration from a PU-based laminate (PET–PU adhesive–metallized PET–PU adhesive–PE) into isooctane and into Tenax® by GC-MS. Peak identification: 1 = undecane, 2 = dodecane, 3 = tridecane, 4 = tetradecane, 5 = octahydro-1,5-dinitroso-1,5-diazocine, 6 = 3-tridecanone, 7 = butylated hydroxytoluene, 8 = unknown adipate, 9 = 1,6-dioxacyclododecane-7,12-dione, 10 = hexadecane, 11 = 1,4,7-trioxacyclotridecane-8,13-dione, 12 = dodecyl cyclopentanoate, 13 = diisobutyl phthalate, 14 = triphenyl phosphate. IS, internal standard (4-*tert*-butylphenol). Reproduced from ref. 8 with permission from Springer, Copyright 2012.

and tetramethyl silicate. Silane compounds are common additives used to improve the chemical and thermal resistance of the adhesives. Other compounds identified were glycols (dipropylene glycol, ethylene glycol monophenyl ether and tripropylene glycol monomethyl ether), often used as chain extenders or cross-linkers; triethylamine, used to neutralize the resin or as an initiator in polymerization reactions; 2-ethyl-4-methyl-1,3-dioxolane, formed by reaction between diols and carbonyl compounds; 2,6-toluene diisocyanate (monomer); butylated hydroxytoluene (BHT), an antioxidant; 2,4-di-*tert*-butylphenol and 2,6-di-*tert*-butylbenzoquinone, derived from Irgafos and Irganox antioxidant degradation; 1,6-dioxacyclododecane-7,12-dione, a by-product of the polymerization reaction; tributyl phosphate and triphenyl phosphate, used as flame retardants; 2-methyl-3(2*H*)-isothiazolone, a biocide; and 1,4-trioxacyclotridecane-8,13-dione. Also, an unknown compound with an adipate structure was among the major compounds in one of the PU adhesives.

Additionally, studies of migration from 14 commercial multilayer packaging materials that contained these PU adhesives and that are used for dry foods such as coffee, for acidic foods such as tomato sauce and for fatty foods such as meat foodstuffs were performed.[8] Only three of the migrants quantified came from the adhesive: an unknown adipate, triphenyl phosphate and 1,4,7-trioxacyclotridecane-8,13-dione.

Athenstädt *et al.*[14] reported the use of GC-MS with either EI or chemical ionization (CI) to detect others NIASs such as cyclic esters formed as by-products during the reaction of adipic and isophthalic acid with monoethylene glycol and diethylene glycol. These four compounds integrate the polyol component of the PU adhesive. Cyclic ester oligomers, which are stable and do not possess free hydroxy groups to react with the diisocyanate, are potential migrants. A cyclic ester of isophthalic acid and tetraethylene glycol and a product of the reaction of isophorone diisocyanate with methanol were identified in the migration tests carried out.

4.3.1.2 Volatile and Semivolatile Compounds from Acrylic Adhesives

Identification of volatile and semivolatile compounds in acrylic-based adhesives used in FCMs using GC-MS has been reported in several publications.[3,13,15] Canellas *et al.*[13] detected 23 compounds in acrylic adhesives using a single-quadrupole MS (QMS) detector. The analysis for screening compounds was carried out by HS-SPME-GC-MS with a PDMS (polydimethylsiloxane) fibre at 80 °C for 15 min and acquisition was performed in scan mode (*m/z* 50–350). In contrast, for quantitative analysis liquid injection and the SIM acquisition mode were used. Two of the detected compounds, with an ester as the main structure, could not be identified. Three compounds were classified as toxicity class III by Cramer rules: 2,4,7,9-tetramethyl-5-decyne-4,7-diol, 2,4,7,9-tetramethyl-5-decyne-4,7-diol ethoxylate and 2-(1-phenylethyl)phenol. In addition, 2-ethyl-1-hexanol and 2-ethylhexyl acetate were found in all acrylic adhesives analysed. These compounds are probably impurities of the methyl methacrylate used to produce the adhesives.

More interesting analytical tools were reported by the same group for identifying migrants from acrylic adhesives.[3,15] On the one hand, GC coupled to a TOF mass spectrometer, based on accurate mass measurements of molecular and main fragment ions, was used.[3] On the other hand, atmospheric pressure GC coupled to quadrupole TOF-MS (APGC-Q-TOF-MS) was explored for the identification of unknowns.[15]

The analysis of acetonitrile extracts of non-cured acrylic adhesives was performed by GC-TOF-MS with CI.[3] This technique allowed the authors to confirm the identification and complete the screening of volatile compounds previously carried out by headspace (HS) and HS-SPME-GC-MS. For the latter, three different SPME fibres were tested: PDMS, PDMS–DVB and polyacrylate. HS-SPME was found to provide more sensitive measurements for compounds present at very low concentrations in the laminate. Thus, the number of compounds detected with the polyacrylate fibre was higher than that detected by HS and the peak area was also higher with SPME.

In the CI mode, pseudomolecular ions $[M + H]^+$ were detected and the identification was based on comparison between the theoretical accurate mass of the proposed compound and the experimental accurate mass obtained. In this way, 14 unknown compounds were identified and 17 compounds previously identified were confirmed. For some compounds, CI provided higher sensitivity than EI.[3] Some of these compounds, such as 2-ethyl-1-hexanol, 2-ethylhexyl *n*-butyrate and 2-methyl-1,2-thiazol-3(2*H*)-one were found in adhesives in which they had not been detected by EI. New compounds were also found by CI, such as 3-methyl-4-heptanone, 1-bromo-2-propanone, 2-ethylhexyl acetate, guanidine, 2-ethylhexyl 2-propenoate, 2-ethyl-2-methylpropylhexanoate, 2,5-pyrrolidinedione, 1-(benzoyloxy)ethanol, 2-(2-butoxyethoxy) ethanol, 2-(2-butoxyethoxy) acetate, 1-dodecanol, 2,4,7,9-tetramethyl-5-decyne-4,7-diol and 2,4-bis(1,1-dimethylethyl)phenol.

APGC ionization is an atmospheric pressure chemical ionization (APCI) process in which the target molecules are ionized by reagent ions generated in a nitrogen plasma. The ionization can occur by charge transfer or protonation and therefore the mass spectra can be dominated by M^+, $[M + H]^+$ or $[M-H]^-$. This soft ionization technique was used in combination with MS/MS for identifying non-targeted compounds in three different acrylic adhesives used in eight three-layer laminates with the structure paper–adhesive–substrate.[15] The substrates were PP, PET or cellulose acetate of different thickness. Samples were extracted by HS-SPME (PDMS fibre, 80 °C, 30 min) and by liquid extraction (acetonitrile for the adhesives and acetone for Tenax®). Migration studies were carried out by covering the plastic surface of the laminate with Tenax®.

In some cases, the technique was used to confirm some of the compounds previously identified by GC-QMS,[3] since the molecular ion was the most abundant in the APGC-Q-TOF-MS spectrum and it provided accurate mass data. It is worth noting that some compounds were detected with APGC-Q-TOF-MS but not by GC-QMS. The chromatograms were acquired in elevated energy (MS^E) mode to try to elucidate new molecules, since it provides exact mass information on precursor and fragment ions in the same run. Figure 4.2 shows the APGC-Q-TOF-MS chromatogram of a liquid extract of one of the acrylic adhesives studied and the high-energy spectra of three compounds [2-methyl-1,2-thiazol-3(2*H*)-one, 5-chloro-2-methyl-1,2-thiazol-3(2*H*)-one and 1,2-benzothiazol-3(2*H*)-one] identified by MS^E. First, the elemental composition of these compounds was obtained from the low-energy exact mass precursor ion spectra: C_4H_5NOS, C_4H_4ClNOS and C_7H_5NOS, respectively. The chemical database provided 65, 19 and 79 possible candidates, respectively, for these elemental compositions. A bibliographic search about adhesives showed that these molecular formulae corresponded to common biocides used in acrylic adhesive formulations. The exact masses of the fragment ions obtained in the high-energy spectra (Figure 4.2) confirmed the identification of the candidates. This finding was relevant, as these compounds are suspected to cause contact allergy in humans and to be cytotoxic.[15]

The migration results showed that only five compounds (2-ethyl-1-hexanol, 2-ethylhexyl acetate, 2-ethylhexyl *n*-butyrate, 2,4,7,9-tetramethyl-5-decyne-4,7-diol and 2,4,7,9-tetramethyl-5-decyne-4,7-diol ethoxylate) migrated above their limits of detection (<10 μg of compound per kilogram of simulant).

APGC-Q-TOF-MS has also been reported by Canellas *et al.*[16] for the identification of volatile migrants from self-stick adhesive labels used for direct food contact. The adhesive used in the label was a pressure-sensitive adhesive commonly used for direct contact on cheese, cold meats and meat derivatives. In this work, it was shown that the adhesive contained several volatile and semivolatile compounds including NIASs and the migration from the label to pork casing filled with isooctane (used as a fat simulant) was studied. In a subsequent study by the same group,[17] the migrants from the self-stick label in fresh sausage was determined by multiple HS-SPME-GC-MS (single quadrupole). The multiple approach allowed volatile compounds to be quantified directly in the solid meat sample.

4.3.1.3 *Volatile and Semivolatile Compounds from Hotmelt Adhesives*

Thirty-nine compounds were detected by GC-MS in two hotmelt adhesives,[13] including several alkanes (from decane to hexacosane) and four compounds classified as toxicity class II by Cramer rules, namely BHT, 3,5-di-*tert*-butylbenzoquinone, *cis*-jasmone and *cis*-calamenene. Also, nine polycyclic aromatic compounds (toxicity class III) were found,[13] namely acenaphthalene, octahydro-1(2*H*)-naphthalenone, octahydrochrysene, decahydrofluorenene, 3,5-di-*tert*-butyl-1,7-dihydroxy-8-methylnaphthalene, 4-(3-hydroxy-2,2,6-trimethyl-cyclohex-1-enyl)pent-3-en-3-one, indene,

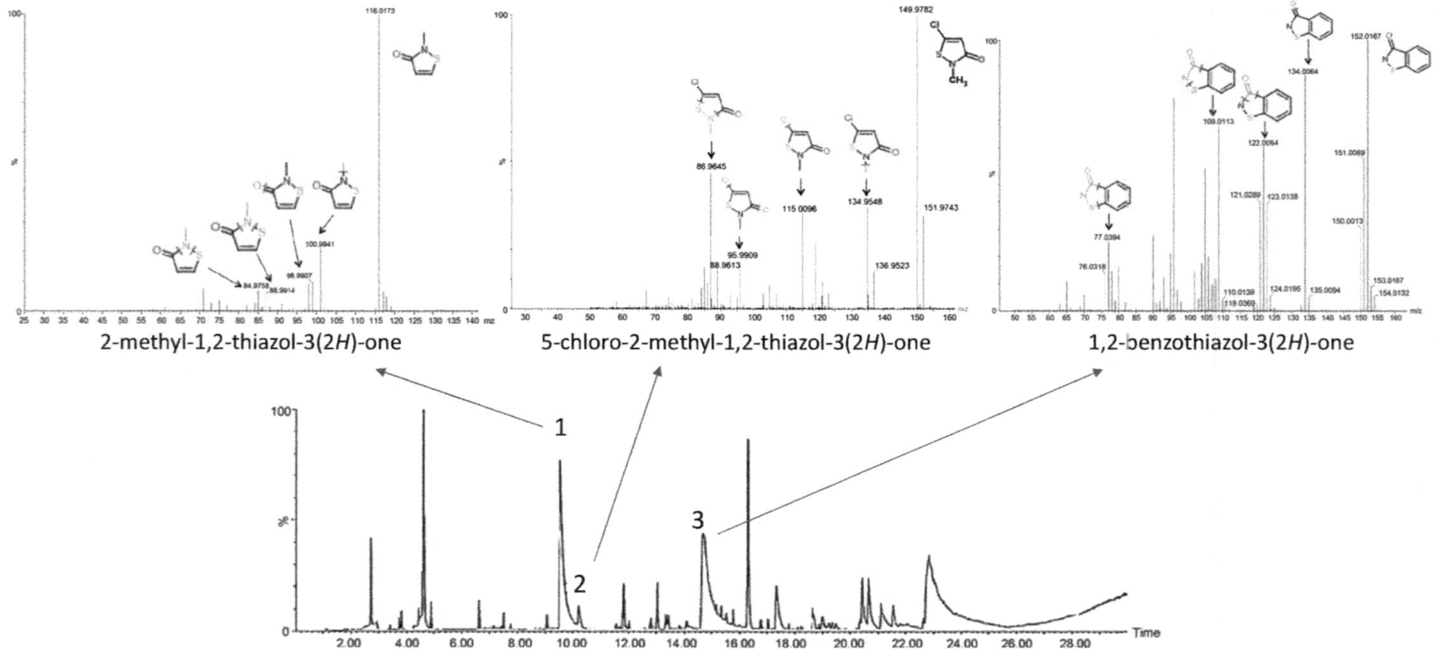

Figure 4.2 Chromatogram of the liquid extract of an acrylic adhesive obtained by APGC-Q-TOF-MS and high-energy spectra of peaks 1–3 identified as 2-methyl-1,2-thiazol-3(2H)-one, 5-chloro-2-methyl-1,2-thiazol-3(2H)-one and 1,2-benzothiazol-3(2H)-one, respectively. Adapted from ref. 15 with permission from Elsevier, Copyright 2012.

2,3-dihydro-1,1,3-trimethyl-3-phenylindole and 2-methyl-7-phenylindole. They probably came from the wax, oil or hydrocarbon resin used in the manufacture of the hotmelt adhesive.

4.3.2 Non-volatile Compounds from Adhesives

To evaluate the migration of polar and non-volatile adhesive components, the selected technique is usually LC followed by MS. Different mass spectrometers have been proposed for migration studies related to this type of compound, depending on the objective, sensitive quantification of target compounds or identification of non-targeted unknown compounds.

4.3.2.1 Non-volatile Compounds from Polyurethane Adhesives

The quantitative determination of primary aromatic amines (PAAs) by LC-MS is another example of the usefulness of MS for migration studies of PU-glued laminates into foods.[4,9,10]

PAAs are highly toxic compounds, suspected to be carcinogenic, which can be formed when the remaining unpolymerized aromatic isocyanates from PU adhesives meet water or humidity after the curing process. In addition, they can also come from recycled paperboard, which contains azo dyes from the printed paper used in the recycling process. According to EU regulations[2,18] for FCMs, the limit of PAAs released by the packaging material must be below 10 μg aniline equivalents per kilogram of food or food simulant.

The colorimetric method usually employed, based on derivatization of PAAs with *N*-(1-naphthyl)ethylene-1,2-diamine dihydrochloride (NEDA), lacks selectivity and does not allow the identification of the individual PAAs present in the food or food simulant. Therefore, several LC-MS methods have been developed. For instance, Lambertini *et al.*[10] carried out the determination by high-performance liquid chromatography–mass spectrometry (HPLC-MS) using a linear ion trap (LIT) instrument equipped with an electrospray ionization (ESI) source and SIM acquisition mode. Aznar *et al.*[9] reported the determination of the same analytes by ultra-performance liquid chromatography (UPLC) coupled to tandem quadrupole MS (triple quadrupole) with ESI in positive ion mode and selected ion recording (SIR) and MRM acquisition modes. They proposed a cation-exchange solid-phase extraction (SPE) step before the LC-MS analysis to concentrate the amines and to eliminate the residual matrix components that could produce an ionic suppression effect during the MS detection. This combination improved the sensitivity and the accuracy of the quantitative method. Table 4.2 summarizes some acquisition parameters of the HPLC-MS and UPLC-MS/MS methods and limits of detection (LODs) reported for PAAs.

Pezo *et al.*[4] reported the results of studies of migration from different multilayer films intended for food packaging, made of different substrates and different PU adhesives. Migration tests were carried out in thermo-sealed

Table 4.2 Determination of primary aromatic amines from PU adhesives by two different LC-MS methods. Data taken from ref. 9 and 10.

		HPLC-(LIT)MS[10]			UPLC-(QQQ)MS/MS[9]			
Amine	CAS No.	MH^+	SIM scan range (m/z)	LOD ($\mu g\ kg^{-1}$)	MH^+	Daughter ion	Collision energy (eV)	LOD_{ANL} (μg ANL equiv. kg^{-1})[a]
p-Phenylenediamine	106-50-3				109.2	91.6	18	0.078
m-Phenylenediamine	108-45-2				109.2	91.6	18	0.003
2,6-Toluenediamine	823-40-5				123.2	108.1	20	0.007
4-Methoxy-*m*-phenylenediamine	615-05-4	139.1	138–140	0.6	139.2	123.3	15	0.012
2,4-Toluenediamine	95-80-7	123.1	122–124	0.4	123.2	108.1	20	0.005
1,5-Diaminonaphthalene	2243-62-1				159.2	141.6	20	0.053
Aniline	62-53-3	94.0	93–95	0.1	94.2	76.4	20	0.300
Benzidine	92-87-5	185.2	184–186	0.5	184.3	155.3	35	0.009
o-Anisidine	90-04-0	124.1	123–125	0.1	124.1	109.2	15	0.014
4,4-Diaminodiphenyl ether	101-80-4	201.1	200–202	0.2	201.3	108.1	20	0.007
o-Toluidine	95-53-4	108.1	107–109	0.1	108.2	90.4	15	0.261
4-Chloroaniline	106-47-8	128.0	127–129	0.2	128.2	92.4	15	0.528
4,4-Methylenedianiline	101-77-9	199.2	198–200	0.4	199.5	106.2	20	0.001
o-Dianisidine	119-90-4	245.2	244–246	0.2	245.3	228.8	20	0.002
2-Methoxy-5-methylaniline	120-71-8	138.1	137–139	0.1	138.2	122.1	20	0.004
3,3′-Dimethylbenzidine	119-93-7	213.3	212–214	0.3	212.5	196.0	20	0.004
2,6-Dimethylaniline	87-62-7				122.2	106.0	20	0.231
4,4′-Thiodianiline	139-65-1	217.1	216–218	0.2	217.2	123.1	20	0.003
2,4-Dimethylaniline	95-68-1				122.2	106.0	20	0.014
2-Naphthylamine	91-59-8	144.1	143–145	0.1	144.3	125.8	30	0.008
4-Chloro-*o*-toluidine	95-69-2	142.1	141–143	0.1				
2,4,5-Trimethylaniline	137-17-7	136.1	135–137	0.1				
4,4-Methylenedi-*o*-toluidine	838-88-0	227.2	226–228	0.3	227.3	119.2	30	0.012
4-Aminobiphenyl	92-67-1	170.2	169–171	0.2	170.2	151.2	30	0.003
4,4′-Methylene-bis(2-chloroaniline)	101-14-4	267.2 + 269.2	266–270	0.9				
3,3′-Dichlorobenzidine	91-94-1	253.2 + 255.2	252–256	1.0				
4-Aminoazobenzene	60-09-3	198.1	197–199	0.1	198.2	105.1	20	
o-Aminoazotoluene	97-56-3	226.1	225–227	0.1				

[a]ANL, aniline.

bags filled with 3% aqueous acetic acid under different time and temperature conditions, depending on the type of food and conditions used in real life. After the migration tests, 39 NIAS impurities and other migrants were identified by UPLC-Q-TOF-MS in the food simulant. Moreover, using the proposed method, the authors showed that the colorimetric method using NEDA overestimates the levels of PAAs. They attributed this to impurities and other NIASs present in the plastic laminates.

Both GC-MS and UPLC-Q-TOF-MS have been proposed for evaluating the migration of NIASs in PU adhesives[19,20] and the role of EVOH as a barrier in laminates. Most of the migrants were identified as adipic-based cyclic compounds[20] derived from this monomer, which is used to obtain the polyester–polyol resins of a bicomponent PU adhesive. One of these NIASs was 1,4,7-trioxacyclotridecane-8,13-dione, a lactone classified as toxicity class III, which can migrate through PE layers very efficiently. This study showed that the presence of EVOH as a barrier layer decreases the migration but it does not inhibit the process completely. In addition, it was observed that this NIAS decomposed in contact with an acidic simulant to yield two degradation products that were also classified as toxicity class III.[19]

Lawson *et al.*[21] reported the use of matrix-assisted laser desorption/ionization mass spectrometry (MALDI-MS) for the rapid screening of three polyether and one polyol components of PU adhesives. They obtained the MALDI-MS fingerprint patterns of these components for reference purposes and carried out studies of migration from cured PET–PU–PE laminates in the form of pouches into distilled water. Only oligomers from the polyol component of the adhesive were detected as migrants in water. However, the use of this technique is limited because the measurements are not quantitative.

Phthalate migration from agglomerated cork stoppers was evaluated by Sendón *et al.*[22] The agglomerated stoppers studied were manufactured from natural cork granules and PU adhesives that contain esters, such as phthalates and adipates, as additives at concentrations of 2–5%. The HPLC-MS/MS method used was validated for dibutyl, diisononyl, diisodecyl and butylbenzyl phthalate. Triple-quadrupole MS/MS with ESI in positive ion mode and MRM acquisition mode was used for quantitative determination, providing quantification limits below the restriction limits. Migration tests were carried out by immersion of cork stoppers in 12% ethanol (as a wine simulant) for 2 h at 70 °C. Diisononyl phthalate was found in the two of the PU adhesives analysed at concentration levels of 2.9 and 10.4%. However, none of the phthalates were detected in the simulant after the migration tests.

4.3.2.2 Non-volatile Compounds from Acrylic Adhesives

Canellas *et al.*[6] identified several non-volatile compounds as migrants from acrylic adhesives using UPLC coupled to two different MS instruments: (a) a TOF-MS instrument that allowed the identification of several components by a combination of exact mass and in-source collision-induced dissociation (CID) and (b) a high-resolution MS instrument that provided an expanded range of fragmentation protocols for structural characterization studies by the combination of IMS and TOF-MS. Moreover, UPLC-Q-TOF-MS has been also proposed for the identification of acrylic adhesive compounds migrated from paper–plastic laminates.[13]

Three different water-based acrylic adhesives were analysed and the migration of adhesive components to Tenax® from different food packaging laminates (PP–adhesive–PE and PP–adhesive–paper) and from a label (made by applying adhesive on Al and PE) was studied.[6] The adhesives were extracted with acetonitrile and analysed by both UPLC-TOF-MS and UPLC-HDMS. The qualitative migration tests were carried out by covering both sides of the packaging with a monolayer of Tenax® in independent experiments. Then, the Tenax® was extracted with methanol and analysed by UPLC-TOF-MS. The MS range acquired was 100–1000 Da because it was considered that it is unlikely that masses above 1000 Da would migrate from the adhesive to the food. An ESI probe was selected because the compounds were supposed to be polar species as they were in a water-based acrylic adhesive, which is considered a polar medium. Positive and negative ion ESI modes were tested and no compounds were identified with the negative ion mode.

UPLC-TOF-MS analysis of the three acrylic adhesives studied showed that there was no common profile of compounds for the pure adhesives, even though they belong to the same type of adhesive. Moreover, the chromatograms showed a series of peaks that could be polymers and some wide peaks that could correspond to more than one compound but it could not be separated. However, the use of high-resolution MS allowed the identification of individual compounds in the spectra even though the chromatographic separation was not perfect. From the data provided by UPLC-TOF-MS analysis (retention time, exact mass and elemental composition), different components were identified: a series of polymers such as poly(ethylene glycol) (PEG), poly(propylene oxide) (PPO) and octylphenol ethoxylate (OPEO) and nonylphenol ethoxylate (NPEO) non-ionic surfactants.

The identification of some polymers could be made based on the exact mass and elemental composition obtained from the difference between each polymer mass. For instance, from mass 173.0790 to 965.5523 Da, corresponding to the first and 19th peaks of the series, respectively, it was found that the elemental composition ranged from $C_6H_{14}O_4Na$ to $C_{42}H_{86}O_{22}Na$, with a difference between the measured accurate mass of each polymer of the series of 44.0262 Da, corresponding to one unit of C_2H_4O. Based on this information, it was concluded that this series of polymers corresponded to PEG, which, with a molecular weight of about 200–800 Da, is usually used as a plasticizer in pressure-sensitive adhesives such as acrylic adhesives.[6]

In other adhesive, the elemental composition of a polymeric series of 11 peaks showed a measured mass difference of 58.0419 Da, corresponding to one C_3H_6O unit between each polymer of the series, and it was therefore identified as propylene oxide, the repeated unit in PPO.

OPEO and NPEO were other non-ionic surfactant polymers identified in the acrylic adhesives studied but, in this case, each polymer could not be separated by UPLC and the identification was made through their spectra. The spectrum of NPEO is shown in Figure 4.3 (top). As can be seen, the typical shape of a polymer spectrum is observed, with measured masses of 331.2254 (n = 2), 375.2527 (n = 3), 419.2787 (n = 4), 463.302 (n = 5), 507.3285 (n = 6), 551.3554 (n = 7), 595.3833 (n = 8), 639.4107 (n = 9), 683.4367 (n = 10), 727.4637 (n = 11), 771.4838 (n = 12), 815.5173 (n = 13), 859.5384 (n = 14), 903.5705 (n = 15), 947.5958 (n = 16) and 991.6204 Da (n = 17).

NPEO is suspected to have endocrine-disrupting properties and to cause harmful effects, including feminization and carcinogenesis, in various organisms. Therefore, the migration to food from packaging containing these adhesives must be studied. The spectrum obtained for the methanol extract of Tenax® after the migration test showed that only small polymers corresponding to low masses (331–551 Da) migrated to the Tenax® (Figure 4.3, bottom). This finding confirmed that migration is mainly governed by the molecular mass and that the smaller molecules can easily cross the materials and be dissolved in or adsorbed on the food.

Despite the capability of UPLC-TOF-MS for identifying unknown compounds, it is difficult to confirm the identity of a compound with only the exact mass information. That is why these adhesives were also analysed by UPLC-HDMS, which provided accurate masses of fragments to elucidate the identity of the known compound among the possible candidates for the elemental composition corresponding to the exact mass of a precursor ion. Thirteen new compounds (see Table 4.3) were identified by UPLC-HDMS, most of them esters that could be used as plasticizers added to the acrylic adhesive formulation.[6]

A three-phase hollow-fibre liquid-phase microextraction (HF-LPME) combined with an HPLC-UV method has been reported for the determination of two isothiazolinone biocides, 2-methyl-4-isothiazolin-3-one and 5-chloro-2-methyl-4-isothiazolin-3-one, in water-based acrylic adhesives used for food packaging materials.[23] The procedure, aimed at enhancing the sensitivity of the determination, involved three-phase HF-LPME using a semipermeable PP membrane, which contained 1-octanol as organic phase in the pores of the membrane. The donor and acceptor phases were aqueous acidic and alkaline media, respectively. The procedure was compared with direct analysis using UPLC-Q-TOF-MS, where the identification of the compounds and the quantification values were confirmed. Although the HF-LPME-HPLC-UV method is less sensitive (with LODs 10–20 times higher), it may be an interesting alternative to the UPLC-Q-TOF-MS method because it is easy to use, sensitive and affordable for any laboratory.

One approach to improving the barrier properties of polymers is the use of nanoparticles (NPs) into the food packaging material. In this context, the incorporation of selenium NPs (SeNPs), which act as free radical scavengers, together with the adhesive in a multilayer structure has been proposed. Palomo-Siguero *et al.*[24] reported the use of asymmetric flow

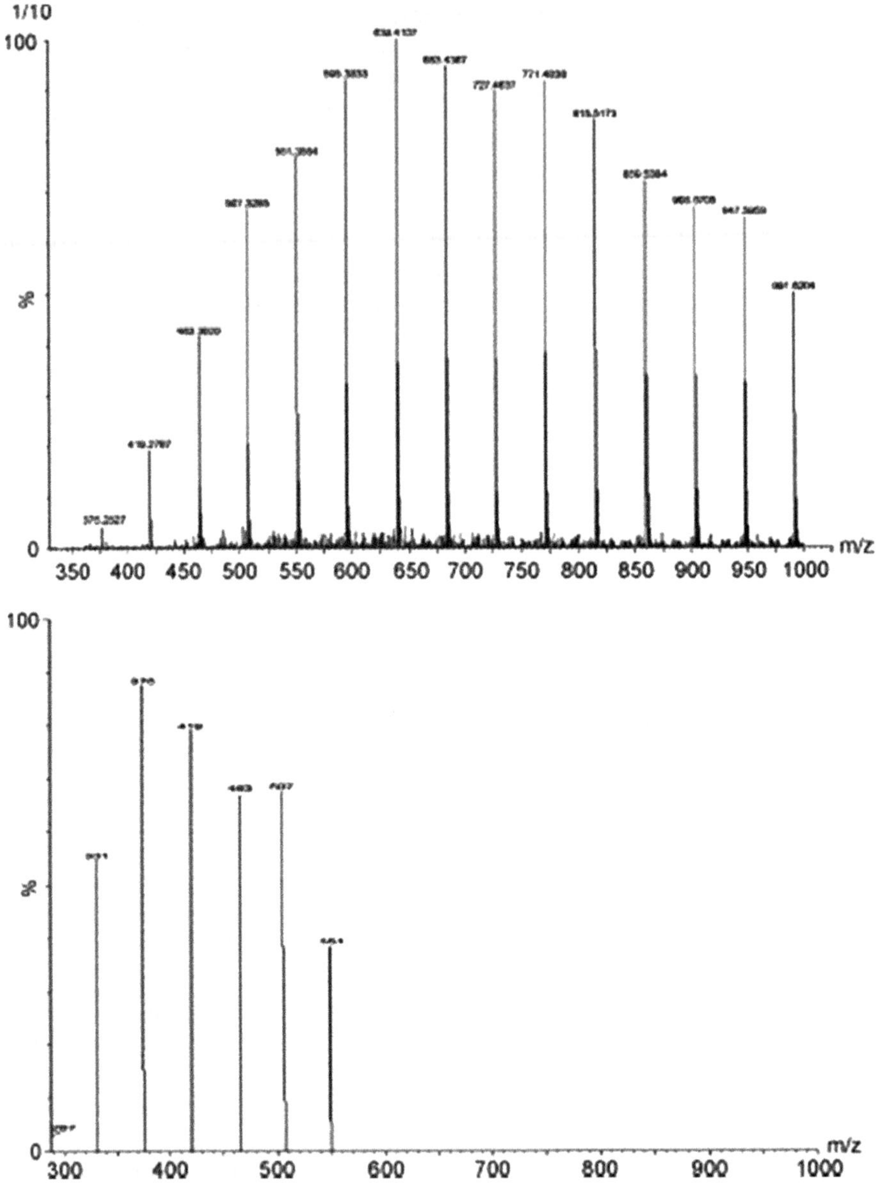

Figure 4.3 ESI-TOF MS spectra of NPEO in an acrylic adhesive (top) and NPEO migrated from the adhesive to Tenax® (bottom). Reproduced from ref. 6 with permission from Elsevier, Copyright 2010.

field-flow fractionation (AF^4) coupled with diode-array detection and inductively coupled plasma mass spectrometry (AF^4-UV-ICP-MS) to characterize SeNPs in an aqueous acrylic adhesive to be used in a multilayer food packaging material.

Table 4.3 Compounds identified in acrylic adhesives by UPLC-ESI-HDMS. Data taken from ref. 6.

Compound	Measured mass (Da)	Theoretical mass (Da)	Formula
Diethylene glycol diacetate	213.0727	213.0739	$C_8H_{14}O_5Na$
2-[2-(2-Ethoxyethoxy)ethoxy]ethyl acetate	243.1195	243.1208	$C_{10}H_{20}O_5Na$
2-(2-Butoxyethoxy)ethanol	185.1154	185.1154	$C_8H_{18}O_3Na$
Bis(2-ethylhexyl)-(*Z*)-but-2-enedioate	363.2527	363.2511	$C_{20}H_{36}O_4Na$
2-(2-butoxyethoxy)ethyl acetate	227.1261	227.1259	$C_{10}H_{20}O_4Na$
2-[2-(2-undecoxyethoxy)ethoxy]ethanol	327.2505	327.2512	$C_{17}H_{36}O_4Na$
2-(2-Octanoyloxyethoxy)ethyl octanoate	381.2630	381.2617	$C_{20}H_{38}O_5Na$
2-[2-(2-Octanoyloxyethoxy)ethoxy]ethyl octanoate	425.2899	425.2879	$C_{22}H_{42}O_6Na$
2-{2-[2-(2-Octanoyloxy)ethoxy]ethoxy}ethyl octanoate	469.3157	469.3141	$C_{24}H_{46}O_7Na$
2-(2-{2-[2-(2-Octanoyloxy)ethoxy]ethoxy} ethoxy)ethyl octanoate	513.3441	513.3403	$C_{26}H_{50}O_8Na$
2-[2-(2-{2-[2-(2-Octanoyloxy)ethoxy]ethoxy} ethoxy)ethoxy]ethyl octanoate	557.3680	557.3666	$C_{28}H_{58}O_9Na$
2-{2-[2-(2-{2-[2-(2-Octanoyloxy)ethoxy] ethoxy}ethoxy)ethoxy]ethoxy}ethyl octanoate	601.3943	601.3928	$C_{30}H_{58}O_{10}Na$
5-[2-(5-Hydroxy-4,4-dimethylpentoxy) ethoxy]-2,2-dimethylpentan-1-ol	313.2338	313.2355	$C_{16}H_{34}O_4Na$

4.3.2.3 Non-volatile Compounds from Hotmelt Adhesives

UPLC-Q-TOF has also been shown to be a powerful and successful tool for identifying and quantifying non-targeted migrant compounds from HM adhesives used in food contact laminates.[5,13]

HM adhesives were studied individually and as a component of laminates with substrate–adhesive–substrate structures, which are commonly used for packaging dry foods such as salt, sugar, cakes, *etc.* Up to 24 compounds were identified in two different HM adhesives, some of them with high toxicity levels according to the theoretical model of Cramer, such as 9,10-dihydroanthracene and retene.[25] One of the adhesives was based on ethylene vinyl acetate and the other on an amorphous poly-α-olefin enriched in propene, and both adhesives contained tackifiers and an antioxidant. Migration tests were carried out with Tenax®, which was extracted with methanol and analysed by UPLC-Q-TOF-MS with positive and negative ion ESI modes and MS^E acquisition mode. As a result, seven migrants were detected and five of them were successfully identified.[5] Their measured masses, the proposed formulae and the theoretical masses are listed in Table 4.4. The identification of the migrants listed in Table 4.4 was confirmed with pure standards. For the unknown 1, the molecular formula selected was $C_{16}H_{31}NO$ and the candidate for this compound was 1-dodecyl-2-pyrrolidinone because it had similar fragmentation spectra to unknown 1, but the identification could not be confirmed because of the lack of a standard. The candidate proposed for

Table 4.4 Non-volatile compounds identified by UPLC-Q-TOF-MS as migrants from hotmelt adhesives used in food packaging materials. Data taken from ref. 5.

Identified compound	Ion	Measure mass (Da)	Proposed formula	Theoretical mass (Da)
Tetradecanamide	$[M+H]^+$	228.2325	$C_{14}H_{30}NO$	228.2307
Unknown 1	$[M+Na]^+$	276.2303	$C_{16}H_{31}NONa$	276.2303
(9*Z*,12*Z*)-Octadecadienamide	$[M+Na]^+$	302.2457	$C_{18}H_{33}NONa$	302.2460
Hexadecanamide	$[M+H]^+$	256.2639	$C_{16}H_{34}NO$	256.2640
(9*Z*)-Octadecanamide	$[M+H]^+$	282.2804	$C_{18}H_{36}NO$	282.2797
3,5-Di-*tert*-butyl-4-hydroxybenz-aldehyde	$[M-H]^-$	233.1542	$C_{15}H_{21}O_2$	233.1542
Unknown 2	$[M-H]^-$	325.1849	$C_{14}H_{29}O_8$	325.1862

unknown 2 was heptaethylene glycol, but it was confirmed that the unknown compound did not correspond to this compound.

The identified migrated compounds from an amorphous poly-α-polyolefin-based HM adhesive were four amides (tetradecanamide, 9,12-octadecadienamide, hexadecanamide and 9-octadecenamide, known as oleamide) that are slip agents and anti-blocking additives used in polyolefin films to reduce friction resistance. Finally, the other compound identified, 3,5-di-*tert*-butyl-4-hydroxybenzaldehyde, could be derived from the degradation of Irganox 1010, which is an antioxidant commonly used in polyolefins. However, the quantitative analysis carried out in this work[5] revealed that none of the migration values exceeded the recommended Cramer exposure values.

4.3.2.4 Non-volatile Compounds from Other Types of Adhesives

Canellas *et al.*[26] reported the identification and quantification of NIASs coming from a water-based biodegradable adhesive in biodegradable food packaging by UPLC-ESI-Q-TOF-MS^E and GC-Q-MS. Thirteen compounds were identified in the adhesive and five of them were NIASs. The migration of the compounds through four different biodegradable materials was studied, showing that three of the identified compounds migrated from the adhesive to the solid food simulant (Tenax®): 2,4,7,9-tetramethyl-5-decyne-4,7-diol, 1,6-dioxacyclododecane-1,12-dione and 1,6,13,18-tetraoxacyclotetracosane-7,12,19,24-tetraone.

Pérez-Palacios *et al.*[27] reported a method for determining four bisphenol-type endocrine-disrupting compounds in food-contact recycled-paper materials by focused ultrasonic solid–liquid extraction and UPLC coupled to high-resolution MS. The target analytes were bisphenol A, bisphenol F and their derivatives bisphenol A diglycidyl ether (BADGE) and bisphenol F diglycidyl ether (BFDGE). These compounds are used in the production of epoxy resins that have many applications, such as epoxy adhesives and printing inks. In addition, Terasaki *et al.*[28] identified several structurally related impurities in a commercial epoxy adhesive component BADGE by NMR spectroscopy, fast atom bombardment mass spectrometry (FAB-MS) and GC-MS. The BADGE sample was fractionated by HPLC and the fractions were analysed by NMR spectroscopy to obtain structural information. Molecular formulae were derived from high resolution FAB-MS and low-resolution GC-MS after trimethylsilyl derivatization.

4.4 Mass Spectrometry in Delamination Studies

Delamination is an undesirable process that may occur because of the migration of packed product ingredients into a multilayer material. As a result, the adhesion strength between layers decreases and the laminate loses its barrier and appearance properties.[29–31]

The interaction between packaging and product has been widely studied because the transfer of additives, monomers, adhesive compounds, *etc.* to the product (particularly in the case of food products) may have human health implications. However, another phenomenon that occurs is the absorption of compounds of the product in packaging. For instance, flavours and aroma-responsible compounds are absorbed by polymers, causing the loss of essential product characteristics. Moreover, this absorption may become an important problem when the absorbed compounds can migrate through the inner layer (contact layer) to the junction with the next layer. In such cases, the adhesive joining the layers may be damaged and adhesion between layers decreases below the minimum quality value.

Delamination of flexible multilayer packaging intended for food, pharmaceutical and cosmetic products was studied by Ortiz and co-workers.[29,30] Sachets, made up of the laminate, the structure of which was PES–PU adhesive–Al–PU adhesive–LLDPE (linear low-density polyethylene), were filled with several aggressive cosmetics such as shampoo, lubricating gel and exfoliating cream and stored in an oven at 40 °C for 10 days. Since volatile and low molecular mass compounds are more prone to migrate through the contact layer (LLDPE), HS-SPME-GC-MS was the analytical technique selected to identify the compounds that had migrated to the packaging. After the migration period, sachets were manually delaminated without opening to avoid contamination of the external layer by the product, obtaining the LLDPE film with product and adhesive remains, on the one hand, and a two-layer PES–Al sample with adhesive remains on the Al side. After opening the sachet and discarding the product, the LLDPE layer was washed with distilled water before being analysed to remove product remains. Then samples were analysed by HS-SPME-GC-MS using a carboxen–polydimethylsiloxane (CAR–PDMS) fibre. MS detection was performed with an ion trap instrument with EI and scan acquisition mode; identification was carried out by a NIST library search. Several volatile compounds used in the cosmetics industry as perfumes, fixing agents and preservatives, such as menthol, dihydromyrcenol and 2-phenoxyethanol, were detected in the Al–PES delaminated layer. In a subsequent study,[31] the effect of the individual compounds on the packaging stability was studied. Compounds suspected of causing delamination,[29,30] such as 2-phenoxyethanol, phenylethanol, menthol, benzyl alcohol, *p*-propenylanisole, dihydromyrcenol, (*R*)-(+)-limonene, α-terpinene, methylparaben, ethylparaben, propylparaben, (*R*)-(+)-carvone, α-terpineol, α-pinene and 3,7-dimethyl-3-octanol, were selected for the study. Each cosmetic ingredient was added to a cream base (NeoPCL® Autoemulsionable O/W) at two concentration levels (1 and 10% m/m) and these formulations were used to fill the sachets, which were subjected to the delamination process. In addition to the above-mentioned PES–Al–LLDPE prepared with a two-component solvent-based PU adhesive, delamination of a PES–Al–LLDPE laminate obtained by an extrusion–coating process with PE film between layers was also studied. Cosmetic ingredients were detected in the inner layers of the laminate (Al–PES) by HS-SPME-GC-MS. The study also revealed that the presence of a phenyl and a hydroxyl group in the compound led to an important loss of adhesion in the adhesive joint laminate. The extrusion-coated laminate was the most prone to delamination by the compounds studied.

To elucidate the delamination mechanisms, different surface techniques, such as X-ray photoelectron spectroscopy, TOF secondary ion MS and attenuated total reflection Fourier transform infrared spectroscopy, were used to analyse the delaminated surfaces.[32] According to the surface analysis results, the PU adhesive remained on the PE surface when the laminate was either mechanically (by a T-peel process) or chemically delaminated (by migrated compounds), but adhesive degradation products were found only on the chemically delaminated Al surface. Therefore, the migrated chemicals seem to interrupt the adhesive cohesive interactions of PU molecules, whereas the mechanical separation of the layers seems to be produced by the rupture of the silane bonds between Al and PU. Conversely, chemical delamination of the extrusion-coated laminate is produced by the breaking of van der Waals bindings inside the PE, and as a result the PE surface was less oxidized and the Al surface contained higher molecular mass polyolefin fragments. In the case of mechanical delamination, the failure of the Al–PE joint was caused by the breaking of H-bonds between polar functional groups on oxidized Al and PE surfaces.

4.5 Conclusions and Trends

HRMS and MS/MS have been shown to be powerful tools for identifying and quantifying adhesive components and NIASs in different types of adhesives used in the manufacture of food contact packaging. Moreover, MS is essential in developing sensitive and selective methods for determining the amounts of adhesive compounds that have migrated from packaging into food simulants and thus to ensure regulatory compliance so as to not endanger human health. A new challenge in this field is the determination of adhesive residues in recycled materials intended for food contact applications. Migration studies will be required when these materials were used, even if they were not in direct contact with the food, because they may be a source of adhesive migrants. In this scenario, more sensitive and more selective analytical methods will be needed.

Adhesive formulations are in continuous evolution to improve their performance, for instance with the inclusion of new ingredients such as NPs. Also, in the frame of the circular economy, the production of new types of materials for biodegradable or easy-to-recycle packaging is demanded. For these reasons, much research on new adhesives for food packaging and on the migration of their components into foods is expected in the future.

References

1. European Parliament and Council Regulation (EU) 2004/1935, *Off. J. Eur. Union*, 2004, **L338**, 4.
2. Commission Directive 2002/72/EC, *Off. J. Eur. Commun.*, 2002, **L220**, 18.
3. C. Nerín, E. Canellas, M. Aznar and P. Silcock, *Food Addit. Contam., Part A*, 2009, **26**, 1592.
4. D. Pezo, M. Fedeli, O. Bosetti and C. Nerín, *Anal. Chim. Acta*, 2012, **756**, 49.
5. P. Vera, E. Canellas and C. Nerín, *Anal. Bioanal. Chem.*, 2013, **405**, 4747.
6. E. Canellas, C. Nerín, R. Moore and P. Silcock, *Anal. Chim. Acta*, 2010, **666**, 62.
7. M. Aznar, P. Vera, E. Canellas, C. Nerín, P. Mercea and A. Störmer, *J. Mater. Chem.*, 2011, **21**, 4358.
8. J. S. Félix, F. Isella, O. Bosetti and C. Nerín, *Anal. Bioanal. Chem.*, 2012, **403**, 2869.
9. M. Aznar, E. Canellas and C. Nerín, *J. Chromatogr. A*, 2009, **1216**, 5176.
10. F. Lambertini, V. Di Lallo, D. Catellani, M. Mattarozzi, M. Careri and M. Suman, *J. Mass Spectrom.*, 2014, **49**, 870.
11. P. Vera, B. Uliaque, E. Canellas, A. Escudero and C. Nerín, *Anal. Chim. Acta*, 2012, **745**, 53.
12. P. Vera, E. Canellas and C. Nerín, *Food Chem.*, 2014, **145**, 237.
13. E. Canellas, P. Vera and C. Nerín, *Food Chem. Toxicol.*, 2015, **75**, 79.
14. B. Athenstädt, M. Fünfrocken and T. C. Schmidt, *Rapid Commun. Mass Spectrom.*, 2012, **26**, 1810.
15. E. Canellas, P. Vera, C. Domeño, P. Alfaro and C. Nerín, *J. Chromatogr. A*, 2012, **1235**, 141.
16. E. Canellas, P. Vera and C. Nerín, *J. Mass Spectrom.*, 2014, **49**, 1181.
17. E. Canellas, P. Vera and C. Nerín, *Food Chem.*, 2016, **197**, 24.
18. Commission Directive 2007/19/EC, *Off. J. Eur. Union*, 2007, **L97**, 50.
19. D. Carrizo, A. Maccagnan, J. S. Félix, C. Nerín and O. Bosetti, *Packag. Technol. Sci.*, 2015, **28**, 1039.
20. F. Isella, E. Canellas, O. Bosetti and C. Nerín, *J. Mass Spectrom.*, 2013, **48**, 430.
21. G. Lawson, S. Bartram, S. Fitchner and E. D. Woodland, *Analyst*, 2000, **125**, 115.

22. R. Sendón, A. Sanches-Silva, J. Bustos, P. Martín, N. Martínez and M. E. Cirugeda, *J. Sep. Sci.*, 2012, **35**, 1319.
23. M. Rosero-Moreano, E. Canellas and C. Nerín, *J. Sep. Sci.*, 2014, **37**, 272.
24. M. Palomo-Siguero, P. Vera, Y. Echegoyen, C. Nerín, C. Cámara and Y. Madrid, *Spectrochim. Acta, Part B*, 2017, **132**, 19.
25. P. Vera, M. Aznar, P. Mercea and C. Nerín, *J. Mater. Chem.*, 2011, **21**, 420.
26. E. Canellas, P. Vera and C. Nerín, *Anal. Bioanal. Chem.*, 2015, **407**, 6781.
27. D. Pérez-Palacios, M. A. Fernández-Recio, C. Moreta and M. T. Tena, *Talanta*, 2012, **99**, 167.
28. M. Terasaki, T. Kazama, F. Shiraishi and M. Makino, *Chemosphere*, 2006, **65**, 873.
29. G. Ortiz and M. T. Tena, *J. Chromatogr. A*, 2006, **1101**, 32.
30. G. Ortiz, A. Garrido-López and M. T. Tena, *Packag. Technol. Sci.*, 2007, **20**, 173.
31. A. Garrido-López, A. Santa-Cruz, E. Moreno, J. Cornago, M. C. Cañas and M. T. Tena, *Packag. Technol. Sci.*, 2009, **22**, 415.
32. A. Garrido-López and M. T. Tena, *Appl. Surf. Sci.*, 2010, **256**, 3799.

CHAPTER 5

Volatile Compounds Through and from Packaging

LUDWIG GRUBER*[a] AND JONATHAN BEAUCHAMP[b]

[a]Department of Product Safety and Analytics, Fraunhofer Institute for Process Engineering and Packaging IVV, Giggenhauser Strasse 35, 85354 Freising, Germany; [b]Department of Sensory Analytics, Fraunhofer Institute for Process Engineering and Packaging IVV, Giggenhauser Strasse 35, 85354 Freising, Germany
*E-mail: ludwig.gruber@ivv.fraunhofer.de

5.1 Introduction

5.1.1 Volatile Organic Compounds

Volatile organic compounds (VOCs) are a group of chemicals that are characterized as having high vapour pressure at room temperature. According to EC Directive 1999/13/EC (Solvent Emissions Directive), VOCs are defined as "any organic compound having an initial boiling point less than or equal to 250 °C".[1] As such, VOCs are often present in the gas-phase, but owing to their typically low molecular weight – and depending on their physicochemical properties – they can adsorb on and migrate through surfaces, as can occur in food contact materials (FCMs), *e.g.* in packaging films. In this context, the analysis of VOCs in relation to the interactions between packaging materials and foods focuses primarily on substances that migrate.

Food Chemistry, Function and Analysis No.10
Food Contact Materials Analysis: Mass Spectrometry Techniques
Edited by Michele Suman
© The Royal Society of Chemistry 2019
Published by the Royal Society of Chemistry, www.rsc.org

This chapter covers the migration of volatile substances in FCM, detailing migration and permeation phenomena, sampling techniques and analytical instrumentation.

5.1.2 Modelling Substance Migration in Food Contact Materials

Migration, permeation and – to a lesser extent – absorption are the principal mechanisms that determine the degree of interaction of volatiles with contact materials. These processes are influenced by several factors, primarily temperature, storage period, geometry and several other environmental parameters.

Food packaging can comprise different materials, including cellulose (paperboard), polymers (films) and metal (cans, films) and composites thereof, and can exhibit varying degrees of complexity, from single components to multilayer packaging systems. Both of these aspects (material and complexity) depend on the food that is packaged and the explicit intention of the packaging system, *e.g.* whether it is merely for logistical convenience along the supply chain or whether it protects the food or extends its shelf-life. In terms of "contamination" of foods due to migration, the packaging material in use may itself act as a source of migrating substances that can influence the quality or safety of packed foods, yet conversely, many packaging materials are designed with the specific intention of protecting the packaged goods from substances that migrate from outside sources, *e.g.* as is often the case with the use of secondary packaging materials.

In general, the mass transfer of migrating substances – or migrants – from FCMs obeys Fick's law of diffusion and can be predicted *via* a theoretical model, *e.g.* for polymeric materials.[2] Typical theoretical concentration curves of a migrant that migrates from an FCM (concentration decrease) into a food (corresponding concentration increase) is presented in Figure 5.1.

For single-layer materials, there are three key parameters that characterize migration behaviour:

1. the initial concentration, $c_{P,0}$, of the migrating substance within the packaging;
2. the diffusion coefficient(s) of the migrating substance in the packaging material, D_P;
3. the partition coefficient, $K_{P,F}$, between the food contact polymer and food.

The migration of compounds from food packaging materials into the packaged food is strictly regulated in European countries by both national and European directives. Regulations typically distinguish between overall migration limits (OMLs) and specific migration limits (SMLs). In the EU, many compounds have SMLs that can be determined *via* established analytical methods that permit systematic compliance testing. SMLs from food

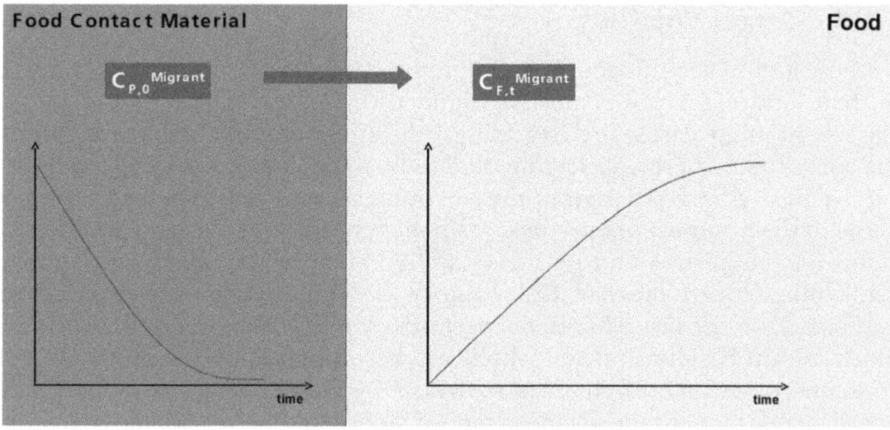

Figure 5.1 Theoretical decrease in the concentration of a migrant in a polymer (left) *versus* its corresponding concentration increase in the food (right) over time in an FCM model. © Fraunhofer IVV.

packaging materials, which range from "none detectable" to 60 mg kg^{-1}, are a fundamental requirement that is dictated by Article 3 of Framework Regulation EC (No.) 1935/2004 and testing schemes for the purposes of verifying compliance with both SMLs and OMLs are provided in Commission Regulation (EU) No. 10/2011 (Plastics Implementing Measure; PIM).[3] Substances that are currently restricted by SMLs include plasticizers, stabilizers and antistatic agents, among others.

As in many areas of VOC analysis, extraction or migration tests on migrating substances often lead to the detection of unknown compounds (in addition to those being targeted). As a consequence, analytical approaches must ideally be sufficiently robust to cater for unknown compounds in addition to non-intentionally added substances (NIAS) during compliance testing. As a general convention, an SML threshold of 10 µg kg^{-1} (10 ppbv) is often applied for NIAS in food during evaluation; this level is specified in EU Regulation No. 10/2011 for migration through a functional barrier (although there are specific exceptions for the use of unpermitted but intentionally added substances in FCM plastics when these are present behind a functional barrier and provided they do not migrate into the packaged food at levels above 10 µg kg^{-1}).

Taking the aforementioned aspects into account, different considerations in the analysis of these compounds must be made, namely:

- the analysis of the initial concentration within the packaging to investigate the migration potential from the packaging material;
- the implementation of specific migration tests, especially for FCMs, in order to perform compliance testing;
- the analysis of the permeation behaviour through the packaging material in order to understand the behaviour as a functional barrier.

5.1.3 Target Volatiles

As indicated in the Introduction, migrating substances are typically volatile in their nature, *i.e.* VOCs, and are highly diverse in terms of their chemical classes. In many cases, the migrating substances are directly dependent on the constituents of the packaging materials. In polymer packaging, migrants can include unreacted monomers, oligomers, non-polymerizable components of the original charge stock, residual solvents and reaction products of stabilizers, adhesives and plasticizers.[4,5] Frequently, complex non-polymerizable mixtures are present. The majority of volatiles present in virgin paper and board, on the other hand, are derivatives of autoxidation of unsaturated lipids within residual resins, which are accompanied by resinous flavour components and short-chain fatty acids.[6] By contrast, recycled paper and board regularly contain a wide range of volatile constituents derived from diverse sources, including printing inks, adhesives and other contaminations. Recently, the problem of mineral oil hydrocarbons (MOHs), comprising mineral oil saturated hydrocarbons (MOSHs) and mineral oil aromatic hydrocarbons (MOAHs), migrating into foodstuffs from packaging containing recycled fibres has emerged[7] and they are of particular concern owing to the suspected carcinogenic nature of some components.

In addition to the packaging material itself being a source of migrating substances, additional components, such as printing inks on the outside of packaging materials, might also be a possible direct source of volatile constituents – including MOHs – in packaging. Two migration mechanisms are feasible in relation to contamination *via* printing ink constituents: on the one hand, the printing inks can migrate (permeate) through the packaging into the food; on the other, the pre-packaging storage of the packaging materials (*e.g.* on rolls or in stacks) results in direct contact of the printed side of the packaging with the food contact side, leading to a direct transfer of the printing ink constituents to the food contact side, which can be subsequently transferred to the food itself after packaging. This latter migration mechanism is known as "set-off." All of these transfer processes are determined by the partition coefficients (between the layers) and the diffusion coefficients (within the layers).[8]

Other target compounds for migration testing might include odour-active substances and by-products that can lead to off-flavour in the packaged food. These might include oxidation products, including aldehydes such as hexanal, and also other off-flavour components, depending on the packaging materials (*cf.* Section 5.4.5).

5.1.4 Analytical Instrumentation

Choosing an appropriate analytical technique(s) to address suitably the question at hand in migration analysis is always a crucial step. Several analytical methods for determining the quality and quantity of specific migrants in food simulants are described in the European Standard EN 13130 published by the European Committee for Standardization.[9] Further advice for testing FCMs is given by the EU's Community Reference Laboratory (CRL) of the Joint Research Centre (JRC) *via* its website[10] or other published guidelines.

The analytical techniques in widespread use for analysing VOCs in relation to migration are based on gas chromatography (GC) coupled with either flame ionization detection (FID) or mass spectrometry (MS).

5.1.4.1 Gas Chromatography

GC exploits the physicochemical properties of target volatiles to separate them physically as they are flushed through the chromatographic column to the detector (*e.g.* FID or MS). The elution time of a target compound through the column depends on several additional factors, such as the properties (polarity) of the column, the flow rate of the carrier gas (typically helium) and the temperature program of the GC oven. In the first stage, however, the constituent volatile compounds of a sample must be transferred from the extract onto the GC column.[11] The sample extract can be in different physical states, such as gaseous, dissolved in a liquid, trapped cryogenically or adsorbed on a solid material. Headspace sampling techniques (*cf.* Section 5.2.2), including headspace-trapping, are very robust methods, as is cryogenic trapping. One particular challenge of the latter, however, when analysing samples containing water (as is the case for most foods), is to avoid blockage of the trap through crystallization, *i.e.* ice. This can be overcome by water removal steps, typically by cooling, such as by using Peltier water traps (for polar and non-polar substances) or membrane water traps (for non-polar substances), but this presents a further challenge for VOC recovery, especially for polar substances, and therefore must be accounted for and checked.

Injection of liquid extracts onto the GC column is also possible and is typically carried out by using syringes for on-column (OC) or split/splitless (SSL) injection. Techniques such as programmed temperature vaporization (PTV) and large-volume (LV) injectors cater for the direct analysis of solvent extracts, although both of these techniques have limitations due to minimal differences in the volatility between the solvent and the constituent analytes of the extract.

As indicated above, the column type and the temperature program are also of critical importance in GC analysis. Columns are available with differing properties relating to selectivity, thermal stability and inertness and the choice of column should be matched with the requirements of the volatiles being targeted, specifically their polarity. For samples that contain both polar and non-polar compounds, the use of only one type of stationary phase may lead to preferential separation of one kind of component at the expense of other classes of components. WAX- and DB-624-type columns are highly selective but their thermal stability is usually poor, whereas DB-1- and DB-5-type columns are both selective and thermally stable. Additionally, comprehensive libraries of compound-related retention indices are available for the latter compounds, which aid in compound identification.

5.1.4.2 Mass Spectrometry

Mass spectrometers are available in different configurations, including ion traps (IT-MS), time-of-flight systems (TOF-MS) and quadrupole technologies to varying degrees of complexity (QMS, tandem or triple quadrupole). The most prevalent type of mass spectrometer used with GC analyses is the quadrupole-based system, typically equipped with a direct-injection inlet.

In order to make target compounds detectable, they must first be ionized. Mass spectrometers operate in either the electron ionization (EI) or chemical ionization (CI) mode, with the former mode being the most commonly used. Ionization in the EI mode is standardized at 70 eV; this produces characteristic – and often unique – mass spectra of compounds that allow them to be identified *via* library-searchable matches in many cases.

The process of fragmentation during EI follows predictable reaction pathways whereby the ions that are formed reflect the most stable cations and radical cations in the gas phase. The highest molecular weight observed in a spectrum is typically that of the parent molecule (minus an electron), which is termed the molecular ion (M^+). Generally, smaller $[M+1]^+$ ions are observed above the calculated monoisotopic mass due to the natural isotopic abundances of predominant elements, such as ^{13}C, ^{2}H, *etc.* The EI mode of operation is useful for compound identification and/or structural elucidation purposes.

By comparison, CI is a softer ionization process that exploits ion–molecule reactions.[12] CI utilizes a similar ion source to EI, albeit with one notable exception, namely that the CI ion source housing is almost closed, *i.e.* its apertures are much smaller than in an EI source, leading to higher pressures in the source (*ca.* 0.1–100 Pa for QMS). As a result, reagent ions react with the analyte (target) molecules to form ions by either proton or hydride transfer. Methane is typically the gas of choice for CI mode because it produces a certain degree of fragmentation and is also preferred for instrumental reasons. In this configuration, a large excess of methane is channelled into the ion source and collisions with energetic electrons produce CH_4^+ and CH_3^+, which further react with methane to form CH_5^+ and $C_2H_5^+$. Subsequently, these ions transfer protons to the sample molecule to form $[M + H]^+$ quasi-molecular ions. The molecular weight information gained with this method aids in compound identification and quantitation. Semiquantitation without standards is usually difficult owing to the highly variable relative response factors of the ionization efficiency, although this depends on the technique in question. Whereas almost all neutral analytes form positive ions, negative chemical ionization (NCI) is especially used for the analysis of compounds that contain acidic groups or electronegative elements (especially halogens).[13]

Mass spectrometers can be operated in one of two modes, either the scan or the selected ion monitoring (SIM) mode; the latter is especially useful for recording the mass traces of target compounds that are known *a priori*, since the characteristic ions of a compound can be monitored based on unique branching ratios, thereby allowing the detection of specific analytes with increased sensitivity relative to the full-scan mode.

The coupling of other mass spectrometers to GC, such as TOF or IT instruments, has advantages when operating in the scan mode, although the linearity ranges are typically not as broad as with QMS. The high scan rate of TOF instruments is a notable advantage of this type of instrument, especially in the field of fast GC and comprehensive two-dimensional gas chromatography–mass spectrometry (GC×GC-MS or 2D-GC-MS).[14] One of the key advantages of GC-IT-MS systems is their ability to perform multiple MS/MS fragmentation steps on a single precursor ion, which adds an additional parameter for compound identification, although this feature is of limited value for many volatiles owing to the lack of suitable fragments achievable in performing such collision-induced dissociation reactions.

More recent developments in mass spectrometry include higher resolution instruments based on techniques such Orbitrap or quadrupole time-of-flight (Q-TOF) instrumentation, but these are costly and generate large volumes of data that make data processing challenging. Nevertheless, exact mass measurements for small molecules delivered by such techniques allow for improved confidence in the selection of elemental compositions for unknown compounds.[13]

The typical instrumentation and techniques used in relation to varying scenarios for specific target compounds are summarized in Table 5.1.

Table 5.1 Typical instrumentation used for substance migration analyses in different scenarios.

Type of substance	Example	Predominant technique
Volatile organic compounds (bp < 150 °C)	Monomers, residual solvents, aroma compounds	Headspace, SPME, purge-and-trap combined with GC-MS and/or GC-FID
Semivolatile organic compounds (bp < 300 °C)	Plasticizers, glycols, additives, photoinitiators and oligomers with MW < 400 Da	Liquid injection GC (split, splitless, PTV, on-column. *etc.*) coupled with FID or MS
Thermally unstable compounds	PAAs, some unsaturated organic compounds, natural fragrances	On-column or PTV injection GC coupled with FID or MS, HPLC-MS
Mixtures of hydrocarbons	MOSHs, MOAHs	LC-GC[48]

5.2 Analysis of Volatile Constituents of Food Contact Materials

5.2.1 Establishing Initial Conditions

A prerequisite for characterizing a migration process is knowledge of the initial concentrations of the volatile migrant in the packaging material ($c_{P,0}$). Several analytical approaches can be used to quantify volatiles in packaging materials and the sampling of these volatiles plays a critical role in achieving accurate results. Sampling of volatile migrants can be achieved *via* either the gas phase (headspace extraction) or the liquid phase (solvent extraction). Although the extraction efficiency plays an important role, other factors must be considered, such as the use of appropriate (and not harsh) conditions to avoid dramatic losses during sample cleanup or disruptive matrix effects during GC-MS analysis. It should also be ensured that the solvent used for the extraction does not interfere with the target compounds on the GC column.

5.2.2 Headspace Sampling

Headspace sampling is a separation technique in which volatile materials may be extracted from a sample matrix *via* the gas phase. Headspace analysis is based on the phenomenon of partitioning of volatiles from the (solid- or liquid-phase) sample into the gas phase (headspace) above it. The higher the volatility of a compound, the higher will be its gas-phase concentration in the sample headspace. Conversely, the less volatile components, which typically represent the bulk of the sample, will tend to remain in the solid or liquid phase. Furthermore, the migration of a compound into the headspace depends not only on its volatility but also its affinity to the original sample phase, in addition to other factors such as temperature, pH and duration.

Headspace sampling techniques can generally be divided into three methods, namely static headspace, dynamic headspace and purge-and-trap (P&T), as represented schematically in Figure 5.2.

5.2.2.1 Static Headspace

In static headspace sampling, the sample is placed in a closed system, for example a glass vial, typically with a large headspace-to-sample ratio. The sample is then allowed to equilibrate with the headspace gas in the enclosed system under controlled conditions (*e.g.* temperature, duration) and sometimes with additional specific parameters, *e.g.* agitation or stirring (in the case of liquid samples). After a defined period of equilibration, the headspace gas is extracted, either *via* a syringe or *via* other extraction techniques, after which the extract is transferred to the analytical system (*e.g.* injection onto a GC column).

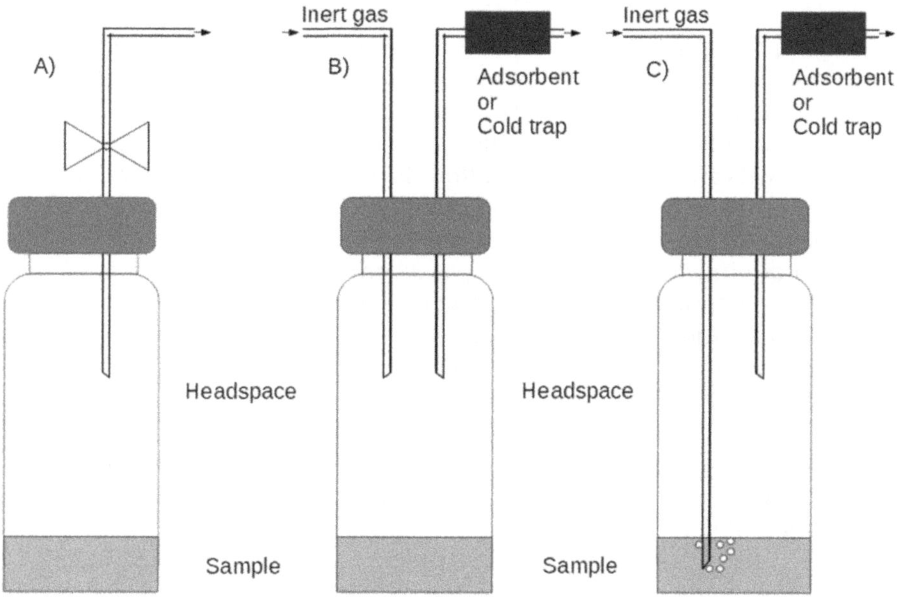

Figure 5.2 Comparison of sampling techniques for extracting volatile organic compounds from a sample: (A) static headspace, (B) dynamic headspace and (C) purge-and-trap (P&T). © Fraunhofer IVV.

5.2.2.2 *Dynamic Headspace*

Dynamic headspace sampling is similar to static headspace sampling, but with sampling achieved by purging the headspace gas out of the enclosed chamber and into a sampling system. Headspace flushing can commence either immediately after placing the sample in the receptacle or after achieving equilibrium between the solid (or liquid) phases and the headspace, as is performed in static headspace analysis. Sample extraction is then performed by flushing the headspace gas through an adsorbent or a cryogenic trap.

5.2.2.3 *Purge-and-trap*

Purge-and-trap (P&T) sampling is similar to dynamic headspace sampling, but with the purge gas being flushed through the liquid matrix to promote the release of volatiles into the gas phase (headspace). Sample extraction is performed in a similar manner to that in dynamic headspace analysis.

5.2.3 Gas-phase Extraction Techniques

5.2.3.1 *Solid-phase Microextraction*

Solid-phase microextraction (SPME) is a sampling technique that involves the use of a fibre coating for extraction.[15] Different fibre coatings are available, such as Carboxen, polydimethylsiloxane (PDMS), divinylbenzene and

Carbowax or mixtures thereof, and the choice of fibre depends on the target analyte type, *e.g.* low molecular weight compounds, amines and nitroaromatic compounds or polar compounds, amongst others. SPME with GC-MS analysis (SPME-GC-MS) is widely used to identify and quantify the volatile compounds in liquid foods, especially alcoholic beverages such as beer and wine.[16] SPME can be used either in liquids or in the gas phase for headspace extraction, whereby the sampling sensitivity in the latter is best when the headspace volume is small. After sample extraction, the SPME fibre is transferred to the injection port of the gas chromatograph, where it is typically heated to promote thermal desorption of the analytes captured on the fibre. SPME is a fast and simple extraction technique that does not require the use of solvents. The accuracy of SPME-GC-MS is affected by matrix effects relating to the compounds present in the extracted sample. The amount of an analyte extracted by the fibre is proportional to its concentration in the sample and the partition coefficient between the sample and fibre, hence it is important to achieve equilibrium during extraction, which can be supported by agitation of the sample.

5.2.3.2 Stir-bar Sorptive Extraction

Stir-bar sorptive extraction (SBSE) is a technique for extracting organic compounds from aqueous matrices.[17] SBSE is typically achieved by using a PDMS-coated magnetic stirring rod that is placed in the sample under defined conditions and for a specific period. After sample extraction, the SBSE rod is transferred to a thermal desorption unit and the volatiles are injected onto the GC column in a similar manner to that in SPME sampling.

Since SBSE rods have a larger capacity to concentrate analytes from a much larger sample volume than SPME, the method is typically very sensitive (for liquid-phase sampling). The extraction efficiencies of both of these sampling methods can be predicted by the octanol–water partition coefficient, log K_{ow}, and apolar compounds with a higher log K_{ow} may show higher relative responses in both methods. Figure 5.3 displays a schematic comparison of the two sampling systems, SPME and SBE.

The use of SBSE in combination with a solvent-assisted approach – SA-SBSE – is a new technique that is treated as a case study (Case Study 1) in relation to a comprehensive analysis of volatile flavour compounds in beverages.[18]

5.2.3.3 Thermal Desorption

Thermal desorption (TD) (also referred to as thermodesorption) is a method for transferring extracted volatiles into an analytical system, typically GC. The technique is often performed on SBSE or P&T extracts. Furthermore, it is in use in many areas of applications, especially in the field of air monitoring.[19] The general principle of TD is always the same: VOCs are collected onto a sorbent that is subsequently heated in a flow of gas that is directly applied to the gas chromatograph.

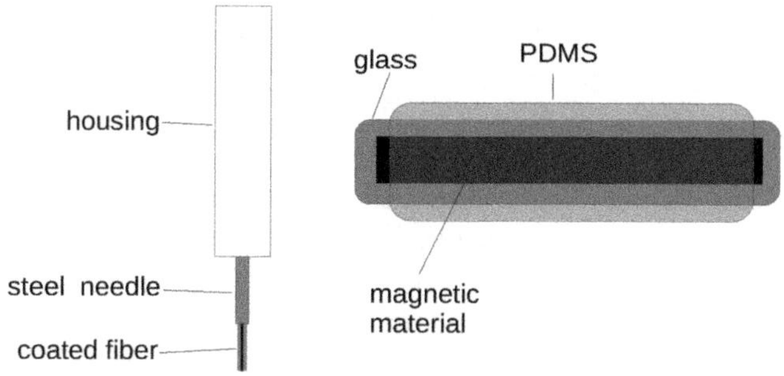

Figure 5.3 Schematic of the extraction apparatus used in solid-phase microextraction (SPME) (left) and stir-bar sorptive extraction (SBSE) (right). © Fraunhofer IVV.

Case Study 1 – Extraction Techniques: SBSE *vs.* SA-SBSE

Aim	Analysis of wide range of aroma compounds with varying log K_{ow} ranging from 1.05 (vanillin) to 7.10 [(*E*,*E*)-farnesene]	
Method	GC column	Type DB-Wax ultra-inert column, 20 m × 0.18 mm id, 0.30 μm film thickness
	Temperature program	40 °C (3 min), 5 °C min^{-1} to 240 °C (7 min), backflush at 240 °C (10 min)
	Sample injection	Twister back-extraction with acetone, 30 min at 800 rpm Large-volume injection, MPS/TDU-ATEX/CIS injection: 200 μL at 0.85 μL s^{-1} TDU temperature 30 °C (30 s), 140 °C min^{-1} to 80 °C (7 min). Desorption at 100 mL min^{-1}, 25 kPa (splitless) CIS 4 liner type Tenax TA®, pneumatic mode low-split 1:3, temperature 20 °C (30 s), 12 °C s^{-1} to 240 °C (hold)
Results	General	Relative responses vary, especially for compounds with low log K_{ow} (*e.g.* fatty acids, phenolic compounds such as chavicol)
	SBSE	Higher relative responses for some apolar compounds (*e.g.* 4-decanolide, hexyl acetate, β-ionone) Presence of salt in conventional SBSE reduces extraction efficiencies of more hydrophobic (apolar) solutes[50]
	SA-SBSE:	Factor of 10 higher relative response factors for the highest log K_{ow} [*e.g.* (*E*,*E*)-farnesene]. SA-SBSE conditions compensate for effect of salt addition. Sixteen (polar) aroma compounds (log K_{ow} range −0.36 to 2.06) identified only by SA-SBSE
Summary	The variation in results demonstrates that although SA-SBSE shows promise as a fast and simple technique, further validation work is necessary, especially in relation to quantitation	

5.2.4 Liquid-phase Extraction Techniques

In contrast to gas-phase (headspace) extraction, liquid-phase extraction procedures allow for sampling directly within the (liquid) sample matrix. Liquid extraction procedures include static extraction, accelerated solvent extraction (ASE), Soxhlet extraction and supercritical fluid extraction (SFE). With solvent-based extraction procedures it is essential to choose solvents that undergo adequate separation from the volatile compounds on the GC column. An additional challenge with liquid extraction procedures is that any necessary cleanup steps must take into account the specific nature of the target compounds and the matrix itself.

5.2.4.1 Static Extraction and Sonication

The targeted analysis of volatile compounds in FCMs is routinely performed by traditional solvent extraction methods that are supported by mechanical shaking and sonication. An alternative to sonication is microwave-assisted extraction, whereby the solvent in contact with the solid sample is heated *via* microwave energy.

Despite the advantages of solvent extraction, in many cases complete recovery of the extracted analytes is not achievable and recovery testing of extraction processes is often complex. Typically this involves the use of an internal standard and it is important to ensure that the extents of recovery of both the target analyte and the internal standard are consistent, precise and reproducible; this is achieved by comparing the analytical results for spiked sample extracts (within the concentration range of interest) with those for non-extracted standards that represent 100% recovery.[20]

5.2.4.2 Soxhlet Extraction

Soxhlet extraction – a reflux technique named after its inventor, Franz von Soxhlet[21] – is preferentially used when the target compound has limited solubility in a solvent and the sample matrix is insoluble in that solvent. The basic components of a Soxhlet apparatus are the condenser, which cools (and condenses) the solvent vapour, the porous container, which holds the solid sample and allows the condensed solvent to saturate and pass through, and the distilling pot, the function of which is to retain the solvent. Traditional Soxhlet extraction – and also related reflux techniques – is employed only if there is an express need to improve the extraction efficiency, since the technique is relatively labour intensive (>24 h per sample) and consumes large quantities of solvent (>150 mL per sample).

5.2.4.3 Accelerated Solvent Extraction

Accelerated solvent extraction (ASE) is a reflux technique similar to Soxhlet extraction that uses organic and aqueous solvents or acids and bases at high temperatures to increase the extraction efficiency of target analytes. For this reason, and because it allows for easier automation, ASE is often a better choice for the extraction of volatiles from FCMs. In particular, ASE is suitable for the extraction of non-volatile compounds and for extracting compounds from polymeric materials. The use of high pressures in the ASE procedure ensures that the solvents remain in a liquid state, thus allowing rapid compound extraction from solid and semi-solid samples using only a small solvent volume (and in an automated manner). An additional advantage of ASE is its closed system, which eliminates contamination issues. The extraction of polyolefins such as polypropylene (PP) and polyethylene (PE) *via* ASE should be conducted with caution, since the relatively harsh extraction conditions might lead to the blockage of capillaries by polymeric precipitates in the non-heated part of the equipment.

5.2.4.4 Solvent-assisted Flavour Evaporation

The technique of solvent-assisted flavour evaporation (SAFE) is in widespread use in the food and flavour industry as a means for the gentle extraction of labile, odour-active compounds.[22] During SAFE distillation, a liquid sample or solvent extract (in dichloromethane) is placed in the glass distillation apparatus that is kept under vacuum, typically *ca.* 5×10^{-3} Pa. The sample is successively added in small amounts to a flash-evaporation chamber that is heated to 80 °C. The gas-phase volatiles that vaporize in the chamber are subsequently transferred *via* a condenser to a cooled chamber, wherein they are frozen out (*via* a liquid nitrogen bath). The resulting extract is then further enriched by Vigreux and micro-distillation to approximately 100 µL.[23] The advantage of this technique over other methods such as high-vacuum transfer is its higher yields of solvent and lipid matrices. Furthermore, it allows for volatile extraction directly from liquid foods.

5.2.4.5 Liquid–Liquid Extraction

Liquid–liquid extraction (LLE) is a technique that takes advantage of the variable solubility of compounds and their differing partitioning between two non-miscible liquid solvents. A simple static LLE procedure involves placing the target sample (*e.g.* for paper products containing lipophilic substances) in a solvent mixture containing equal parts of two different solvents (*e.g.* ethanol–*n*-hexane, 1 : 1 v/v) in a closed vial for a defined period (*e.g.* 24 h). A typical ratio of sample to solvent is 1 g in 10 mL. After the extraction period, water (*e.g.* 20 mL) is added to the extract and the mixture is agitated for several minutes. After a brief phase separation period (*ca.* 5 min), an aliquot of the upper *n*-hexane phase is extracted and used for GC analyses. In addition to sample extraction, LLE may also be utilized to enrich specific aqueous solutions or for sample cleanup of packaging or solid food samples prior to analysis.

5.2.5 Sample Cleanup

Liquid-phase extraction techniques typically require a subsequent sample cleanup stage to reduce contaminants introduced during extraction. The major challenge during sample cleanup strategies is to avoid losses of the volatile target substances. A critical phase in the procedure is the evaporation process, whereby analyte losses for each step should be minimized; possible analyte losses can be ascertained by rigorous testing. A further influencing factor is the choice of solvent, which should be free of contamination and should also elute separately from the analytes on the GC-column to avoid masking the analyte signals with those of the solvent. The general recommendation is that cleanup steps be kept as simple as possible, as they are susceptible to introducing unnecessary analytical drawbacks.

5.2.5.1 Size-exclusion Chromatography

Size-exclusion chromatography (SEC) – or gel-permeation chromatography – separates molecules by size *via* filtration through a gel. SEC is particularly suitable for removing the high molecular weight fraction of a matrix, *e.g.* in plastics or oily and fatty food samples. Different gels with different properties are available for SEC and the choice of gel depends on the application. A common resin in use is Bio-Rad Bio-Beads S-X3 (styrene–divinylbenzene copolymer) with a 200–400 mesh for 30–90 µm particles and an exclusion limit of 2000 Da; its typical filling mass is 50 g using solvents such as azeotropic ethyl acetate–/cyclohexane (1 : 1 v/v) or dichloromethane.

5.2.5.2 Solid-phase Extraction

Solid-phase extraction (SPE) is a versatile method for enriching and purifying samples prior to analysis. It is used especially for real food samples, but the methods must be tailored towards the specific analyte and the matrix in question. The most common SPE technique in use is normal-phase chromatography, typically involving a polar analyte, a mid- to non-polar solvent and a polar stationary phase. By comparison, reversed-phase chromatography involves a polar (usually aqueous) or moderately polar sample matrix (mobile phase) and a non-polar stationary phase. The analyte of interest is typically mid- to non-polar.

5.3 Analysis of Volatile Substance Migration into FCMs

5.3.1 Specific Migration Limits

FCMs may come into contact with foods at various stages along the food chain, during either manufacturing, packaging, preparation or storage. SMLs (*cf.* Section 5.1.2) for individual substances relevant to FCMs are typically based on observations from toxicological studies. In general, food simulants – rather than foodstuffs – are used for SML compliance testing and an appropriate choice of food simulants is essential. For plastic materials, for example, the amounts of substances migrating into foodstuffs with high fat contents are usually higher than those into aqueous foodstuffs, which in most cases is due to the higher solubility of the migrating organic compounds in fat compared with water.

Remaining with plastics as FCMs, both EU Regulation 10/2011 and EU Directive 97/48/EC provide guidelines for migration testing, including simulant categories for specific food types (Table 5.2) and test conditions for extended contact times (Table 5.3). In this context, it is important to note that the test conditions stipulated by these regulations cannot be easily transferred – or even applied – as test conditions for paper and board.

Table 5.2 Recommended food simulants of EU Regulation No 10/2011.

Type of food	Simulant[a]	EU Regulation No. 10/2011
Watery (pH > 4.5)	A	10% v/v ethanol
Sour (pH < 4.5)	B	3% w/v acetic acid
Alcoholic	C	≤20% v/v ethanol
Half fatty	D1	50% v/v ethanol
Fatty	D2	Plant oil
Dry	E	MPPO[b]

[a]Food simulant categories relating to the type of food stated in the first column.
[b]MPPO (Tenax®) is an alternative food simulant that can be used in a number of cases, *e.g.* for packaging made of paper and board or as a simulant for fatty food where plant oils are not applicable.

Table 5.3 Summary of test conditions according to EU Regulation No. 10/2011 for specific migration testing of plastics for contact times longer than 30 days at and below room temperature.

Test conditions	Procedure
10 days at 20 °C	All storage times under frozen conditions
10 days at 40 °C	All storage times under refrigerated and frozen conditions including heating at 70 °C for up to 2 h or heating at 100 °C for up to 15 min
10 days at 50 °C	All storage times under refrigerated and frozen conditions including heating at 70 °C for up to 2 h or heating at 100 °C for up to 15 min and storage times of up to 6 months at room temperature
10 days at 60 °C	Long-term storage for longer than 6 months at room temperature and below including heating at 70 °C for up to 2 h or heating at 100 °C for up to 15 min. Also maximum conditions for accelerated tests
10 days at 40 °C	If there is scientific evidence that migration of the respective substance in the polymer has reached equilibration under this test condition

Migration tests are performed either by filling, immersion or using migration cells. The choice of method depends on the packaging sample being investigated; for example, migration tests of bottles are almost exclusively performed by filling, as this is relatively trivial to carry out and best reflects real-life scenarios. Immersion tests using liquid food simulants are used to assess the migration from utensils and objects for which both sides may come into contact with the food. Migration cells (see Figure 5.4) are primarily used for testing plastic films (typically of thickness <0.5 mm) for which one side is exposed to the food. For practical reasons, modified poly(phenylene oxide) (MPPO or Tenax®) is particularly suited as an alternative food simulant for high-temperature tests and especially paper and board.

Analytical techniques that are used to identify the presence of these substances in foods (and also the food simulants) are mostly GC-based methods, with either FID or MS (*cf.* Section 5.1.4).

5.3.2 Examples of Migration Tests for Specific Targets and Matrices

Migration testing varies according to the target volatiles, the food matrix and the type of packaging material. A series of case studies presented in this section (all from internal trials at Fraunhofer IVV and mostly unpublished data) aims to offer a diverse overview of different scenarios to highlight the complexities and challenges faced, whereby the conditions for testing must always be adapted to the individual circumstances.

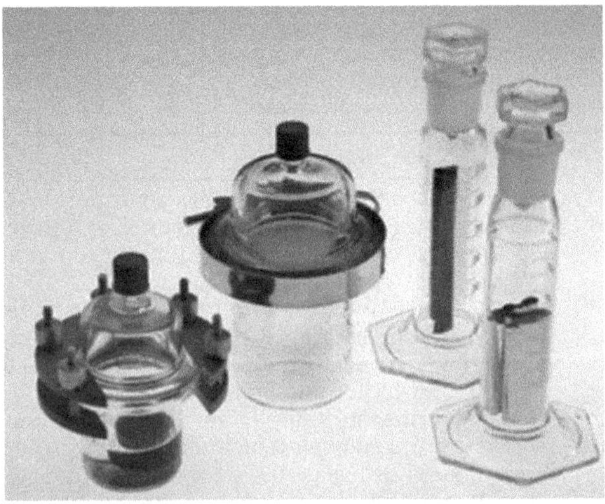

Figure 5.4 Various migration cells including a cell for testing bottle caps (left) and single-sided migration (centre) and glass cylinders for total immersion tests (right). © Fraunhofer IVV.

In the first scenario (Case Study 2), the use of Tenax® as a simulant for characterizing the migration behaviour of polycyclic aromatic hydrocarbons (PAHs) from polymer packaging materials is assessed. The procedure is based on EN 1186-13b but has been modified to cater for the analysis of PAHs.

Plastic packaging is also the subject of the next scenario (Case Study 3) and relates to EN 13628-1:2002, which outlines the use of static headspace analysis with multiple headspace extraction to quantify residual solvents in plastic packaging by GC. The approach outlined in this case study is generally used when the chemical identity of the residual solvents is known *a priori*: almost all technical polyolefins contain groups of volatile oligomers that show characteristic patterns in GC analysis. In multiple headspace extraction, the equilibrium between the sample and its headspace must be re-established between successive extractions. If the period between sample extractions is too brief, the chromatographic area of the analytes will decrease for an increasing number of injections owing to depletion in the sample headspace.[24]

According to theory,[25] the logarithm of the peak area is linearly correlated with the number of injections. From this regression, the total amount of the analyte in the headspace vial can be calculated independently of the polymer type and the vapour pressure of the analyte. Using a multi-analyte standard (*e.g.* *n*-hexane) in combination with the total evaporation technique, the total

Case Study 2 – Migration Test: PAHs from Plastic *via* Tenax

Aim	Assessment of Tenax® as a useful simulant for characterizing the migration behaviour of PAHs from plastic materials	
Method	Specification	European Norm EN 1186-13b, modified for PAHs
	Simulant	Modified poly(phenylene oxide) (MPPO, Tenax®)
	Procedure	The plastic samples are placed in Petri dishes; the surface of the specimen is completely covered with Tenax® (contact area/volume: 4 g Tenax® dm^{-2}) The Petri dishes are covered and stored for 10 days at 60 °C. After storage (and after cooling to ambient temperature), the Tenax® is extracted with *n*-hexane with the addition of an isotopically labelled PAH standard mixture (internal standard) (24 h at 40 °C). SPE clean-up prior to GC-MS analysis; use of a specific SPE column, *e.g.* Strata SI-1 silica, 55 µm, 70 Å (Phenomenex) Quantitative analysis of PAHs performed *via* GC-QMS in SIM mode. Quantitation performed using deuterated internal standards. All analyses are performed in triplicate
Summary	The technique using Tenax® as a simulant offers efficient extraction of PAHs	

Case Study 3 – Migration Test: Residual Solvents in Plastic Packaging

Aim	Estimation of the release of substances from plastic materials by direct skin contact	
Method	Sample	1.0 g polymer or 1.0 dm^2 film, sealed in a 22 mL headspace vial
	GC column	Type ZB 1, 30 m × 0.25 mm id, 0.25 μm film thickness
	Temperature program	50 °C (4 min), 20 °C min^{-1} to 320 °C (15 min)
	Other parameters	Pressure 50 kPa, helium, split: 10 mL min^{-1} Headspace autosampler (Perkin-Elmer HS 40 XL) Oven temperature 200 °C, needle temperature 210 °C, transfer line 210 °C Equilibration time 1 h, pressurization time 3 min, injection time 0.02 min, withdrawal time 3 min Multiple headspace extraction mode with six injections Integration of FID (or MS total ion count, TIC) GC signals
Summary	Equilibrium between the sample and its headspace must be re-established before further extraction to avoid depletion	

area of the analyte and the standard can be compared. Two examples for the MHE quantification of NIAS are shown in Figure 5.5.

The migration of acetaldehyde from poly(ethylene terephthalate) (PET) bottles into non-carbonated and carbonated mineral water is of special interest because it has a low odour threshold of between 10 and 20 ppbv[26] and is detectable at low concentrations in mineral water as a fruity off-flavour. In this case study (Case Study 4), the determination of trace migration of acetaldehyde is performed using P&T for sample extraction, with subsequent analysis by GC-FID/MS.[27] A plot of acetaldehyde concentrations in two mineral water samples over a 1 year shelf-life is displayed in Figure 5.6. The limits of detection (LODs) for acetaldehyde in GC-FID and GC-MS were determined as 50 and 100 pptv, respectively, which are more than two orders of magnitude lower than the odour threshold limit of acetaldehyde in mineral water.

The analysis of volatile migrants is typically performed using GC, but in some cases liquid chromatography (LC), more specifically high-performance LC (HPLC), is necessary. As an example, benzophenone, which is often found in printing inks and coatings for food packaging, has been found to migrate into packaged foods from PE-coated board, either by direct contact or through the gas phase. This phenomenon has been attributed in part to the fact that PE is fairly permeable to low molecular weight substances and does not offer a good barrier to migration.[28] Benzophenone has been assessed

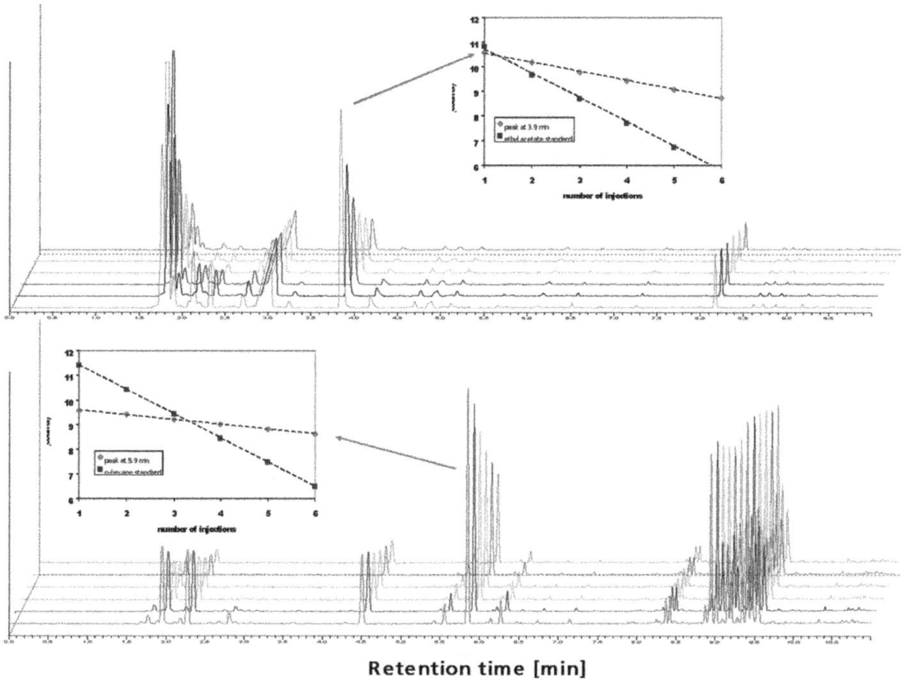

Figure 5.5 Headspace gas chromatograms of residual solvents in printed films (1 dm^2, top) and polypropylene pellets (1.0 g, bottom). Reproduced with kind permission from Mauer and Welle, 2008.[24]

Case Study 4 – Migration Test: Acetaldehyde Migration into Water in PET Bottles

Aim	Trace detection of acetaldehyde in mineral water stored in a PET bottle	
Method	Sampling	10 mL sample volume at 40 °C, vent at 80 °C, transfer line at 200 °C, trap at −35 °C. Purge time 20 min
	GC column	Type Rxi-624, 60 m × 0.32 mm id, 1.8 μm film thickness He carrier gas at 120 kPa
	Temperature program	35 °C (6 min), 5 °C min^{-1} to 90 °C, 10 °C min^{-1} to 260 °C (15 min)
	Detector	FID and MS
Summary	GC-FID and GC-MS exhibited different LODs for acetaldehyde (50 and 100 pptv, respectively)	

for its migration from low-density PE (LDPE) in a series of tests employing different time–temperature conditions, with subsequent analysis by HPLC coupled to diode-array detection (DAD). The experimental data were used to calculate the parameters for migration modelling (diffusion and partition

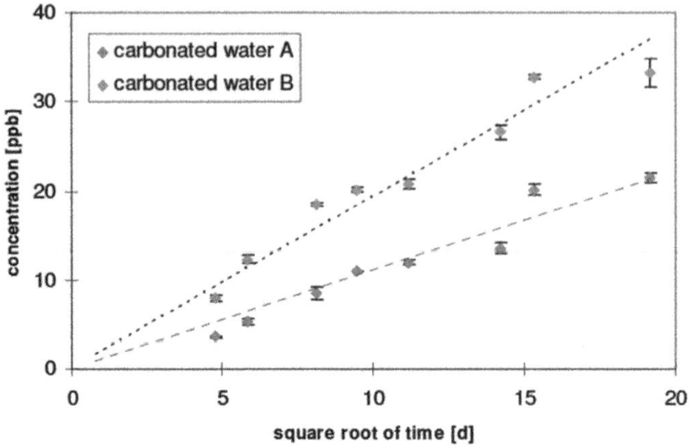

Figure 5.6 Migration kinetics of acetaldehyde from refillable PET bottles into two carbonated mineral water samples held at 23 °C for a shelf-life of 365 days. © Fraunhofer IVV.

coefficients) and the results showed a strong influence of the temperature and physical state of the food on the diffusion coefficient (within both the polymer and the foodstuff). By comparison, the partition coefficient is only significantly affected by the food characteristics, particularly fat content, and not temperature.

In many cases, it is not just a single compound that is of interest. In the following scenario (Case Study 5), P&T extraction with subsequent TD-GC-MS analysis was used to detect a series of volatiles present in water using a reference mixture. The reference mixture was a certified mixture for volatile organic analysis (VOA) containing 73 VOCs (Drinking Water VOA MegaMix™) containing *ca.* 2 mg mL^{-1} of each compound and the procedure is based on the US Environmental Protection Agency (EPA) method EPA 524.4. The analyses revealed that the resolution on the column was acceptable for all volatile substances (see Figure 5.7) and that the detection limit of the method was significantly lower than that achieved with liquid injection.

Similar experiments have been performed *via* P&T for vinyl chloride present in mineral water, where it was possible to reach a trace detection limit of 0.1 ng L^{-1}, which is four orders of magnitudes lower than that with liquid injection (see Figure 5.8). This low detection limit is attributable to the preconcentration step on the cryotrap.

Volatile fluorotelomer alcohols (FTOHs) that might derive from fluorine-treated paper and board have been widely studied as precursors to perfluorocarboxylates, *e.g.* 8 : 2 FTOH degrades to perfluorooctanoic acid (PFOA). In the following case study (Case Study 6), a series of analytes were investigated in fluorine-treated paper and board, specifically four FTOHs (4 : 2 FTOH, 6 : 2 FTOH, 8 : 2 FTOH and 10 : 2 FTOH) and selected perfluorooctanesulfonamidoethanol (FOSE) and perfluorooctanesulfonamide (FOSA)

Case Study 5 – Migration Test: Analysis of 73 Volatiles *via* P&T-TD-GC-MS

Aim	Detection of a series of volatiles from a liquid matrix using purge-and-trap	
Method	Sampling	10 mL sample volume Certified reference mixture: Drinking Water VOA MegaMix™ (Restek Standard #30601). Concentration of each substance *ca.* 2 ppbv 1 µL transfer to a Tenax® tube P&T at 200 °C, valve 200 °C, purge time 10 min at 20 mL min^{-1}, Peltier water trap, transfer line 280 °C, trap −35 °C
	GC column	Type DB-624, 60 m × 0.32 mm id, 1.8 µm film thickness He carrier gas at 120 kPa
	Temperature program	35 °C (6 min), 5 °C min^{-1} to 90 °C, 10 °C min^{-1} to 260 °C (15 min)
	Detector	FID and QMS
Summary	Detection of all volatiles possible, at lower detection limit than *via* liquid injection	

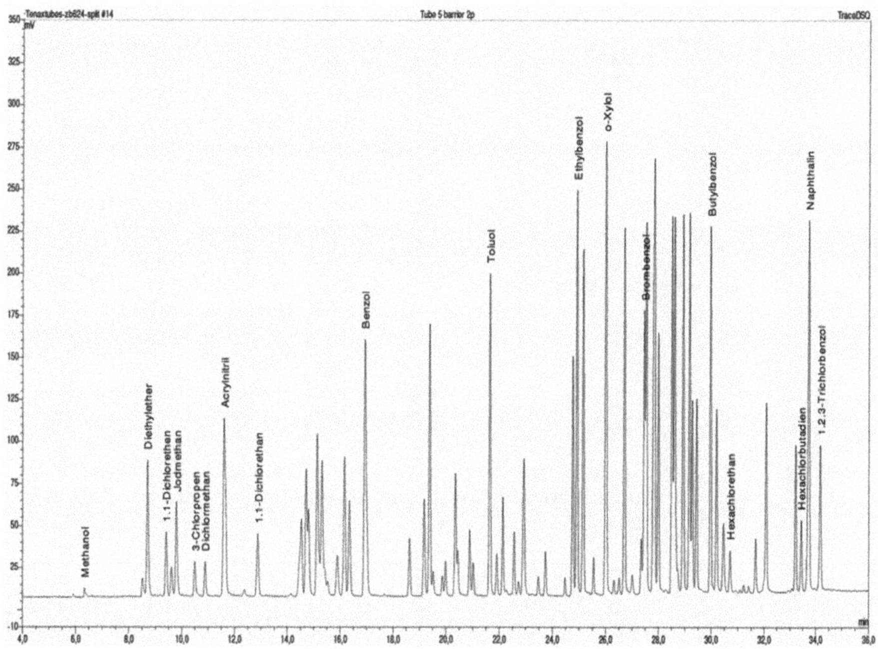

Figure 5.7 Purge-and-trap (TD mode) GC-MS fragmentogram of EPA standard mixture 524.2 Rev 4.10. © Fraunhofer IVV.

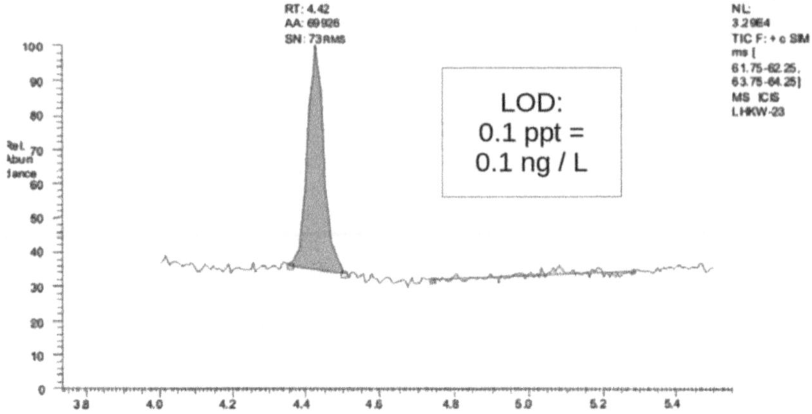

Figure 5.8 Purge-and-trap GC-MS fragmentograms of vinyl chloride at a detection limit of 0.1 ng L^{-1} in mineral water. © Fraunhofer IVV/IMT Innovative Messtechnik GmbH, Vohenstrauss, Germany.

Case Study 6 – Migration Test: Analysis of Volatile Fluorinated Compounds from Treated Paper/Board

Aim	Detection of FTOHs, FOSEs and FOSAs in fluorine-treated paper and board	
Method	Sampling	Static extraction with dichloromethane, 1 g of paper in 10 mL of solvent **Caution**! Evaporation leads to a loss of 4 : 2 FTOH and the potential for perfluoroalkylphosphates (PAPs) to degrade to FTOHs in the GC injector
	GC column	Type DB-35 MS, 30 m × 0.25 mm id, 0.5 µm film thickness He carrier gas, flow rate 1.8 mL min^{-1}
	Temperature program	35 °C (5 min), 70 °C min^{-1} to 60 °C (4 min), 12 °C min^{-1} to 180 °C, 45 °C min^{-1} to 320 °C (1 min), SSL injector temperature 250 °C
	Detector	Q^3MS, EI and CI modes, methane as reactant gas
Summary	Highest selectivity and sensitivity for FTOHs and other volatile fluorinated compounds achieved in positive CI mode (GC-PCI-MS)	

compounds. The best selectivity and sensitivity for FTOHs and other volatile fluorinated compounds was achieved when operating the GC in positive CI mode (GC-PCI-MS), as has been similarly reported previously.[29] This can be explained by the observed mass spectra and signal intensity for 10 : 2 FTOH (see Figures 5.9a and 5.10). In the CI mode, specific ions at *m*/*z* 527.0 and

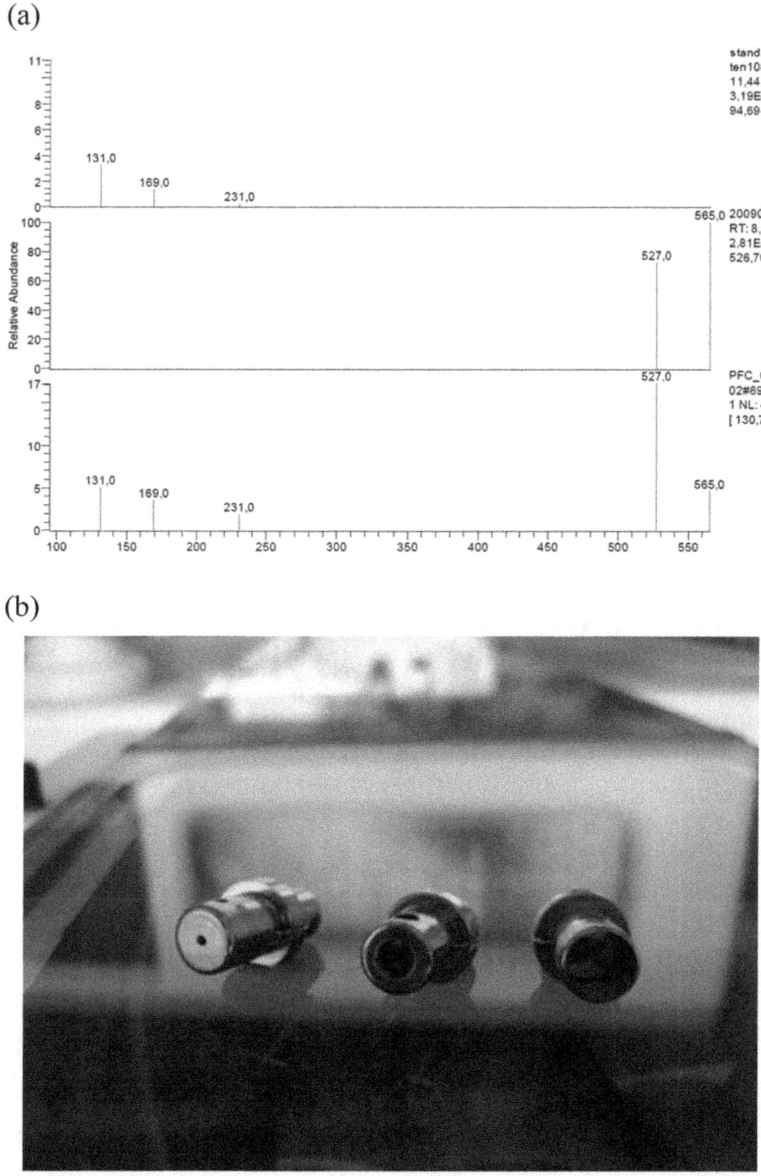

Figure 5.9 (a) Mass spectra of 10 : 2 FTOH and (b) ion volumes for different ionization modes: electron ionization (EI) (right), chemical ionization (CI) (left) and mixed EI–CI (centre). © Fraunhofer IVV.

565.0 are observable, compared with some non-specific fluorocarbon fragments that are present in the EI mode. In addition to the ionization mode, the designs of the CI volume and ion source play a major role in achieving low LODs for FTOHs (see Figure 5.9b), whereby the LOD can vary by several orders of magnitude: in a triple-quadrupole (Q^3MS) system, LODs in

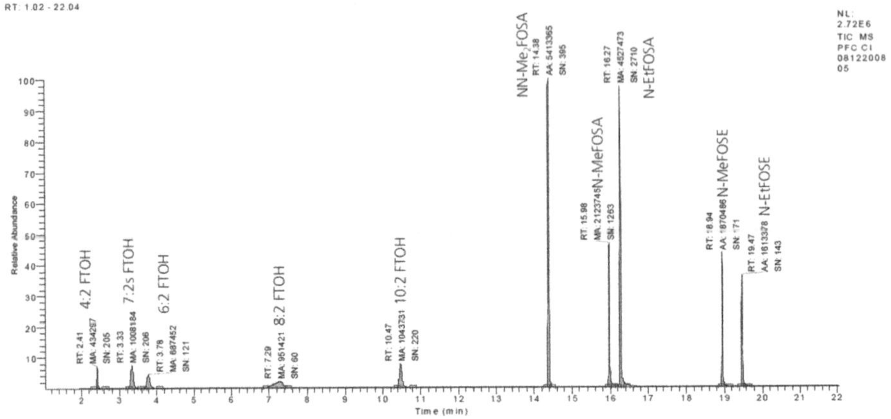

Figure 5.10 Representative chromatogram of a standard mixture detected using the chemical ionization (CI) mode. © Fraunhofer IVV.

the low-ppbv range (*ca.* 1 ng mL^{-1}) are achievable, whereas an alternative CI source design in a single-quadrupole system achieved a much higher (poorer) LOD of 100 ppbv. In some cases a mixed EI–CI ion volume, based on a simple modification of a CI ion source, is useful, such as for screening purposes. In this design, the smaller aperture of the CI ion volume leads to lower pressures in the source, thereby achieving a signal that is a mixture of specific CI-related ions and more non-specific EI-related ions.

The final case study in this section (Case Study 7) involves a case of migration testing beyond food scenarios, namely for estimating the release of substances from plastic materials by direct skin contact.

5.4 Analysis of Permeation Through FCMs

5.4.1 Permeation Parameters

Packaging materials are often designed to actively protect the packaged goods (*e.g.* by use of functional barriers) from substances that could potentially migrate from external sources, *e.g.* from secondary packaging. The permeation of a substance through a functional barrier is influenced by several parameters, primarily the concentration of the permeant in the donor or in the gas phase. Indeed, the initial concentration is a crucial parameter: the higher the initial concentration, the higher is the permeation rate of the permeant through the functional barrier. Figure 5.11 offers a schematic representation of the permeation process and its relevant parameters.

In addition, the distribution equilibria are critical: on the one hand the equilibrium between the gas phase and the functional barrier ($K_{g,b}$) and on the other the equilibrium between the barrier and the food ($K_{b,f}$). The equilibrium constant $K_{g,b}$ is material specific for a given barrier–permeant pair at a specific temperature. $K_{g,b}$ is chiefly determined by the polarities of the

Case Study 7 – Migration Test: PAHs in a Skin Contact Simulation

Aim	Estimation of the release of substances from plastic materials by direct skin contact	
Method	Simulant	20% v/v ethanol in water
	Contact	24 h at 40 °C. Area : volume = 0.16 dm^2 : 22 mL, 20% v/v ethanol–water
	Procedure	Samples fully immersed (full contact) in simulant Each sample is weighed, complete surface area is calculated Migration contact is discontinued after a specified time Migration solutions are extracted (*e.g. via* LLE using *n*-hexane) *n*-Hexane fraction is evaporated to 1 mL Analysis of enriched extract *via* GC-MS
Summary	The technique demonstrated applications beyond FCMs by exploring skin exposure	

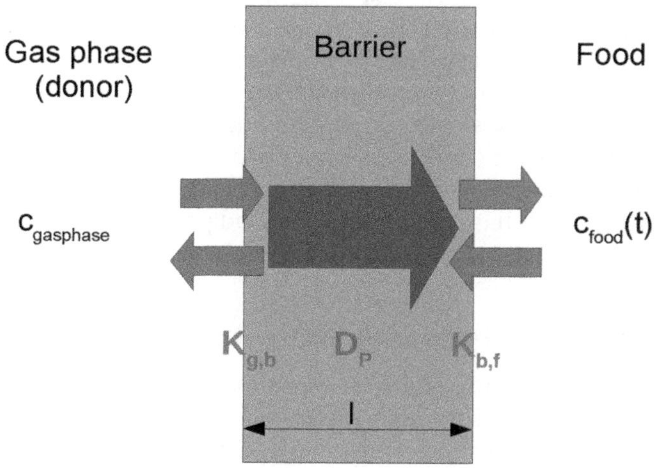

Figure 5.11 Critical parameters influencing permeation through a functional barrier: *c* represents the concentration of the permeant in the gas-phase and food matrix, $K_{g,b}$ is the equilibrium between the gas-phase and the functional barrier, $K_{b,f}$ is the equilibrium between the barrier and the food, D_p is the diffusion coefficient and I is the barrier interface. © Fraunhofer IVV.

polymer and the permeant. The equilibrium constant between barrier and food does not play an important role in good functional barriers, as the equilibrium between the barrier and food is generally not reached within the typical shelf-life of the product.

Not all functional barriers are equally effective. Polymeric and fibre-based materials can be used as functional barriers to prevent contamination of food from the environment or from other packaging components. Inner packaging made of paper or polyolefins delays migration (lag time) but does not prevent it. On the other hand, aluminium, PET or polyamide (PA) barrier layers are considered to be migration-tight barriers that almost completely prevent the migration of volatile substances.

In order for the barrier properties of a film against organic contaminants to be evaluated, it is essential to know the lag times or diffusion coefficients for the specific target permeants; these properties can be determined from empirical data *via* the following methods:[30]

- migration tests for barrier properties;
- permeation tests with static or dynamic acceptor;
- lag-time experiments.

5.4.2 Migration Tests for Barrier Properties

The quality of a functional barrier can be evaluated by performing simple migration tests. The migration test is carried out using real packaging materials in combination with either real food or food simulants, *e.g.* Tenax®, at representative concentrations. In many cases, storage is carried out at a higher temperature in order to shorten the test period. It is important to note that a migration test delivers only a snapshot of a specific packaging and food for the contact conditions under which the test was conducted. As a numerical output, migration tests deliver a value that represents the quantity of a contaminant that has migrated to the food per unit area of packaging (in mg dm^{-2}), as demonstrated by a practical example in the literature.[31]

5.4.3 Permeation Tests with Static or Dynamic Acceptor

Experimentally determined permeation derived from permeation tests with static or dynamic acceptors offers a simple procedure for assessing functional barriers. An explicit premise of such tests is the assumption that permeation commences immediately, at a constant permeation rate (*i.e.* no lag time), which typically overestimates permeation.

If the permeation rate for a specific permeant and barrier is known, the time to reach the SML level, t_{SML}, can be calculated according to the following equation:

$$t_{SML} = \frac{SML \times m}{P \times A} \tag{5.1}$$

where m is the mass of food (in kg), P is the permeation rate (determined experimentally, in μg day^{-1} dm^{-2}) and A is the contact area of the packaging material (in dm^2). The units of t_{SML} and SML are days and μg kg^{-1} food, respectively.

5.4.4 Lag-time Experiments

Lag-time experiments are similar to permeation tests but with a dynamic acceptor. The design of the apparatus in both tests is very similar, but in lag-time tests the permeant is supplied at a constant concentration in the gas phase, rather than spiked directly into the sample matrix. This setup means that the complete system can be conditioned at the experimental target temperature prior to testing. At a predefined time, the measurement is then started either by changing over to a dosed gas flow or by injecting the permeant through a septum, which provides a precise starting point for permeation and thereby allows the lag time to be determined accurately.

The empirically determined lag time is used to determine the diffusion coefficient,[32] D_P, and the partition coefficient, $K_{g,b}$, can be calculated from the permeation curve. As both D_P and $K_{g,b}$ are material constants that are independent of the concentration of the substances used, the permeants can be chosen to cover an optimum concentration range. As in the permeation test with a dynamic acceptor, the value measured in the lag time experiment is the permeation rate. As both the diffusion and partition coefficients represent material constants at a given temperature, these parameters need be determined only once for a barrier polymer, after which the permeation can be calculated for all possible barrier thicknesses and packaging geometries.[33] In the following scenario (Case Study 8), a barrier film test is presented to determine permeation for cardboard. The film is placed in a special permeation cell held at constant temperature (*e.g.* 40 °C), with the spiked cardboard

Case Study 8 – Permeation Test: Barrier Film Test for Spiked Cardboard		
Aim	Permeation rate determination for films using selected compounds spiked on cardboard	
Method	Procedure	Special permeation cells containing test film held at 40 °C Cardboard spiked with permeants placed in part of cell; films in direct contact with cardboard Upper side of cells flushed with N_2 into GC-FID system
	Results	The experimentally determined permeation rates for non-barriers, such as biaxially oriented polypropylene (BOPP, 20 μm) differ from those of barriers, such as biaxially oriented poly(ethylene terephthalate) (BOPET, 12 μm) or metallized EVOH, by several orders of magnitude. In general, barrier layers can be considered to be migration-tight barriers that almost entirely prevent the migration of volatile substances
Summary	Simple and efficient experimental setup for determining the permeation parameters of specific permeants through packaging films	

placed in the lower part of the cell such that it is in direct contact with the film. The upper side of the cell is flushed with a constant flow of pure nitrogen; the nitrogen purges the permeated substances out of the cell and this gas is subsequently analysed. This can be achieved by the use of a multi-port valve, *e.g.* a 16-position valve, which successively passes the gas streams of several permeation cells to a pre-trap with a connected enrichment unit and GC-FID system, as shown schematically in Figure 5.12. The experimentally determined permeation rates for non-barriers, such as biaxially oriented polypropylene (BOPP, 20 μm), differ from those of barriers such as biaxially oriented PET (BOPET, 12 μm) or metallized ethylene–vinyl alcohol (EVOH), by several orders of magnitude (see Table 5.4). In general, barrier layers can be considered to be migration-tight barriers that almost entirely prevent the migration of volatile substances.

5.4.5 Analysis of Odour-active Compounds in FCMs

The sensory properties of food and packaging play a crucial role in consumer acceptance of products. Changes in the sensory properties of packaging (*cf.* Chapter 6) result in undesirable consequences for the packaged foods, leading to off-odours (or off-flavours). Off-odours in packaging result from the presence of odour-active compounds, which can develop for several reasons. Off-odours in relation to packaging materials mostly arise as a result of chemical changes of their constituents, such as oxidation or thermal degradation, but can also be due to contamination.[34–36] The analytical determination of off-odours is typically performed using GC–olfactometry (GC-O), whereby the compounds eluting from the GC column are split into equal portions for

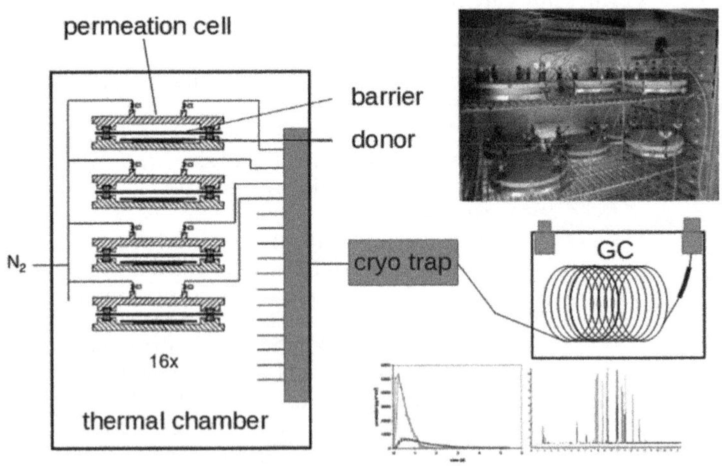

Figure 5.12 Scheme of a permeation test stand and photograph of a multi-port system for success testing of cells. Reproduced with kind permission from Ewender and Welle.[32] © Fraunhofer IVV.

Table 5.4 Experimentally determined permeation rates at 40 °C for some plastic layers and films.[49]

		Permeation rate/µg day^{-1} dm^{-2}					
Compound	Bp/°C	BOPP (1)[a] 20 µm	BOPP (2)[a] 20 µm	EVOH, metallized 20 µm	Acrylic layer 31 µm	PVDC–acrylic 25 µm	BOPET 12 µm
Dodecane	216	4900	1640	0.02	1.43	1.51	0.010
Naphthalene	218	693	95	0.009	0.44	0.39	0.010
1-Methylnaphthalene	243	1590	808	<0.006	0.98	0.79	<0.006
Tetradecane	254	1390	1250	0.014	3.90	4.05	0.008
1-Ethylnaphthalene	260	1000	690	<0.006	1.90	1.57	<0.006
2,7-Diisopropylnaphthalene	279	88.0	79.5	<0.006	1.30	1.45	<0.006
TXIB™[b]	280	244	103	<0.009	0.25	0.26	0.015
Hexadecane	287	238	275	0.008	3.85	6.14	<0.006
Benzophenone	305	68.0	76.5	0.008	0.21	0.26	<0.007
Octadecane	317	33.0	31.5	0.010	2.61	4.02	0.009
4-Methylbenzophenone	326	22.0	22.5	0.017	1.27	0.59	0.017
Phenanthrene	336	32.0	31.5	<0.006	1.37	1.00	<0.006
Eicosane	343	5.51	4.95	<0.006	0.76	1.17	<0.006
Docosane	369	1.02	0.90	<0.007	0.11	0.11	<0.007
Tetracosane	391	<0.03	<0.03	<0.007	<0.007	<0.007	<0.007

[a]BOPP (1) and BOPP (2) are from two different suppliers.
[b]TXIB™, 2,2,4-trimethyl-1,3-pentanediol diisobutyrate.

transfer to the detector (FID or MS) and an olfactory detection port; in the latter, a trained sensory panellist sniffs the eluting gas and notes the odour impressions perceived throughout the GC run. A schematic of this apparatus is depicted in Figure 5.13.

GC-O offers an additional dimension to the data in the form of odour attributes, which are related to the retention time information (retention indices).[37] In the case of compounds with high odour potency (low odour threshold), using the human nose as an additional detector has the advantage that it is often more sensitive than the analytical detection systems employed in conventional GC, meaning that many odour-active compounds are clearly detectable by the human assessor but elicit only a small or negligible peak in the detector. Furthermore, odour attributes can be useful in determining the nature of co-eluting compounds, such as stereoisomers, that are indistinguishable by the detector but elicit distinctive odour impressions in the assessor.

5.4.6 Screening Approaches for Volatiles in FCMs

Recent years have seen legislation and researchers increasingly focusing on NIASs that are present in packaging materials. According to Article 3 of the Commission Regulation (EU) No. 10/2011, a NIAS is defined as "an impurity

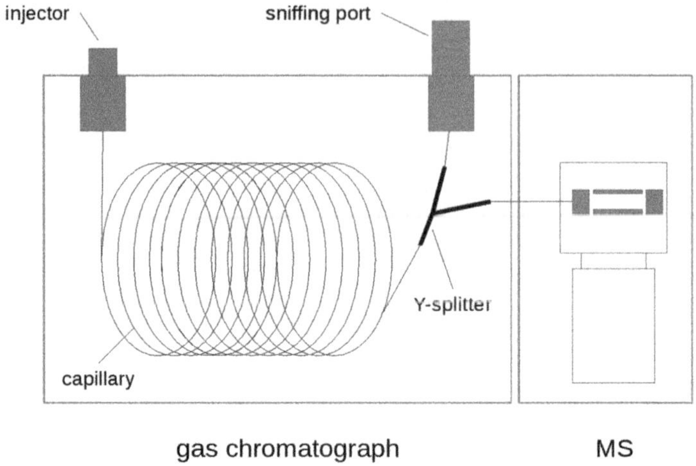

Figure 5.13 Gas chromatography–mass spectrometry–olfactometry (GC-MS-O) for the detection of odour-active VOCs. © Fraunhofer IVV.

in the substances used or a reaction intermediate formed during the production process or a decomposition or reaction product."[3] It is stipulated that NIAS should not exceed a migration rate of 10 µg kg^{-1} food (excluding substances that are carcinogenic, mutagenic or reprotoxic). In practice, this means that there is a need to identify NIAS in FCMs or food samples and examine their migration potential and actual migration by screening approaches. Since the physicochemical properties of NIAS present in a sample are often unknown, a combination of different analytical tools must be used to detect the majority of these substances.

A combined screening approach for volatile NIAS may comprise the following steps:

- A screening of packaging material by means of P&T-GC-MS or headspace GC-MS to investigate the migration potential from the packaging material.
- Specific migration tests with Tenax® and/or appropriate ethanol–water mixtures as food simulants to perform compliance testing. Food simulants such as ethanol–water mixtures can be extracted with LLE (with addition of water) to reach the necessary detection limit of 10 µg kg^{-1} food.

As GC-MS quantification of unknown compounds is hampered by the need for appropriate internal or external standards, an alternative quantification approach is to use GC-FID. Flame ionization detectors are universal detectors that do not exhibit the large substance-specific variations observed in MS detection, thereby reducing possible errors in quantitation.

5.5 Ambient Desorption and Direct Injection Mass Spectrometry Techniques

5.5.1 Direct Analytical Techniques

Migration and permeation testing primarily utilize chromatographic techniques – GC or LC – that require laborious and time-consuming sample preparations prior to analysis. An alternative to these approaches is to use ambient desorption- or direct injection-based techniques, which ionize samples directly before detection in the mass spectrometer and typically do not require sample extraction or enrichment steps. These techniques are currently not in widespread use in packaging-related migration tests, but offer huge potential for screening and high-throughput analysis. The final section of this chapter introduces these techniques and their potential for use in analysing volatiles in FCMs.

5.5.2 Ambient Desorption

Ambient desorption ionization mass spectrometry (ADI-MS) broadly encompasses MS-based systems that utilize surface desorption under ambient conditions in order to ionize and transfer analytes to the mass spectrometer. Regarding packaging applications, ADI-MS techniques permit the analysis of surface treatment and also volatile substances in paper and board. A number of ambient ionization schemes exist, *e.g.* desorption electrospray ionization (DESI)[38] and direct analysis in real time (DART).[39] In both cases, ions are formed in an ion source that is situated external to the MS and no sample preparation or separation is required. Paper spray ionization (PSI)[40] is a similar approach whereby the analyte ions are generated by the application of a high voltage to a paper triangle wetted with a small volume of the target solution. In most cases, ADI-MS techniques deliver sensitive fingerprints but semiquantitative results are often difficult to obtain.

5.5.3 Direct Injection Mass Spectrometry

As an alternative to ambient desorption, direct injection mass spectrometry (DIMS) involves direct transfer of a sample gas (containing the target analytes) into an ionization or reaction region for subsequent detection by MS Since DIMS methods traditionally do not include GC separation, they typically utilize soft CI in order obtain the molecular ion with minimal fragmentation, thereby ensuring simpler mass spectra that are easier to interpret.[41] DIMS techniques include proton–transfer reaction mass spectrometry (PTR-MS),[42] selected ion flow tube mass spectrometry (SIFT-MS)[43] and ion–molecule reaction mass spectrometry (IMR-MS).[44] All three techniques utilize proton–transfer reactions, but charge transfer is also in use, particularly in SIFT-MS.[45] A schematic depiction of a PTR-MS instrument is shown in Figure 5.14.

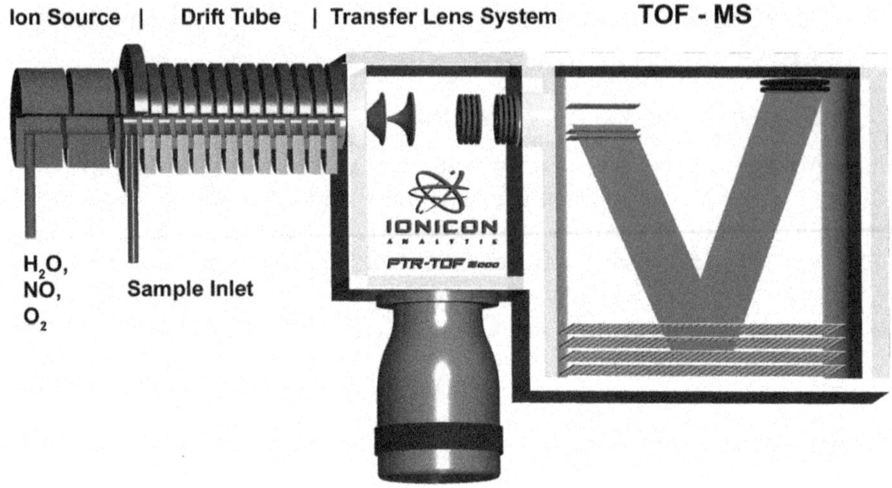

Figure 5.14 Schematic representation of a proton-transfer reaction time-of-flight mass spectrometry (PTR-TOF-MS) instrument comprising a hollow-cathode ion source, a drift tube (reaction chamber), a transfer lens system and a TOF-MS detection system. Reproduced with kind permission, © IONICON Analytik GmbH.

Although all techniques traditionally use quadrupole MS systems, recent advances have seen PTR-MS coupled to TOF-MS, which offers improved mass resolution (separation of isobaric compounds) and fast scan times. Further, a fast GC interface for PTR-TOF-MS offers the possibility of additionally separating some isomeric compounds.[46] PTR-QMS and PTR-TOF-MS have been demonstrated to be suitable for detecting MOHs in spiked paperboard samples, offering a potential rapid screening method *via* direct headspace analysis.[47]

References

1. *Council Directive 1999/13/EC of 11 March 1999 on the Limitation of Emissions of Volatile Organic Compounds Due to the Use of Organic Solvents in Certain Activities and Installations*, 1999.
2. *Plastic Packaging Materials for Food: Barrier Function, Mass Transport, Quality Assurance and Legislation*, ed. O. G. Piringer and A. L. Baner, WILEY-VCH Verlag GmbH, Weilheim, 2007.
3. *COMMISSION REGULATION (EU) No 10/2011 of 14 January 2011 on Plastic Materials and Articles Intended to Come into Contact with Food*, 2011.
4. M. Lommatzsch, M. Biedermann, K. Grob and T. J. Simat, *Food Addit. Contam., Part A*, 2016, **33**, 473–488.
5. A. Störmer and R. Franz, *Food Addit. Contam.*, 2009, **26**, 1581–1591.
6. G. Ziegleder, *Packag. Technol. Sci.*, 1998, **11**, 231–239.
7. M. Biedermann and K. Grob, *J. Chromatogr. A*, 2014, **1255**, 76–99.

8. A. Störmer and E. Pajares Alva, presented in part at the 6th International Symposium on Food Packaging – Scientific Developments Supporting Safety and Innovation, Barcelona, Spain, 16–18 November 2016, 2016.
9. *EN 13130-2-28 Materials and Articles in Contact with Foodstuffs – Plastics Substances Subject to Limitation*, 2004.
10. EURL Food Contact Materials, https://ec.europa.eu/jrc/en/eurl/food-contact-materials, (accessed 31.7.2018), 2018.
11. J. Dewulf and H. Van Langenhove, *Trends Anal. Chem.*, 2002, **21**, 637.
12. A. G. Harrison, *Chemical Ionization Mass Spectrometry*, CRC Press, Boca Raton, Florida, 2nd edn, 1992.
13. E. de Hoffmann and V. Stroobant, *Mass Spectrometry: Principles and Applications*, 3rd edn, 2014.
14. P. Q. Tranchida, F. A. Franchina, P. Dugo and L. Mondello, *Mass Spectrom. Rev.*, 2016, **35**, 524–534.
15. J. Pawliszyn, *J. Chromatogr. Sci.*, 2000, **38**, 270–278.
16. M. W. Heaven and D. Nash, *Food Control*, 2012, **27**, 214–227.
17. E. Baltussen, P. Sandra, F. David and C. Cramers, *J. Microcolumn Sep.*, 1999, **11**, 737–747.
18. K. MacNamara, N. Ochiai, K. Sasamoto and A. Hoffman, Application Note 198: A Combination of Standard (SBSE) and Solvent-assisted (SA-SBSE) Stir Bar Sorptive Extraction for Comprehensive Analysis of Flavor Compounds in Beverages, G. GmbH, 2018.
19. *EPA Method TO-17-monitoring 'Air Toxics' in Ambient Air Using Sorbent Tubes and Automated, Cryogenfree Thermal Desorption*, 1999.
20. D. T. Burns, K. Danzer and A. Townshend, *Pure Appl. Chem.*, 2002, **74**, 2201–2205.
21. F. Soxhlet, *Dingler's Polytech. J. (in German)*, 1879, **232**, 461–465.
22. W. Engel, W. Bahr and P. Schieberle, *Eur. Food Res. Technol.*, 1999, **209**, 237–241.
23. *Review of Isolation and Concentration Techniques*, ed. J. M. H. Bemelmans, Applied Science Publisher, London, UK, 1979.
24. A. Mauer and F. Welle, presented in part at the 4th International Symposium on Food Packaging – Scientific Developments Supporting Safety and Quality, Prague, CZ, 19–21 November 2008, 2008.
25. B. Kolb and L. S. Ettre, *Chromatographia*, 1991, **32**, 505–514.
26. G. Haack and J. Ewender, presented in part at the 2nd International Symposium of Food Packaging, Vienna, AT, 2000.
27. J. Ewender, R. Franz, A. Mauer and F. Welle, *Dtsch. Lebensm.-Rundsch.*, 2003, **99**, 215–221.
28. J. Maia, A. Rodríguez-Bernaldo de Quirós, R. Sendón, J. M. Cruz, A. Seiler, R. Franz, C. Simoneau, L. Castle, M. Driffield, P. Mercea, P. Oldring, V. Tosa and P. Paseiro, *Food Addit. Contam., Part A: Chem., Anal., Control, Exposure Risk Assess.*, 2016, **33**, 715–724.
29. A. Jahnke, L. Ahrens, R. Ebinghaus, U. Berger, J. L. Barber and C. Temme, *Anal. Bioanal. Chem.*, 2007, **387**, 965–975.
30. H. Diehl and F. Welle, *Food Packag. Shelf Life*, 2015, **5**, 41–49.

31. V. Guazzotti, S. Limbo, L. Piergiovanni, R. Fengler, D. Fiedler and L. Gruber, *Food Packag. Shelf Life*, 2015, **3**, 10.
32. J. Ewender and F. Welle, *Packag. Technol. Sci.*, 2014, **27**, 963–974.
33. J. Ewender, R. Fengler, R. Franz, L. Gruber and F. Welle, *DLG Expert Report*, 2016, **10**.
34. A. Strube and A. Buettner, presented in part at the 4th International Symposium on Food Packaging – Scientific Developments Supporting Safety and Quality, Prague, Czech Republic, 19–21 November 2008, 2008.
35. A. Strube, C. Groetzinger and A. Buettner, *Water Sci. Technol.: Water Supply*, 2009, **9**, 299–309.
36. O. Tyapkova, M. Czerny and A. Buettner, *Polym. Degrad. Stab.*, 2009, **94**, 757–769.
37. E. Kováts, *Helv. Chim. Acta*, 1958, **41**, 1915–1932.
38. Z. Takáts, J. M. Wiseman, B. Gologan and R. G. Cooks, *Science*, 2004, **306**, 471–473.
39. R. B. Cody, J. A. Laramée and H. D. Durst, *Anal. Chem.*, 2005, **77**, 2297–2302.
40. J. Liu, H. Wang, N. E. Manicke, J.-M. Lin, R. G. Cooks and Z. Ouyang, *Anal. Chem.*, 2010, **82**, 2463–2471.
41. J. Beauchamp and E. Zardin, in *Springer Handbook of Odor*, ed. A. Buettner, Springer Nature, Cham, Switzerland, 2017, ch. 18, pp. 355–408, DOI: 10.1007/978-3-319-26932-0.
42. A. Hansel, A. Jordan, R. Holzinger, P. Prazeller, W. Vogel and W. Lindinger, *Int. J. Mass Spectrom. Ion Processes*, 1995, **149/150**, 609–619.
43. D. Smith and P. Španěl, *Rapid Commun. Mass Spectrom.*, 1996, **10**, 1183–1198.
44. U. Tegtmeyer, H. P. Weiss and R. Schlögl, *Fresenius. J. Anal. Chem.*, 1993, **347**, 263–268.
45. D. Smith, P. Španěl, J. Herbig and J. Beauchamp, *J. Breath Res.*, 2014, **8**, 027101.
46. J. Beauchamp and J. Herbig, in *The Chemical Sensory Informatics of Food: Measurement, Analysis, Integration*, eds B. Guthrie, J. Beauchamp, A. Buettner and B. K. Lavine, American Chemical Society, Washington, DC, 2015, vol. 1191, ch. 17, pp. 235–251.
47. J. Beauchamp, E. Zardin, M. Grünbaum, R. Fengler, L. Gruber and A. Buettner, Conference contribution: 1st Food Chemistry Conference, Amsterdam, Netherlands, 2016.
48. M. Biedermann, K. Fiselier and K. Grob, *J. Agric. Food Chem.*, 2009, **57**, 8711–8721.
49. J. Ewender, H.-C. Langowski and F. Welle, *Verpack.-Rundsch.*, 2015, 50–51.
50. V. M. Leon, B. Alvarez, M. A. Cobollo, S. Munoz and I. Valor, *J. Chromatogr. A*, 2003, **999**, 91–101.

CHAPTER 6

Migration and Sensory Changes in Packaged Food Products

LUCIANO PIERGIOVANNI*[a] AND LUANA AMOROSO[b]

[a]DeFENS, Department of Food, Environmental and Nutritional Sciences – PackLAB, Università degli Studi di Milano, Via Celoria 2, 20133 Milan, Italy; [b]Di3A, Department of Agriculture, Food and Environment, Università degli Studi di Catania, Via Santa Sofia 100, 95123 Catania, Italy
*E-mail: Luciano.Piergiovanni@unimi.it

6.1 Introduction

The formulation of a comprehensive scientific view of migration phenomena relating to food packaging materials is about 50 years old, which is not a long time for a distinct subject of study. The significance of the topic, moreover, became increasingly relevant owing to the potential adverse consequences of the diffusion of additives, monomers and various contaminants in terms of the danger and toxicity for consumers. The migration phenomena involving packaging materials that lead to sensory changes in foods, cosmetics and other products have not been as widely investigated and monitored as those relevant to safety aspects. The main basic occurrences and the theory behind the migration, however, are substantially the same; the analytical approaches are also similar, even though there are specific differences that need to be emphasized.

Food Chemistry, Function and Analysis No.10
Food Contact Materials Analysis: Mass Spectrometry Techniques
Edited by Michele Suman
© The Royal Society of Chemistry 2019
Published by the Royal Society of Chemistry, www.rsc.org

The different aspects parts of this critical and complex topic are briefly discussed here, in order to provide a framework of knowledge that is useful for striving for a better use of packaging materials and for better food quality.

6.2 Sensory Properties of Food and Beverages

Our contact with food is always a sensorial approach: human sensors, that is, taste, smell, sight, touch and hearing, perform an analysis according to which the product could be found appealing or not for the consumer and, consequently, be accepted or rejected. The behaviour of the sense organs is based on a stimulus, *i.e.* any chemical or physical activator, capable of causing a response in a receptor.[1] Each receptor is specialized in receiving only one class of stimuli, which, after activating it, reach, in the form of impulses, the corresponding sensory areas in the brain, to be interpreted and translated into sensations (Figure 6.1). All sensory modalities are involved in the perception of a food product. These can play different roles at different times, but often interact. Perception is, in fact, a holistic process in which the different modes come into play together, integrating and complementing each other.[2] The different sensations can be distinguished from one another on the basis of the sensory modality involved.

The visual characteristics are related to the overall appearance of the product in terms of colour, shape and materiality of the surface: a particular appearance may suggest to the consumer a certain type of texture.

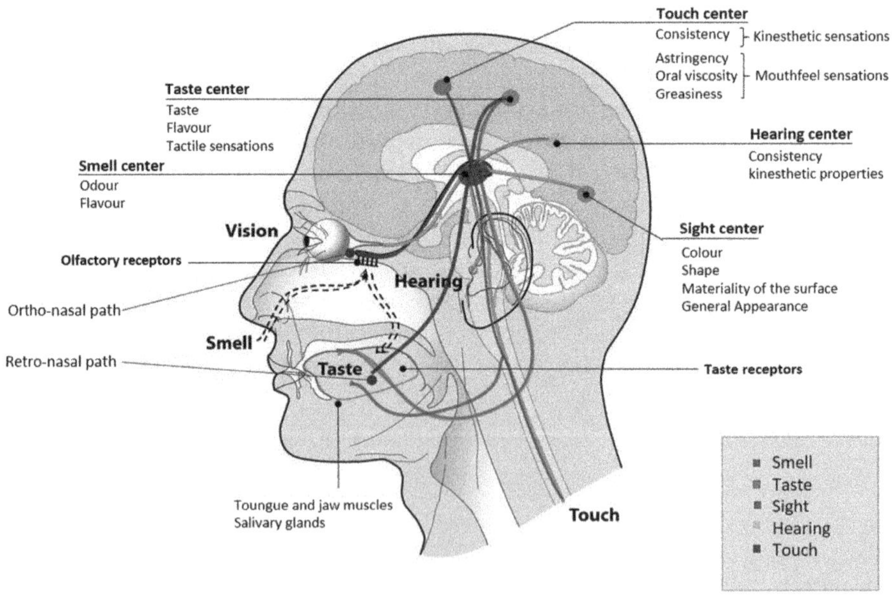

Figure 6.1 The sensations that depend on the sensory apparatus are multiple and have different connections with the brain.

Olfactory perception is distinguished into two types, depending on the channel involved:[2] the term orthonasal perception is used to indicate direct perception through the nose, based on the absorption of volatile substances from olfactory mucosal receptors, whereas the term retronasal perception indicates the odorous sensations in the mouth, perceived as a result of chemical stimulation of the olfactory epithelium, which occurs on introducing the product into the oral cavity. To distinguish these two types of olfactory sensations, we talk about odour descriptors to refer to characteristics gathered in the orthonasal path and flavour descriptors to indicate odorous sensations perceived through the retronasal path. More generally, the term flavour means an integrated set of gustatory, olfactory and chemaesthetic characteristics that are perceived thanks to a chemical stimulation on introducing the food product into the mouth. Olfactory stimuli are very transient and, furthermore, very complex. The odorants that we perceive from the natural daily environment often consist of hundreds of different volatile substances. For example, the odour of banana is considered to consist of more than 350 and the aroma of fresh coffee of more than 800 volatiles.[3] Moreover, individual volatile substances can combine with others and, consequently, increase significantly the number of putative stimuli. More than 400 000 compounds are considered to be odorous to the human olfactory system.[4]

The taste characteristics are related to the stimulation of receptors of different types, mainly located on the tongue. Currently, five basic tastes have been identified: bitter, sweet, salty, acid and umami. The gustatory sensations are always associated with olfactory sensations, determined by the evaporation of volatile substances, which, once released in the mouth, reach the nose *via* the retronasal and tactile sensations determined by the oral interaction between saliva and non-volatile substances (phenols, polysaccharides).

The sensations that depend on the sensory apparatus of touch are multiple and can be divided into different categories, based on the type of stimulation involved. These sensory modalities include those sensations based on a physical stimulation of the mechanoreceptors, located on the fingertips, but also on the face and in the mouth. In the latter case, the term "mouthfeel" is used to indicate specifically tactile sensations in the mouth, such as astringency, oral viscosity and greasiness. In contrast, the term "kinaesthetic" is related to the feelings based on muscle sensitivity: it corresponds to the mechanical movement of the muscles resulting from the stress exerted by the muscles of the hands, the jaw and the tongue when the food is handled or chewed. From this sensory modality derives the perception of the physico-mechanical characteristics linked to the consistency of food (elasticity, hardness, adhesiveness, friability).

The auditory sense comes into play together with the perception of consistency and is therefore closely related to visual and tactile characteristics: the vibrations caused by mastication and swallowing of a food product, reaching the inner ear, communicate the textural characteristics of the ingested matter; typical examples are the crunchiness of potato fries, snacks and some unripe fruits and vegetables. Several studies have demonstrated the strongly influential role that auditory information plays in our perception of food, especially with crunchy foods,[5] which are largely characterized by tactile, mechanical, kinaesthetic and auditory properties.[6]

The sensory profile of foods, drawn up on the basis of the description and quantification of its characteristics of appearance, odour, flavour and texture, is a primary set of information for the optimization, differentiation and innovation of products and it is particularly useful for identifying the key factors that drive product acceptability and consumer preferences. Although a number of intrinsic and extrinsic factors (microbiological integrity, nutritional content, brand, price and marketing considerations, availability, context. *etc.*), contribute simultaneously to the perceived quality of a packaged product, and this makes food choices more complex than a simple result from a blind-labelled acceptance test would suggest, the hedonic reaction to its sensory attributes is often one of the most important leverages for purchasers: it determines the extent and quality of the expectations that we subconsciously forecast for the eating and drinking experience and, thus, influences repeat consumption.

Owing to the important relationship between overall satisfaction and the intensity of various sensory characteristics, it is easily appreciated that food companies place great value on understanding and quantifying the sensorial attributes of products, to achieve eventual success in a well-defined reference market. Such properties can be evaluated using the human senses, in addition to extremely sensitive instrumental methods, but, in both cases, a comprehensive sensory assessment represents a crucially important tool for the food and beverage industry, in order to:

- optimize, differentiate or innovate a product, identifying the drivers of consumer preferences to ensure that the final sensory quality of their products remains consistent, leading to the sale of their goods;
- investigate the relationships between sensory properties and chemical–physical characteristics to define chemical parameters related to the conformity of the products;
- develop targeted formulations through the knowledge of the technological functionality of foods and the interactions between ingredients/components;
- estimate the technological functionality of new ingredients, how they change when subjected to different handling and processes conditions and their relations with perceivable properties;
- communicate the perceivable properties of a product, potentially using them as headlights during a promotional campaign;
- analyse the changes over the storage time and their relation to sensory shelf life;
- study the sensory effects, either positive or negative, of packaging processes, including those related to food contact materials (FCMs).

In this last regard, it is appropriate to keep in mind that an important aspect of packaging is its design and appearance, since the packaging constitutes the interface between consumer and product, and therefore plays a significant role in how packaged food is received by consumers. Again, it should be emphasized how there will always be a certain degree of interaction between packaging, food and the environment surrounding food. In many cases, manufacturers focus on minimizing this interaction, and, therefore, improve packaging with the aim of interacting with the food to such a small extent that it may be considered insignificant in sensory terms. On the other hand, some types of packaging try to utilize and encourage an interaction between packaging and food; this form of intentional and useful interplay is the so-called active packaging.

These few concepts are sufficient to demonstrate that whole sensory profile evaluation of packaged foods is a much more intricate matter than it might appear, owing to the high degree of their physical and chemosensory complexity. Even the simplest product will vary in a number of sensory aspects, each capable of interacting with other attributes at various levels and through several mechanisms. Therefore, a single attribute or ingredient can produce a variety of sensory impressions, with varying impact on the perceived intensities of these components and, certainly, on the degree of overall consumer acceptance of the product.

Addressing issues related to sensory changes in packaged food products, resulting from interactions with packaging materials and from the failure of materials to preserve product integrity or quality, therefore involves multidisciplinary knowledge that ranges from molecular chemistry to cognitive psychology, provided by sensory analysts, materials scientists, packaging manufacturers, food processors and consumer demand specialists.

6.3 Changes of Sensory Properties

The sensory changes of packaged food products occur through several different mechanisms and for various reasons. Not all of them are strictly related to interactions with packaging materials and a classical (but not very often used) distinction is made between "taints" and "off-flavour." The former is the result of contamination by foreign chemicals derived from an external source, whereas the latter is an atypical odour or taste due to intrinsic food product degradation.[7] An atypical sensory change, moreover, can be also determined by the packaging technology used or by an inappropriate selection of packaging material. Vacuum or modified atmosphere packaging, for instance, unavoidably changes the equilibrium of several components of the vapour, liquid and solid phases of the packaged food, and this can affect the odour and the flavour perception. In fresh foods, the physiological processes can proceed differently or be inhibited in a confined environment, particularly if the atmosphere is different from air, influencing the colour, odour or taste of the packaged products. Moreover, a wide range of possible damage of sensory properties can be associated with a wrong packaging design that results in being unable to protect the product quality properly. A package that has a too high gas barrier or, in contrast, that is insufficiently permeable, a material offering a too low a barrier to water vapour transmission, thus leading to the drying or moistening of foods, or a material transparent to wavelengths that affect the food colour, all represent typical examples of mistakes in packaging selection that can determine even serious defects of the sensory properties during the shelf life of the products.

Perhaps the most investigated phenomenon of sensory property changes associated with the atmosphere inside the packages is related to meat colour, and the blooming and darkening of red meat has been extensively studied. The colour of fresh meat is regulated by the relative contents of myoglobin, oxymyoglobin and metmyoglobin at the surface, and all these forms, which have different colours and concentrations, are strictly related to the oxygen partial pressure.[8] The studied cases of crispness or softness losses due to water vapour exchanges across the packaging are well known, certainly with regard to bakery products, and for these products the sensory characteristics related to rheology are an essential part of consumers' appreciation and have been very carefully considered.[9] Finally, the last example of how packaging choices can affect important sensory properties is related to oxidation and colour fading with reference to light transmission. In our laboratories, we studied the effects of the light barrier of different metallized films in delaying chemical degradation reactions, which occur in dry lasagne made with eggs and spinach, and we found a correlation between the transmittance values of the packaging films and a reduction in shelf life.[10]

This premise, aside from the fundamental theme of this chapter, is necessary to emphasize that resolving sensory issues related to packaged products involves a wide and deep analysis, case by case, and a good knowledge of different aspects. Unfortunately, very few reports have approached these aspects of packaging optimization, leaving packaging technologists without adequate and specific "good manufacturing practices" and "guidelines" for a packaging design that takes sensory properties into adequate consideration.[11]

With reference to the changes in sensory properties due to mass transfer phenomena between foods and FCMs (as defined in Section 6.5), the literature seems more extensive and is regularly updated. Taking into account the two possible phenomena, *i.e.* something released from the contact material to the food or taken away from the food and migrated into the packaging material, the consequences for the sensory profile of foods and beverages may involve a change in colour, a change in odour or a change in taste/flavour, the rheological properties being substantially unaffected.

6.3.1 Changes of Colour due to Mass Transfer

The colour of both foods and beverages is, possibly, the first sensory attribute perceived by the consumer. Colours are widely and differently distributed in plant and animal life and any change of colour is assumed to be an alert or a message. It is not just a hedonic aspect of the food choice, because consumers well know that a different colour might mean deterioration, expiration or even falsification. Nevertheless, "colour blindness" affects a large number of individuals: in several European countries as many as 8% of men and 0.4% of women experience congenital colour deficiency. Also known as colour vision deficiency, colour blindness is the decreased ability to see colour or differences in colour.[12]

Changes in food colour due exclusively to the migration of packaging components are rare, even though packaging materials are almost always pigmented or printed with coloured inks over a large part of their surface. For flexible packaging materials, in particular, a potential risk of migration of dye components is conceivable and related to the "set-off" phenomenon. Set-off occurs when printed surfaces of a packaging material come into contact with the food contact side. This is the case during rolling or stacking of FCMs that are printed on the external side.[13] The migration of non-volatile compounds from printing inks in food packaging materials through the set-off is, therefore, a real issue and the development of potential functional barriers to this migration is nowadays a relevant matter of research.[14] Studies related to the set-off issue basically address the aspects of regulatory compliance of packaging materials and it is not possible to find any data about food colour changes due to this kind of interaction.

Paradoxically, somehow, the migration phenomena from FCMs are seen as a means for maintaining or enhancing food colour. Actually, the very first case of food packaging interaction, with positive effects on the food colour, concerned canned food and in particular vegetable products in tinplate containers. Spontaneous corrosion in food cans is characterized by anodic dissolution of the metals with ion formation.[15] In a tinplate food container, it is desirable that tin is the sacrificial member of the pair, leading to protection of the steel substrate. This corrosion is accompanied by atomic hydrogen release (coming from limited steel corrosion), which instead of forming molecular hydrogen (eqn (6.1)), moves in the electrolyte, creating a very reducing environment at the interface, which is favourable for both the colour and flavour of many green and red vegetable food products, and even more effective for common antioxidant additives.

$$Sn^0 \rightarrow Sn^{4+} + 4e^- \rightleftharpoons 4H^+ + 4e^- \rightarrow H^0 \tag{6.1}$$

For completeness, two final points must be noted. First, corrosion of metal containers can also be responsible for undesired colour and taste changes. Sulfur compounds in canned products can lead to unpleasant blackish staining of tinplate surfaces. Sulfur or sulfur dioxide may be reduced to hydrogen sulfide in soft drink and wine cans, giving an offensive smell. Even a very low concentration of sulfur dioxide (1 ppm) is known to speed up corrosion in acidic media, enhancing these effects. Sulfide ions can react with tin and iron to produce blue–black tin sulfide or a black deposit of iron sulfide, respectively, which cause visible staining on the metal surface that can possibly be transferred to the food product.[16] Second, these kinds of interactions, related to corrosion phenomena, are not properly migration events; at least as these events are defined in Section 6.5.

Real diffusion phenomena, *i.e.* migration events, have been investigated many times with the aim of preserving food colour. Silver nanoparticles, which are well known as an effective antimicrobial agent, with a very broad spectrum of inhibitory activities, have recently also been considered as a colour preservative. Their release from nanostructured polyethylene–silver oxide (PE–Ag_2O) bags was used in the preservation of the colour of apple slices and their commercial quality. Ag^+ ions can reduce apple browning by inducing the decomposition of ethylene accumulated in the package. The number of silver nanoparticles that migrated from the nanostructured PE–Ag_2O bags was fairly low in this study, less than 10 μg L^{-1}, a concentration within the requirements of World Health Organization (WHO) guidelines for drinking-water quality, which indicate levels of silver up to 0.1 mg L^{-1} to be safe. The shelf life of freshly cut fruits, evaluated on the basis of various sensory attributes, was extended to 12 days, the normal shelf life being 6 days.[17] The migration of effective antioxidant molecules might also be more directly related to food colour changes. Antioxidants are additives commonly used in polyolefins [low- and high-density polyethylene (LDPE, HDPE), polypropylene (PP)], which are typically the food contact polymer in all composite and multi-layer flexible packaging materials. Many articles have been published concerning the monitoring of the migration of phenolic antioxidants, such as butylated hydroxyanisole (BHA), butylated hydroxytoluene (BHT), pentaerythrityl tetrakis(3,5-di-*tert*-butyl-4-hydroxyphenyl)propionate (Irganox 1010) and tris(2,4-di-*tert*-butylphenyl) phosphite (Irgafos 168), which are the most commonly used.[18–20]

6.3.2 Changes of Odour and Flavour due to Mass Transfer

The odour of food and beverage products is no less important than their colour in consumers' appreciation. Perhaps it comes later (first we look at it, then we smell it), but it certainly has the same significance as colour in terms of rating the product. Similarly to colour, but maybe in a more critical way, odour perception is variable. The ability to identify odour varies among people, decreasing with age and showing gender differences, and also deficits for smokers and those with other behaviours. Chronic problems in odour detection have been reported in a few people who suffer, temporarily or permanently, "anosmia," which is the inability to perceive odour or a lack of functioning olfaction. An important point is that, with some exceptions (synergy and masking of perception with odour mixtures are fairly common in beverages[21]), the odour can be related to a specific substance; this is often the case with food tainting due to interactions with the packaging. Indeed, it is common, in current language, to speak of "plastic odour," "carton odour" and even "metallic odour."

Whereas changes in food colour due to migration from packaging materials are fairly rare, changes in the typical odour with packaging interactions are much more frequent. This is supported by the number of tests and procedures in use to detect potential tainting packaging materials. The most common sensory test is the "Robinson test," considered the most appropriate sensorial analytical method to evaluate the effect of packaging materials able to change the sensory properties of packed food products, described later in Section 6.6.2. Substances that affect the odour and flavour of foods and beverages and coming from packaging material by mass transfer can reach the products both by contacting them and also through the air/vapour phase without any contact. This is discussed further in Section 6.5, but it is readily evident how high the probability is that such an interaction can take place in any stage of the storage, transport and commercial life of a packaged product. Again concerning printed flexible packaging, several issues have been found concerning the solvents for printing inks, for which purpose ethyl acetate, ethoxypropanol, toluene and methyl ethyl ketone have been and still are widely used. All these solvents have high vapour pressures at room temperature and very low odour thresholds, varying from 1 mg L^{-1} (ethyl acetate) to 160 mg L^{-1} (ethoxypropanol), and can have serious adverse effects on the sensory profile of packaged foods. In the past, mesityl oxide (4-methylpent-3-en-2-one) was also used as a solvent for paints and lacquer coatings and its reaction with hydrogen sulfide, occasionally but naturally present in some foods, leads to the production of 4-mercapto-4-methylpentan-2-one, which has a unpleasant catty odour.[7]

Finally, it seems opportune to mention several studies carried out in recent years with the aim of developing devices that are able to release amounts of fragrance to counteract the possible decline of sensory properties. The optimum setup of such active packaging concepts requires the understanding of the diffusion properties of the fragrance in order to ensure an adequate delivery of positive odour-active components during the use or commercial life of the product; at the same time, these devices represent examples of the addressing of migration events towards a possible enhancement of the sensory profile.[22,23]

In addition to the release of flavouring substances, based on the same principles of controlled release already mentioned for colour control, smart materials can be used to influence the migration of antioxidants to obtain desirable outcomes in terms of odour preservation. The release of BHT from LDPE film, for instance, was investigated to assess its potential in reducing the oxidation of cheese during its shelf life[24] and, in another study, the release of BHT from polyolefin film into whole milk powder was determined during 30 days at 30 °C in order to control powder oxidation, which is a basic criterion for maintaining the quality of milk and dairy products.[25]

6.4 Food Packaging Materials

The materials used in the field of food packaging belong to four very different groups of solids, namely plastics, cellulosic materials, ceramic materials and metals, where a wide variability of different properties comes from even small chemical and technological factors.[26] Therefore, their final properties can be very different with respect to many aspects, included those pertinent to the topic of this chapter. A brief description of the four groups may be useful in addressing properly the attention required in designing and selecting packaging for the best performance in terms of preserving the sensory properties of foods and beverages.

6.4.1 Metal Packaging Materials

Several fundamental properties, common to all metals, make them useful for packaging applications. They are tough (*i.e.* resistant to stress and abuse), malleable (*i.e.* it is possible to mould metals to obtain almost any shape) and have high thermal conductivity (*i.e.* any thermal treatment is possible and recycling is feasible). The property most relevant to the present topic, however, is the compactness of their molecular structure, responsible for their high density, which prevents any diffusion of light, vapour or gas, even through very small thicknesses, leading to an absolute barrier. Aluminium alloys, tin plate, tin-free steels, polymer-coated steels and stainless steels can occasionally be responsible for food tainting and changes in some important sensory attributes (colour, odour or taste), but this may occur just with corrosion phenomena or leaching events and not for real diffusion mechanisms. Nevertheless, it must be remembered that in many cases, the metal surface is protected by a layer of plastic or coated for sealing purposes and, in this instance, migration of the organic components with possible sensory damage is possible, even in the form of small oligomeric components.[27]

6.4.2 Ceramic Packaging Materials

The term ceramic refers to all materials defined as inorganic and non-metallic whose chemistry is centred on silicon structures, such as china, porcelain, earthenware and glasses, which are the most important and used for food packaging. According to the most accurate definition, glass is "an amorphous, inorganic product of fusion that has been cooled to a rigid condition without crystallizing."[27] Actually, the term applies more to a physical state than to a chemical structure and also some special organic materials are denoted glasses or glassy. The inorganic ingredients used in the glass-making process are all crystalline but, owing to the high temperature of the process (1450–1500 °C), any crystalline structure is lost and the fundamental structural units rearrange in the amorphous state. The ability of the Si–O–Si bonds to display angles from 120° to 180° makes possible the randomized structure from which derive the main properties of glass containers: their inertness and compactness which, as occurs with metals, prevent any diffusion phenomena. Once again, however, it is worth noting that glass containers always need plastic, rubber or other materials for closing purposes and is not possible to exclude interactions that lead to unwanted sensory changes. A typical issue related to cork closures for wine bottles, for instance, is the so-called "cork taint," to which several compounds are thought to contribute. Chloroanisoles, in particular for their low sensory threshold, have been extensively studied[28] and identified as microbial degradation products of chlorophenols or chlorinated solutions used to decolorize cork. A more serious issue with respect to glass container closure is related to the degradation of epoxidized soybean oil (ESBO), used as a plasticizer in PVC gaskets in lids for glass jars. Its oxidation leads to the production of short-chain aldehydes that gives a musty and rancid odour to contact foods.

6.4.3 Cellulosic Packaging Materials

Cellulose is, by far, the most abundant and widespread natural polymer, with a yearly production of more than 75 billion tons,[29] and being, worldwide, the largest proportion of the total packaging materials (about 40%). Cellulosic packaging materials, however, are always heterogeneous solids in which the cellulose content can be lower than 50% by weight. In addition to the fundamental constituents of the primary source (mainly wood), *i.e.* lignin and hemicellulose, a wide variety of so-called non-fibrous additives are present. Many chemical additives are regularly used in the manufacture of cellulosic packaging materials, *e.g.* fillers, pigments, moisture-proof agents, oil-proof agents, adhesives, antifoaming agents, draining and floating agents, chelating agents, plasticizers, biocides and others. Among these additives, biocides play an important role. Chlorophenols have been widely used as fungicides and biocides, also in wood and wooden articles (*e.g.* pallets), and we have already mentioned (cork taint) that they can be methylated to chloroanisoles by numerous microorganisms, with serious consequences for the sensory properties of contact foods.[30] Even though it is not very common for there to be direct contact between cellulosic materials and food products, occasionally some sensorial and potential safety problems with packaged foods are related to the transfer of such chemicals when recycled pulps are used, since they may contain various short-chain aldehydes that have an unpleasant odour and also traces of saturated and aromatic hydrocarbons derived from mineral oils.[31]

6.4.4 Plastic Packaging Materials

Although "plastics" and "polymers" are often used as synonyms, they have different meanings. Not all polymers exhibit plastic properties (*e.g.* many biopolymers such as chitosan, cellulose, proteins and complex lipids are polymers but are not plastic) and plastics contain various amounts of additives that are not polymers, *e.g.* plasticizers, antioxidants, pigments, antistatics, fillers and many others. These are fundamental in order to fulfil the expected final properties but they have low molecular weights and are not polymers at all. Taking into account the most recent developments in research and commercial practice, this brief presentation on plastic packaging materials will focus on two arbitrary categories: synthetic oil-derived polymers and bioplastics. The last category currently constitutes less than 1% of the total worldwide plastic production (about 320 million tons per year) but there are serious prospects for fast growth. Recently, the European Bioplastics Organization proposed to define a plastic as a bioplastic if it is either bio-based or biodegradable or shows both of these features. Bioplastics correspond to three groups of different materials: (1) bio-based but non-biodegradable plastics such as polyethylene (bioPE), polypropylene (bioPP), poly(vinyl chloride) (bioPVC) and polyester (bioPET) that are synthesized starting from bio-based raw materials but identical with the current polymers based on fossil resources; (2) plastics that are both bio-based and biodegradable, such as poly(lactic acid) (PLA), polyhydroxyalkanoate (PHA), thermoplastic starch (TPS) and others; (3) plastics that are obtained starting from oil products but, for their intrinsic characteristics, are biodegradable, such as poly(butylene succinate) (PBS), polycaprolactone (PCL), poly(vinyl alcohol) (PVOH) and a few others. The very large category of oil-based polymers include the very common polyolefins polyethylene, used in food packaging as high-density polyethylene (HDPE), low-density polyethylene (LDPE) and linear low-density polyethylene (LLDPE), and polypropylene (PP), which together account for about 75% of the total plastic packaging materials, poly(ethylene terephthalate) (PET) (about 7%), polystyrene (PS) (about 7%), poly(vinyl chloride) (PVC), polyamide (PA) and other minor polymers. The specific properties of all of these polymers are readily available from many sources;[26] however, some issues related to the subject deserve to be mentioned. HDPE is commonly used to produce screw caps for plastic bottles, which are treated with erucamide as lubricant that can release mono- and di-unsaturated carbonyl compounds that can give a plastic-like off-flavour to mineral water.[32] Concerning all the polyolefins and PP in particular, it is worth mentioning the possible thermal degradation products in the form of oligomers but also, and much more critical, short- and medium-chain odorous aldehydes. A very small aldehyde, acetaldehyde, is the thermal degradation product of PET and its production and sensory effects have been studied in detail because its very low odour threshold (0.02 mg L^{-1}) created many problems for PET bottle producers.

For flexible packaging, in particular, rarely is one material used alone in the manufacture of the final packages, and various other materials are included in order to assemble a structure that can fulfil all the targets needed for the packaging and distribution. Therefore, the complexity of packaging is augmented from a chemical point of view. The most common processes for producing multi-layer structures are co-extrusion, adhesive lamination, coating, metallization, multi-injection moulding and multi-layered cardboards forming. All of them add components (adhesives, coatings, compatibilizers for co-extrusion, additives, *etc.*) that can be responsible for migration phenomena and potential sensory changes. Adhesives, in particular, cause many migration problems that are not just sensory related. In 1999, Schweitzer *et al.*[33] investigated the reaction between the monomer of a polyester adhesive and an impurity in propylene glycol (propionaldehyde), giving rise to 2-ethyl-5,5-dimethyl-1,3-dioxane (2-EDD), a very pleasant fragrance (fruity blueberry) with a very low odour threshold in water of about 0.01 mg L^{-1}.

6.5 Migration Phenomena

Although the term "migration" is currently used with different meanings and with reference to various different phenomena, in the food packaging field it is always intended to mean a kind of interaction between food and contact materials, with negative consequences for consumers. However, a rigorous description of the unwanted event and quantification of its consequences are possible only if a precise definition is given to the term. Migration should be intended as a mass transfer from an external source into food or even from food into the contacting material. It occurs by submicroscopic processes and, in particular, by diffusion processes. Properly defined in this way, it has been possible for many years to investigate these events in depth and to find useful tools for predictions and assessments. With reference to migration from plastic packaging materials into food products, two relevant milestones in this particular part of packaging science deserve to be mentioned: the first application of Fick's laws to the explanation of residual monomer migration and the development of prediction models (and the appearance of related software for their practical use and to be able to represent the phenomena concerned dynamically and quantitatively).[34–36]

The paper that is still currently mentioned as the "pioneering paper"[36] that first applied the diffusion laws to migration phenomena was related to the migration of styrene (vinylbenzene, styrol).[37] This simple, aromatic molecule (Figure 6.2) is the monomer of PS (polystyrene), a polymer widely used in many packaging applications; it is not very toxic but is certainly a threat to the sensory properties of contact foods and beverages.

The toxicology of styrene has been extensively studied because it remains as an unreacted monomer in many items of styrene polymers and copolymers in a measurable amount (even percentage levels). Its acute toxicity,

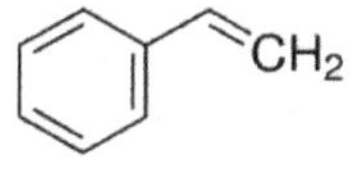

Figure 6.2 Styrene (vinylbenzene, styrol) was the first migrant to be studied in detail for its odour and migration behaviour.

expressed as the LD_{50} value for rats, is about 5 g kg^{-1} for oral administration and 2.4 g kg^{-1} for intraperitoneal administration, and the lethal concentration (LC_{50}) for 2 h of exposure is 5000 ppm for mice, hence it is not really a very dangerous substance.[38,39]

However, styrene is a colourless–yellowish viscous liquid with a pungent and unpleasant odour; it has been said that styrene use is self-limiting in packages: no one, in fact, would use heavily contaminated material, owing to the odour. The odour threshold for styrene is actually very low in water (0.01 mg L^{-1}) and also in air (70 μg m^{-3}), where its half-life is also short and estimated to be 2 h. In air, in fact, it is easily chemically transformed into benzaldehyde and formaldehyde, both of which are odorous air pollutants.[40] Very often, it is the styrene odour that is referred by consumers as "plastic odour" and in the 1960s this was a critical issue, leading to a wide investigation into its possible migration phenomena. In this way, both physics and chemistry and also mathematical modelling became involved in migration issues.

6.5.1 Migration Models

Predictive mathematical models for migration estimation are essentially based on diffusion theory and considerations of partitioning effects. Therefore, the diffusion coefficient of the migrant into the plastic (D_P, $cm^2\ s^{-1}$), for estimating the dynamics of the mass transfer (eqn (6.2)), and the partition coefficient ($K_{P,F}$) (eqn (6.3)), for establishing the equilibrium level of the transfer, have to be known or properly estimated in order to apply migration models to the solution of migration problems.

$$\frac{\partial C_{P_{x,t}}}{\partial t} = D_P \frac{\partial^2 C_{P_{x,t}}}{\partial x^2} \tag{6.2}$$

$$K_{P,F} = \frac{C_{P,\infty}\rho_P}{C_{F,\infty}\rho_F} \tag{6.3}$$

where $C_{P_{x,t}}$ (v/v) is the migrant concentration at point x of the thickness and at time t, D_P is its diffusion coefficient in the thickness of the packaging material, $K_{P,F}$ is the partition coefficient (packaging/food) of the migrant at equilibrium, $C_{P,\infty}$ and ρ_P are the migrant concentration (v/v) in the plastic at equilibrium and the migrant density, respectively, and $C_{F,\infty}$ and ρ_F are the corresponding parameters relating to food.

All the currently used models are deterministic, *i.e.* they are not stochastic; in fact, they assume that, for a given polymer, migrant, contact food (or simulant) and temperature, the amount migrated at time t always has one specific value. All the currently used models are also prudentially overestimating in nature, aiming at the "worst case prediction," and are based on several general requirements/constraints. Some examples of such constraints are the following: it is assumed that the migrant is uniformly distributed in the thickness of the FCM; the total sum of the migrant amount in packaging and in the food is always constant during migration; no boundary resistance is assumed for the transfer of the migrant; and the migration proceeds only from contact material towards food or *vice versa* in special cases. The general requirements/constraints can be differently adopted and chosen according to the model used, but models based on diffusion and partition have been applied to all the various formalities of the phenomenon of mass transfer that we call migration.

6.5.2 Phenomenal Description of Migration

The migration may occur in a wide range of different modalities and many authors have proposed schemes for the possible classification of migration phenomena.[41] A detailed description is beyond the scope of this chapter, but a short presentation on this topic seems useful because of the connections with potential changes of sensory profiles.

6.5.2.1 Negative Migration – Flavour Scalping

A first important distinction concerns the direction of the mass transfer. The majority of migration studies have concerned transfer from packaging materials to food, but it is not uncommon that food components can move into the contact material, seriously affecting the sensory properties. A standard classification of these opposing migration phenomena denotes the migration from packaging to food as "positive migration", whereas the adsorption of food components by the contact material is termed "negative migration" (or reverse migration).[41] Flavour scalping, a phenomenon governed by the same physical principles as positive migration, is a typical case of negative migration. Both volatile and non-volatile food components, responsible for the blend of taste and smell sensations evoked by a substance in the mouth, can have diffusional and partitioning characteristics adequate for mass transfer into packaging materials. Limonene and other aroma compounds from fruit juices have been extensively studied with regard to flavour scalping by LDPE, which is the sealing layer in almost all composite structures and has been proved to be very prone to the sorption of such molecules[42,43]. Aroma scalping also depends on the aroma solvent used: the aroma polar components were more adsorbed into polymers if their solvent was non-polar and *vice versa*.[44] To reduce the extent of scalping and its negative effects, suitable polarity and morphology of the polymers may be chosen.

The transfer of flavour compounds into polymer packages may also indirectly affect food quality by causing delamination of multi-layer packages or by altering the barrier and mechanical properties of plastic packaging materials,[43,45] both important factors for the shelf life of many packed foods. Hirose *et al.*[46] reported that the oxygen permeability of LDPE and two types of ionomer increased due to the presence of absorbed D-limonene. Johansson and Leufvén[45] investigated the effect of rapeseed oil on the oxygen barrier properties of different polymer packaging materials and found that, unlike amorphous PET, which remained an excellent oxygen barrier even after storage in rapeseed oil for 40 days, polyolefins (PP and HDPE) showed an increased oxygen transmission rate (OTR) after the same time of contact. This influence on oxygen barrier properties, resulting from flavour absorption, was attributed to swelling of the polymer matrix, which changes the polymeric network involved in permeability phenomena. Sadler and Braddock[47] showed that the oxygen permeability of LDPE was proportional to the mass of absorbed limonene. An increased oxygen permeability of LDPE indicated that absorption of volatiles must be responsible for structural changes in the polymer.

6.5.2.2 Positive Migration Modalities

Volatile behaviour and the diffusion coefficient are the key parameters for the complete description of three different modalities of positive migration:

1. Volatile substances do not require contact between the packaging and the food in order to lead to migration, even though they may be affected by the contact. Typically, this type of interaction applies to dry solid foods with poor direct contact with the package. Migration is limited to volatile compounds that have relatively high vapour pressures at room temperature and, obviously, can change the sensory properties. The migration phenomenon is controlled mainly by the diffusion in the packaging material and vapour pressure, and much less by the characteristics of the food phase. Volatile compounds in the packaging materials may also evaporate into the outside environment without causing migration towards food but leading to odorous packages.
2. Non-volatile substances, having high diffusion coefficients in the packaging walls, require contact between the packaging and the food in order to lead to migration. This is very likely the most common case of migration from plastic packaging and, depending on the potential migrant and its affinity for the contact food, can cause possible changes to the food and beverage sensory profile.
3. Non-volatile substances, having low or very low diffusion coefficients, should be ideally preserved from mass transfer phenomena. However, in a dual interactive mode, migrants with low diffusion in the plastic at the initial contact with food may attain a high diffusion coefficient after suitable food components have been adsorbed and diffused into the packaging walls, *i.e.* by a negative migration phenomenon. This dual–mode interaction of swelling of plastic material subsequently accelerates the diffusion of the migrant and can have even more serious organoleptic consequences, a double effect on sensory properties: a loss of food components plus a gain of plastic components.

6.6 Evidence-based Assessment

Food companies normally perform sensory tests on their final products and also packaging companies often address the potential risk of provoking, *via* interaction phenomena, tainting of the packaged products. This is not just good manufacturing practice because the mandatory regulations concerning FCMs all around the world impose specific attention to this kind of issue. In the European Union, for instance, Framework Regulation 1935/2004, Art. 3, states that, "... Materials and articles, including active and intelligent materials and articles, shall be manufactured in compliance with good manufacturing practice so that, under normal or foreseeable conditions of use, they do not transfer their constituents to food in quantities which could: (a) endanger human health or (b) bring about an unacceptable change in the composition of the food or (c) *bring about a deterioration in the organoleptic characteristics thereof* ..." (our italics). The respect of such a constraint is not an easy task to achieve in advance without a good knowledge of the possible phenomena responsible and, moreover, in the absence of appropriate tools and procedures. Once a risk of threat to the sensory profile has been identified, a mandatory step is to obtain, in the absence of a specific limit of migration established by law, the threshold values for its perception, in order to estimate the risk of tainting. This, obviously, is achieved by means of a sensory analysis that will be discussed in more detail in Section 6.6.2. Some chemicals can be detected at extremely low concentrations and individuals, as already noticed, may be more or less sensitive to certain odours and compounds.[48]

Odour thresholds have been defined and described in different ways. The "detection threshold" is the concentration that can be detected without any requirements on identification or recognition of the stimulus, whereas the "recognition threshold" is the minimum concentration at which a stimulus can be identified or recognized. These are absolute thresholds, whereas the relative thresholds are the difference thresholds that state the smallest change in concentration of a compound required to give a distinguishable change in perception. There are different units for the threshold values, according to the media, the most common being mg m^{-3} in air, mg L^{-1} in water and mg kg^{-1} in other media. Generally, they refer to a probability of detection equal to 0.5, *i.e.* 50% of the population will detect it at that level.

The detection threshold is always concentration dependent and when the perception sensitivity is plotted against the logarithm of concentration, generally a logistic curve (sigmoidal curve) is obtained, as shown in Figure 6.3.

There is a fairly large literature on this subject and some compilations of odour thresholds, with more than 10 000 values reported, are available.[49] The solving of sensory issues with packaged food products often involves a dual approach, since the evidence-based assessment is obtained by both instrumental and sensory analysis.

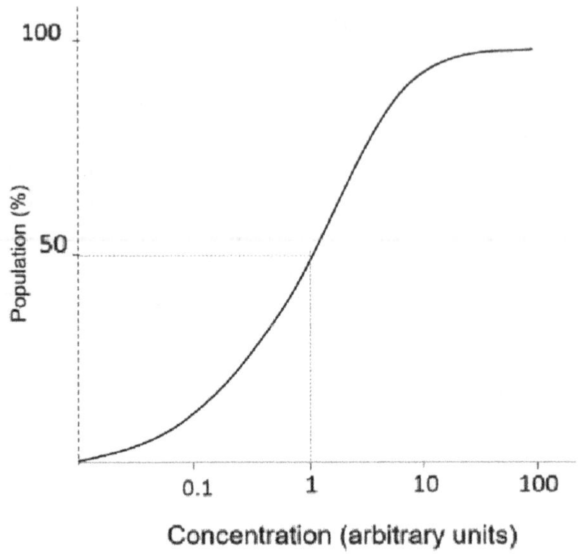

Figure 6.3 Odour and taste thresholds vary in the population according to a sigmoidal (logistic) function.

6.6.1 Instrumental Evaluation

All the most sensitive and high-resolution analytical techniques have been used in the assessment of various forms of taints. In a few cases, high-performance liquid chromatography (HPLC) was used in conjunction with appropriate detection systems. Fluorescence detection was used to detect and quantify substances responsible for sensory changes, such as skatole and indole in meat.[50] Electrospray ion-trap tandem mass spectrometry was used for the determination of ESBO in foods.[51] ESBO is a pool of organic compounds derived from the epoxidation of soybean oil, used as a plasticizer and stabilizer in PVC gaskets of metal lids used to seal glass jars and bottles. It has a mild vegetable odour, and the potential for migration and sensory changes exists as for all oxidized products. Mass spectrometry (MS) techniques has been found to be very effective in detection and problem solving with regard to FCMs. However, the majority of the instrumental analyses have been performed with gas chromatography (GC), sometimes coupled with flame ionization detection (FID) and, much more often, coupled with MS. FID provides an easier quantification step and has been used alone, but also together with other detection systems. Obenland *et al.*, for instance, were able to demonstrate how different commercial packagings caused reductions in the flavour quality of navel oranges *via* scalping phenomena.[52] MS detection systems have been extensively used for investigating changes in the volatile profiles of several foods, but it is outside the scope of this chapter to provide an exhaustive review of this topic; as a significant example, this technique provided results well correlated with sensory analysis for whole pasteurized milk packaged in different containers.[53] The sensitivity of such a detection system made it possible to detect less than 0.5 μg mL^{-1} of many volatile and odorous substances (*e.g.* ethyl acetate, 3-methyl-2-butanone, hexanal, dimethyl disulfide), coming both from migration phenomena from the packaging and from autoxidation reactions promoted by the transparency of the container.

In any case, the key point for effective and accurate instrumental analysis in all food–packaging interactions with sensory damage is the sample preparation, with the aim of increasing the concentration of analytes.[48] Several methods have been used for the extraction and concentration of volatiles and semivolatile compounds from packaging materials and foods, *e.g.* liquid–liquid extraction, simultaneous steam distillation–solvent extraction (SDE), static headspace, dynamic headspace, direct thermal desorption, solid-phase microextraction (SPME) and, more recently, headspace sorptive extraction (HSSE), stir-bar sorptive extraction (SBSE) and closed-loop stripping analysis (CLSA), widely reported in the literature. The choice of extraction method, however, depends on various specific factors related both to the matrix and to the causes of sensory changes.

In many circumstances, the GC instrument (equipped with an FID or MS detection system) is coupled with an "olfactometer" (CG-O); here a "sniffer port" is combined with the detector to allow a panellist to identify the odour of a substance as it leaves the column (Figure 6.4). The carrier gas must be mixed with humid air before the trained analyst records the time and intensity and describes the odour.[54] Reviews of this technique are available.[55]

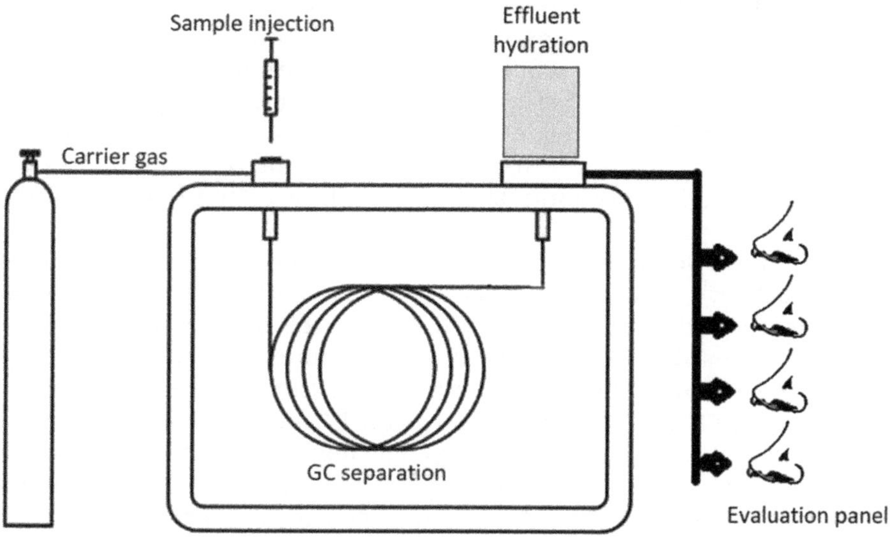

Figure 6.4 Schematic diagram of an olfactometry test, connecting GC analysis with sensory perception.

Instrumental tools for assessing sensory changes include the modern techniques of the electronic nose (E-nose) and electronic tongue (E-tongue).[56] E-nose instruments are designed to mimic the human nose; they contain arrays of sensors, relatively selective, that interact with volatile molecules to produce some sort of signal that can be recorded, whereas E-tongue devices react to non-volatile compounds present in a liquid. The core of the instruments, however, is the connected software system, which uses multivariate statistics for pattern recognition. This makes it possible to determine if one sample is similar to or different from another, based on a rapid analysis of headspace or solvent, and thus to assess differences in volatile and non-volatile substances. In our laboratory we used the E-nose technique to investigate the odour properties of 25 homo- and copolymers of ethylene and five homo- and copolymers of propylene, intended for food packaging uses, in comparison with conventional sensory panel tests. The instrumental results were well correlated with the panel answers and the application of cluster analysis made it possible to differentiate polymer odours into strong, medium and weak.[57]

6.6.2 Sensory Analysis

Sensory analysis, very far from degustation, is a scientific discipline based on principles of experimental design and statistical analysis that, by proper use of the human senses, makes possible an objective and quantifiable evaluation of the sensory profile of consumer products. Therefore, the same principles and techniques apply to foods, beverages and packaging materials. As we have already stated, the mandatory regulations concerning FCMs imposed constraints on migration (limits of global and specific migration) and on possible sensory tainting. However, strangely, the tests for verifying the compliance of FCMs with migration limits are described in official law, whereas sensory assessment covered only by voluntary standards. Fortunately, at least in Europe, very good standards are readily available. The discriminants for the selection of standards are the type of material, direct or indirect contact and the effects of odour or flavour on taste.

The European Standard EN 1230-1, "Paper and Board Intended to Come into Contact with Foodstuffs," specifies the test method for assessment of the odour released by a paper or board sample. It is applicable to all kinds of paper and board, including coated and/or printed material, intended to come into direct or indirect contact with foodstuffs. The second part of that standard, EN 1230-2, "Off-flavour (Taint)," specifies whether a paper or board sample contains substances that may be transmitted through the air space to a test substance and affect its taste. It is applicable to all kinds of paper and board, including coated and printed material, intended to come into contact with foodstuffs. Other food packaging materials should be subjected to a German standard, DIN 10955, which specifies test methods for the sensory analysis of packaging materials

Figure 6.5 Schematic diagram of a Robinson test. The packaging material is confined in a closed container but not in contact with grated milk chocolate.

that are intended for wrapping of foodstuffs that come in direct contact with the packaging material.

The already mentioned European Standard EN 1230-2 includes the procedure most known and applied in the packaging sector, namely the Robinson test, developed by the Office International du Cacao et du Chocolat. In this test, a panel assesses possible taste differences of milk chocolate caused by the packaging material positioned very close but not in contact with the material (Figure 6.5). Given the fatty matrix, the high affinity to migration components and its well-known and delicate odour and flavour, milk chocolate is an ideal test medium. It is worth remembering that flavour is always defined as the blend of taste and smell sensations evoked by a substance, volatile or non-volatile, in the mouth.

All of these procedures rely on groups of evaluators (the panellists) who are trained and expert. Actually, they do not perform a "consumer test" but are prepared according to specific standards and act differently, according to the type of sensory test needed and stated in the objectives of the sensory analysis. Three main methods of sensory testing are in use: Discrimination, Descriptive Analysis and Acceptance. The sensory test objectives will determine the appropriate method to be used. Discrimination tests, the simplest, are based on a comparison of samples and ask, "Is there a difference"? Discrimination tests are used to compare differences among products or between a current and a new formulation or packaging. The triangle test is most commonly used to distinguish if there is a difference between products. It is statistically effective and significant results can be obtained using small panels. Descriptive analysis provides the description of the sensory qualities of food. It pertains to the sensing and describing both qualitative and quantitative sensory attributes. Quantitative attributes reflect the degree of the characteristic and is expressed by a scale value. Acceptance tests evaluate liking for foods and are consumer or affective tests and, normally, are not used for the purpose of assessing damage to the sensory profile due to packaging interactions.

6.7 Conclusion

The interaction phenomena between foods or beverages and contact packaging materials can induce unwanted changes to various sensory properties. Mainly odour, colour and taste/flavour can be affected by the migration of constituents, coming from the contact material's components or from non-intentionally added substances that can be generated during the manufacture of the packaging. However, these possible phenomena are not the most frequent cause of sensory damage to packaged food products. Very likely, the majority of sensory issues, apart those due to the food's own characteristics, comes from inadequate packaging design or selection. Examination of the research literature, however, reveals that the diffusion of specific substances from packaging materials to a food is often seen as a potential tool for preserving or enhancing the sensory properties, throughout the so-called active packaging devices or concepts.

In conclusion, the possible events of interactions with detrimental effects on the sensory profile of foods and beverages need to be better investigated, understood and monitored with more attention and commitment, in particular because it is evident that the theory behind these unwanted phenomena, and also the evaluation procedures for detecting them, are known and adequately advanced.

References

1. E. Pagliarini, *Valutazione sensoriale: aspetti teorici, pratici e metodologici*, Hoepli, Milano, 2002.
2. Società Italiana di Scienze sensoriali, *Atlante sensoriale dei prodotti alimentari*, Tecniche Nuove, 1 edn, 2013.
3. R. Hudson, *J. Comp. Physiol.*, 1999, **185**, 297.
4. U. Stockhorst and R. Pietrowsky, *Physiol. Behav.*, 2004, **83**, 3.
5. B. K. Drake, *J. Food Sci.*, 1963, **28**, 233.
6. Z. M. Vickers and S. S. Wasserman, *J. Texture Stud.*, 1980, **10**, 319.
7. D. S. Mottram, *Int. J. Food Sci. Technol.*, 1998, **33**, 19.
8. K. Krzywicki, *Meat Sci.*, 1979, **3**, 1.
9. N. L. Chin and P. J. Martin, in *Bakery Products Science and Technology*, ed. W. Zhou, Y. H. Hui, I. De Leyn, M. A. Pagani, C. M. Rosell, J. D. Selman and N. Therdthai, Wiley-Blackwell, 2 edn, 2014, ch. 26, pp. 453–472.
10. L. Piergiovanni and S. Limbo, *Packag. Technol. Sci.*, 2004, **17**, 155.
11. S. E. Duncan and J. B. Webster, *Adv. Food Nutr. Res.*, 2009, **56**, 17.
12. X. B. V. Chan, S. M. S. Goh and N. C. Tan, *Asia Pac. Fam. Med.*, 2014, **13**, 10.
13. K. Bentayeb, L. K. Ackerman, T. Lord and T. H. Begley, *Food Addit. Contam., Part A*, 2013, **30**, 750.
14. M. Aznar, C. Domeño, C. Nerín and O. Bosetti, *Dyes Pigm.*, 2015, **114**, 85.
15. G. W. Patrick, *Anti-Corros. Methods Mater.*, 1976, **23**, 9.

16. C. Mannheim, in *Food Product-package Compatibility*, ed. J. I. Gary, B. R. Harte and J. Miltz, Technomic Publishing, Lancaster, 1986, pp. 105–133.
17. L. Zhou, S. Lv, G. He, Q. He and B. Shi, *J. Food Qual.*, 2011, **34**, 171.
18. B. Marcato, S. Guerra, M. Vianello and S. Scalia, *Int. J. Pharm.*, 2003, **257**, 217.
19. M. S. Dopico-García, J. M. López-Vilariñó and M. V. González-Rodríguez, *J. Agric. Food Chem.*, 2007, **55**, 3225.
20. K. Kyungmo, C. Yoonjee, C. J. Chun, P. Se-Jong and H. Jaejoon, *J. Food Sci.*, 2018, **83**, 1005.
21. H. T. Lawless, *J. Sens. Stud.*, 1986, **1**, 259.
22. B. P. F. Day, in *Smart Packaging Technologies for Fast Moving Consumer Goods*, ed. J. Kerry and P. Butler, Wiley, 2008, ch. 1, pp. 1–8.
23. L. Vermeiren, F. Devlieghere, M. Van Beest, N. De Kruijf and J. Debevere, *Trends Food Sci. Technol.*, 1999, **10**, 77.
24. C. D. Soto-Cantú, A. Z. Graciano-Verdugo, E. Peralta, A. R. Islas-Rubio, A. González-Córdova, A. González-León and H. Soto-Valdez, *J. Dairy Sci.*, 2008, **91**, 11.
25. D. Granda-Restrepo, E. Peralta, R. Troncoso-Rojas and H. Soto-Valdez, *Int. Dairy J.*, 2009, **19**, 481.
26. L. Piergiovanni and S. Limbo, *Food Packaging Materials*, Springer, 2016.
27. J. Bott, A. Störmer and P. Albers, *Food Packag. Shelf Life*, 2018, **16**, 112.
28. L. J. Parr, M. G. Gee, D. G. Land, D. Robinson and R. F. Curtis, *J. Sci. Food Agric.*, 1974, **25**, 835.
29. Y. Habibi, *Chem. Soc. Rev.*, 2014, **43**, 1519.
30. T. Lord, in *Taints and Off-flavours in Foods*, ed. B. Baigrie, Woodhead, Cambridge, 2003, ch. 4, pp. 64–111.
31. R. Lorenzini, K. Fiselier, M. Biedermann, M. Barbanera, I. Braschi and K. Grob, *Food Addit. Contam., Part A*, 2010, **27**, 1765.
32. A. Strube, A. Buettner and C. Groetzinger, *Water Sci. Technol.: Water Supply*, 2009, **9**, 299.
33. L. Schweitzer, J. Noblet, Q. Ye, E. Ruth and I. H. Suffet, *Water Sci. Technol.*, 1999, **40**, 217.
34. J. Crank, *The Mathematics of Diffusion*, Oxford University Press, 1979.
35. A. Baner, J. Brandsch, R. Franz and O. Piringer, *Food Addit. Contam.*, 1996, **13**, 587.
36. J. A. L. Wan, P. C. Chatwin and L. L. Katan, *Phil. Trans. R Soc. Lond. A*, 1995, **350**, 379.
37. T. Garlanda and M. Masoero, *Chim. Ind.*, 1966, **48**, 936.
38. J. A. Bond and H. M. Bolt, *Crit. Rev. Toxicol.*, 1989, **19**, 227.
39. H. Härkönen, *Scand. J. Work, Environ. Health*, 1978, **4**, 104.
40. World Health Organization Regional Office, *WHO Regional Publications, European Series*, 1987, **23**, 118.
41. P. C. Chatwin, in *Migration from Food Contact Materials*, ed. L. L. Katan, Springer US, Boston, MA, 1996, pp. 26–50.
42. T. Nielsen and M. Jägerstad, *Trends Food Sci. Technol.*, 1994, **5**, 353.

43. S. C. Fayoux, A. M. Seuvre and A. J. Voilley, *Packag. Technol. Sci.*, 1997, **10**, 145.
44. P. Hernandez-MuÑoz, C. AtalÁ and R. Gavara, *Food Addit. Contam.*, 2001, **18**, 673.
45. F. Johansson and A. Leufvén, *J. Food Sci.*, 1997, **62**, 355.
46. K. Hirose, B. R. Harte, J. R. Giacin, J. Miltz and C. Stine, in *Food and Packaging Interactions*, ed. J. H. Hotchkiss, American Chemical Society, 1988, ch. 3, pp. 28–41.
47. G. D. Sadler and R. J. Braddock, *J. Food Sci.*, 1990, **55**, 587.
48. K. Ridgway, S. P. D. Lalljie and R. M. Smith, *Food Addit. Contam.*, 2010, **27**, 146.
49. L. J. van Gemert, *Odour Thresholds: Compilations of Odour Threshold Values in Air, Water and Other Media*, Oliemans Punter, 2011.
50. J. A. Garcia-Regueiro and I. Diaz, *Meat Sci.*, 1989, **25**, 307.
51. M. Suman, S. La Tegola, D. Catellani and U. Bersellini, *J. Agric. Food Chem.*, 2005, **53**, 9879.
52. D. Obenland, S. Collin, J. Sievert, K. Fjeld, J. Doctor and M. L. Arpaia, *Postharvest Biol. Technol.*, 2008, **47**, 159.
53. A. E. Karatapanis, A. V. Badeka, K. A. Riganakos, I. N. Savvaidis and M. G. Kontominas, *Int. Dairy J.*, 2006, **16**, 750.
54. A. Mounchili, J. J. Wichtel, J. O. Bosset, I. R. Dohoo, M. Imhof, D. Altieri, S. Mallia and H. Stryhn, *Int. Dairy J.*, 2005, **15**, 1203.
55. B. Plutowska and W. Wardencki, *Food Chem.*, 2008, **107**, 449.
56. E. A. Baldwin, J. Bai, A. Plotto and S. Dea, *Sensors*, 2011, **11**, 4744.
57. L. Torri, L. Piergiovanni and E. Caldiroli, *Food Addit. Contam.*, 2008, **25**, 490.

CHAPTER 7

Risk Assessment of Plastic Packaging for Food Applications

MAGDALENA WRONA AND CRISTINA NERIN*

Grupo GUIA, Departamento de Química Analítica, Universidad de Zaragoza, C/ María de Luna 3, edificio Torres Quevedo, 50018 Zaragoza, Spain
*E-mail: cnerin@unizar.es

7.1 Introduction

In the contemporary world, food packaging plays a vital role in preserving food quality and it has become an integral part of our everyday life. Food packaging can be defined as an element with a specific design created to contain food products and to protect them from the environment during transport and storage.[1] Three main types of protection of food can be distinguished: physical, chemical and biological. Physical protection allows the food to be protected from high pressures, vibrations, heat, *etc.*, whereas chemical protection provides food with a barrier against oxygen, water vapour and dust. Biological protection ensures a barrier to microorganisms, insects, rodents, *etc.*, preventing disease and spoilage. Moreover, application of a biological barrier maintains the conditions inside the packaging to control ripening and ageing. However, packaging not only protects food from external contamination and damage but also serves a range of other functions such as dosage

Food Chemistry, Function and Analysis No.10
Food Contact Materials Analysis: Mass Spectrometry Techniques
Edited by Michele Suman
© The Royal Society of Chemistry 2019
Published by the Royal Society of Chemistry, www.rsc.org

control, provision of nutrition information, decreasing food waste and minimization of the application of food additives.[1] Therefore, the design of packaging plays a crucial role in the protection of the shelf life of a food product. The proper selection of packaging materials and technologies maintains the quality and freshness of the product during distribution and storage.

Food packaging materials can be described as a group of materials implemented in food applications. Packaging encompasses a wide range of material types across plastics, glass, metal and paper. Owing to their chemical structures and properties, the most commonly applied materials in the food industry are plastics.[2,3]

7.2 Migration

As already mentioned, food packaging protects foodstuffs, but it can also be a source of chemical contamination of food due to migration processes. Migration can be defined as the mass transfer between the packaging material and the packaged food under specific conditions.[4] The most concern has grown about the migration of hazardous compounds from plastic materials.

The consequences of migration come primarily from the interactions between food and packaging, which result in changes in the composition of the food due to the incorporation of migrants coming from the environment or the packaging itself. These substances are considered in some cases to be toxic to humans. Moreover, the stability of the packaging itself can be affected by alterations to the physicochemical and mechanical characteristics of the packaging material during the storage of the packaged product caused by a loss of packaging components.[5]

In practice, there are two types of migrations: global migration and specific migration. Global migration represents the total amount of components, known or not, that have been transferred from the packaging material to food during preparation and/or storage. Global migration is a measure of the inertness of the packaging material, whereas specific migration refers to the amount of a specific and identifiable substance that is transferred to the food.[6]

Migration does not occur equally in all packaging materials with food contact. In polymeric materials, including plastics and cellulose derivatives, the interaction occurs through a mechanism based on mass transfer through the polymer matrix, whereby the components, additives and residues present in the material are incorporated into the packaged product.[7,8] Figure 7.1 presents a scheme of a migration process in a polymer–food system.

The migration of additive components or residues in polymeric packaging materials to food is a phenomenon of mass transfer, as a consequence of the tendency of all chemical systems to approach equilibrium, and, as such, it is controlled by two mechanisms: diffusion and sorption. The overall process is developed in three stages: diffusion of the migrants into the polymer matrix, solvation or dissolution of the migrants in the food–packaging interface and dispersion of the migrants in the food.[9]

Migrants are compounds of low molecular weight that are present in the polymer matrix. Possible chemical migrants include plasticizers,

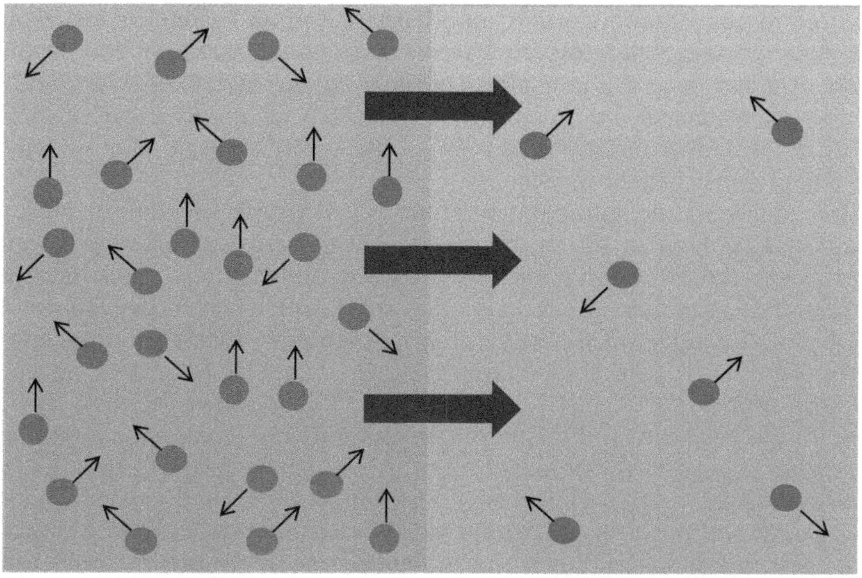

Figure 7.1 Scheme of a migration process in a polymer–food system.

antioxidants, non-intentionally added substances (NIAS), thermal stabilizers, slip compounds, monomers and oligomers. When these compounds reach a specified limit in the food, quality and safety may be jeopardized. Therefore, the study of the migration of food packaging compounds on exposing food or food-simulating liquids (or solid) to storage conditions is of great importance.[10,11]

Undoubtedly, migration must be considered in the context of different types of packaging. The knowledge of novel solutions for plastic packaging applications that are commercially available is crucial because it allows the prediction of potential harmful compounds present in food contact materials (FCMs). Moreover, an appropriate method of analysis can be chosen to detect and quantify possible migrants. Modern society requires modern solutions, especially in the area of food safety and hence food packaging. Consequently, a wide range of different types of plastic packaging can be distinguished, some of which are briefly described in the following sections. Moreover, hazardous migrants are discussed in terms of risk assessment.

7.3 Multilayer Packaging

The application of multilayer materials is driven by the necessity to design a single packaging structure that combines different properties and functions. In many cases it is not possible to meet all the packaging design requirements using a single monolayer material. For example, the high barrier properties required in many food packaging applications cannot be achieved with only one polymer layer. Therefore, multilayer material composed of two or more layers of materials, each consisting of plastics, and also aluminium and/or cardboard, which are bound together by means of adhesives or extrusion coatings, provide a final material able to be used in modified atmosphere packaging (MAP), vacuum packs, *etc.*

The adhesives are complex mixtures of chemical substances such as monomers, polymers, fillers, pigments, antioxidants, tackifiers, solvents, plasticizers and adhesion promoters, which vary widely in their chemical and physical properties. In the case of multilayer materials, they are usually applied on the full surface of two or more different substrates (*e.g.* polymers, paper, aluminium) forming the laminates. They can also be applied on a partial area, forming pouches (*e.g.* bags for chips) or windows (bags for bread), or sealing aluminium foil lids with plastic food containers called easy-peel packaging (*e.g.* yogurt). Moreover, they are used in self-stick labels, *e.g.* for fresh sausages.[12]

Although in most applications adhesives are not in direct contact with food, it has been shown that once incorporated into the packaging their constituent compounds may diffuse through the material and may establish surface contact with the packaged food. The results of investigations of different types of adhesives (acrylic, vinyl, hotmelt, rubber, polyurethane) revealed the migration of both volatile and non-volatile compounds.[13–21] Among them, the most hazardous are those compounds which do not appear in the positive list of compounds in European legislation and they belong to toxicity class III.[22] During the risk assessment of adhesives, it is necessary to recalculate the migration values using the real proportion of the surface covered by the adhesive with respect to the volume of food contained in the packaging. In many applications adhesives are used to give the geometric shape to pouches and boxes, where the adhesive is applied on a small area (less than 5%). Moreover, target analysis can be carried out of specific toxic compounds, *e.g.* primary aromatic amines (PAAs) that are reaction products from residual isocyanates in polyurethane adhesives.[23]

Adhesives are also manufactured with different chemical substances that constitute a source of off-flavours that could affect packaged food, causing consumer complaints, higher production costs or even a possible loss of brand confidence and market share.[15] Because of this, it is important to perform migration assays connected with olfactometry studies.[19,24]

7.4 Oxo-biodegradable Packaging

In recent years, the increasing demand for the production and application of plastics has created a global waste disposal problem. Polymers cause serious environmental pollution since their degradability is very slow, having a negative impact on wildlife.[25] In order to reduce environmental contamination, materials with faster degradation processes are being investigated and developed as substitutes for the traditional polymers.[26] Such polymers can be divided into three main classes: photodegradable, biodegradable and oxo-biodegradable. A photodegradable polymer upon exposure to sunlight undergoes photooxidative degradation. A biodegradable polymer, by microbial activity in the presence (aerobic degradation) or absence (anaerobic degradation) of oxygen, converts its organic compounds to carbon dioxide, water, inorganic compounds, methane and biomas.[27] To accomplish the biodegradation of polymers, the microorganisms first need to cleave the polymer chains in order to reduce its molecular weight, making feasible its transportation into the cells, where most of the biochemical processes take place.[28] Finally, an oxo-biodegradable polymer has pro-oxidant additives that catalyse the degradation process to accelerate it, leading to the formation of short-chain organic chemicals that will then biodegrade.[29] Knowledge of the performance of oxo-biodegradable materials allows the prediction of migrating compounds characteristic of particular types of polymers. We can distinguish antioxidants and stabilizers, developed to increase the durability of polyolefins, and transition metal complexes used as pro-oxidants. Polyolefins have relatively high molar mass values; therefore, oligomers can migrate from their principal chain matrix. Moreover, oxidation products of oxo-biodegradable materials are small molecule compounds such as acids, aldehydes, ketones and alcohols.[30–32]

7.5 Packaging from Recycled Polymers

One of the alternatives for reducing the environmental impact of plastic waste is recycling. As a result, the demand for recycled post-consumer packaging materials continues to grow and a clear increase in the development of numerous procedures for recycling of polymers to ensure their safety can be observed. Nevertheless, the recycling of post-consumer plastics into materials for direct food contact applications requires a detailed knowledge of possible contaminations in order to assess the risk to consumers' health. Emphasis should be placed on the chemical substances intentionally added to all plastic products to enhance their physicochemical properties and simultaneously extend their shelf life. These additives can potentially migrate from FCMs created from recycled polymers and lead to undesirable effects on human health. Moreover, they can be released from plastics during the various recycling steps and recovery processes, thus increasing their concentration. Therefore, safe recycling means that the emission of substances of high concern and contamination of recycled products are avoided.[33]

The efficiency of decontamination in recycling technology must be checked by a challenge test that has been adopted by the European Commission. The test involves contamination of the polymer with a high concentration of substances called surrogates, that mimic a broad range of potential contaminants that may be present in the post-consumer material due to consumer packaging misuse, under standardized conditions of time and temperature. In the next step, the fortified polymer is recycled by thermomechanical reprocessing, where processes such as sorting, washing, grinding, melting and pelletizing can be performed. Subsequent analysis of the contaminated polymer and the recycled material will provide evidence on the efficiency of the recycling process.[34,35]

It should be highlighted that, in the case of having the recycled material only before manufacturing the final product for food packaging, it is not possible to perform migration studies. In that event, extractability tests can be applied. Powdered material fortified by surrogates represents the worst-case scenario for safety issues. It can be assumed that the results of migration assays of manufactured recycled packaging will always lower be than the results of the analysis of direct extracts from the raw material. Extraction results can provide the required information about the feasibility of the recycled material for food contact applications, hence these tests can be applied instead of migration assays.[36]

One of the important health concerns regarding contaminants and migrants from recycled pellets are mineral oil hydrocarbons (MOHs). These are mixtures of alkanes (linear hydrocarbons) and naphthenes (cyclic hydrocarbons) called mineral oil saturated hydrocarbons (MOSHs) and highly alkylated mono- and/or polyaromatic hydrocarbons called mineral oil aromatic hydrocarbons (MOAHs). Hydrocarbons having at least one aromatic ring are considered as MOAHs, even though they may consist predominantly of saturated carbons. Mineral oils are by-products made from petroleum and are mainly used as lubricants for engines and technical instruments.[37] Mineral oils appear as contamination appears as a hump in the chromatographic trace, as can be seen in Figure 7.2.

MOSHs in the range C_{16}–C_{35} may accumulate in human tissue and could have harmful health effects. However, the most dangerous type of mineral oils are MOAHs, which include the potential carcinogenic aromatic hydrocarbons with their ability to alter genetic material and disrupt the hormonal system. Since no level of consumption of MOAHs can be considered safe, the European Food Safety Authority (EFSA) established that any exposure to them through food represents a potential health risk. Therefore, the analysis of mineral oil mixtures in recycled packaging is essential.[38–40] Other sources of mineral oils also connected with packaging can be polyolefin-based hot-melt adhesives and offset printing inks. The range of alkanes to be considered is C_{16}–C_{50}.

7.5.1 Printing Inks

Inks are commonly used in food packaging materials and, therefore, migration of ink components to food must be studied. They can be defined as a coloured fluid or paste used for writing, drawing or printing, and are mainly composed of a pigment or dye, suspended or dissolved in a solvent.

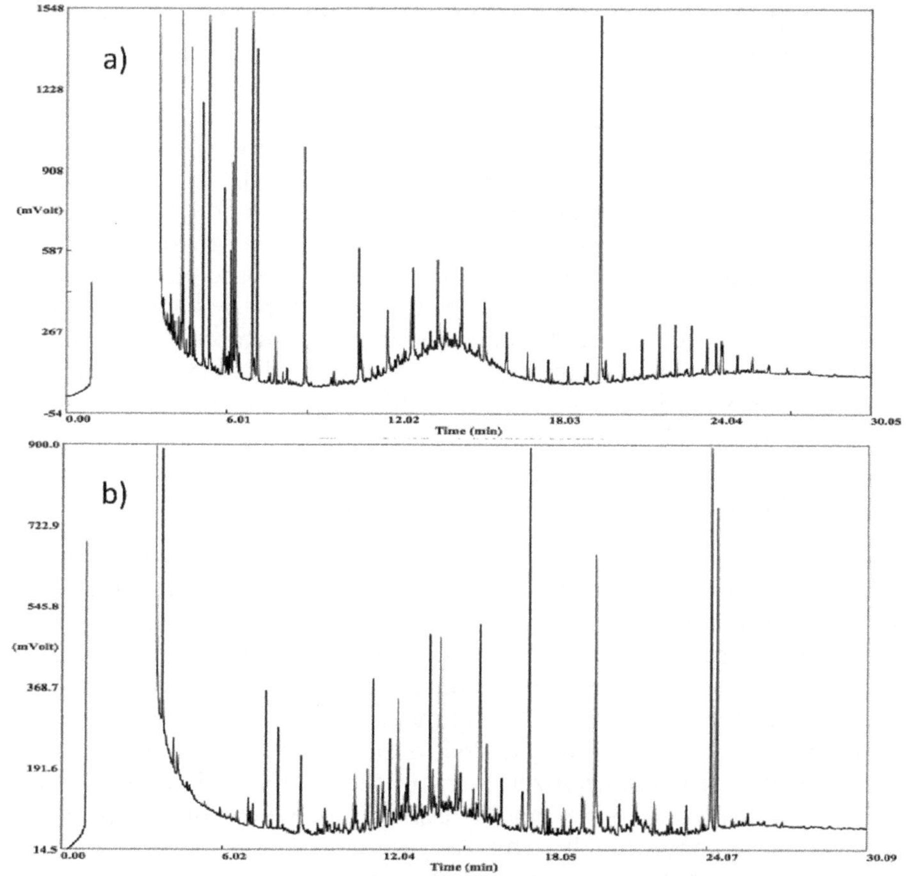

Figure 7.2 Chromatograms showing humps of (a) MOSHs and (b) MOAHs extracted from recycled poly(ethylene terephthalate) (PET) pellets, separated on silica gel with silver nitrate and analysed by gas chromatography with flame ionization detection (GC-FID).

When multilayer materials are used in food packaging applications, migration can take place, not only from the food contact surface, which is the internal side of the packaging, but also from external layers due to diffusion and partition processes. The possibility of migration of ink components into food increases when the materials are produced and stored in rolls. The transfer of ink components from the external printed surface of food packaging to the food contact surface is called a set-off phenomenon. The European Printing Ink Association (EuPIA) addresses the most important questions about printing inks and recommends a list of compounds that can be used for producing printing inks for food packaging. During the manufacture of food packaging inks, a wide range of chemical compounds are applied: additives, colorants, pigment additives, polymeric resins, surfactants, solvents and photoinitiators.

The results of studies of the migration of different multilayer materials for food packaging with different printing technologies showed the presence of different non-volatile ink components such as colorants (Solvent Red 49), plasticizers (dimethyl sebacate, tributyl *O*-acetylcitrate) and surfactants (Schercodine M, triethylene glycol caprylate). An oxidation product of an ink additive (triphenylphosphine oxide) was also detected. The importance of cut-off selection for migration experiments should be highlighted since inks are not homogeneously distributed over the surface and therefore different compounds can be found in migration depending on the piece of material analysed.[41] The number of migrants can be decreased by application of an external lacquer. Nevertheless, some new migrants can appear following the reaction between ink and lacquer.[42] Even though only authorized chemical compounds are used, migration studies are necessary owing to the possibility of migration of NIAS.

7.5.2 Non-intentionally Added Substances

NIAS are chemical compounds that migrate from FCMs but they were not added on purpose to the FCM. They may originate from degradation of the material components, impurities in raw materials applied for processing of food packaging and mixtures of chemical compounds that can react with each other and form new unwanted compounds. They could even come from the interaction between the FCM and the food. However, as these chemical compounds were not originally added to the material they may not appear in the positive list of authorized starting substances and additives. Some examples of NIAS determined and identified in polymers for food packaging applications are presented in Table 7.1.

It should be highlighted that any potential health hazard caused by NIAS coming from the final material should definitely be assessed by the manufacturer according to scientific bases of risk assessment that are internationally recognized.[53] However, the identification of NIAS is a very challenging task, for numerous reasons. First, the real composition of plastic materials, additives and adhesives is unknown. FCMs can include complex multilayer structures with up to seven or more different layers, either coextruded or laminated. They can be manufactured using different adhesives and different polymers or paper as substrates, which can also be coated with varnishes and printing inks. Further, the detailed compositions of plastic materials and adhesives are in general highly confidential, and only the major components of the packaging are usually revealed. In addition, there are several different participants involved in packaging manufacture, such as the polymer producer, the manufacturer, the converter and the producers of adhesives, printing inks and varnishes. In addition, the final function and application of the packaging produced have an influence on food–packaging interactions.[11]

Second, there are series of compounds such as the stabilizers, UV absorbers, plasticizers and antioxidants that were created specially to confer

Table 7.1 NIAS detected in migration from different food packaging materials, their origin and the analytical method of detection.[a]

FCM	NIAS	Analytical method	Reference
Active packaging	N^2-Dodecanoyl-L-arginine (LAS)	UPLC®-Q-TOF-MS^E	43
Active packaging	$C_{10}H_{16}O_2$	UPLC®-Q-TOF-MS^E	17
Adhesives	Primary aromatic amines (PAASs)	UPLC®-MS/MS UPLC®-Q-TOF-MS^E	23
Adhesives	Abietic acid derivatives	GC-MS	44
Polyethylene	(*Z*)-9-Octadecenamide	GC-MS	45
Polyethylene	2,4-Di-*tert*-butyl-6-nitrophenol	GC-MS	45
Poly(ethylene terephthalate)	Nonylphenol	GC-MS	46
Poly(ethylene terephthalate)	Acetaldehyde	GC-MS	47
Polypropylene	Dimethylbenzaldehyde	ESI-MS GC-MS	48
Polypropylene	Diphenyl phosphate	SPME-GC-MS	49
Printing ink	Benzophenone	HPLC-UV LC-MS/MS	50
Printing ink	Bisphenol A (BPA)	HPLC-UV LC-MS/MS	50
PVC, plasticized	ESBO chlorohydrins	UPLC®-Q-TOF-MS^E	51
Recycled PE	Diethylhexyl phthalate	GC-MS	52
Recycled PP	Aza-2,2,6,6-tetramethyl-3-cyclohexene	GC-MS	52

[a]ESBO, epoxidized soybean oil; PE, polyethylene; PP, polypropylene; PVC, poly(vinyl chloride); GC, gas chromatography; LC, liquid chromatography; UPLC®, ultra-high-performance liquid chromatography; Q, quadrupole; TOF, time-of-flight; MS, mass spectrometry; MS^E, elevated energy mass spectrometry; ESI, electrospray ionization; SPME, solid-phase microextraction.

additional functions on the plastic matrix, but there is little knowledge about their formulae.[54] In addition, adhesives, varnishes and printing inks are not regulated by EU law,[11] even though producers should release specific "statement of composition" documents.

Third, quantification of NIAS is difficult because of the presence of numerous chemical classes of substances and their different analytical limits of detection and response factors in chemical and analytical detectors. NIAS have to be detected and unequivocally identified in food simulants after the migration tests, especially when risk assessment is involved. In addition, it is very important to know the origin of the NIAS in order to avoid their presence in packaged food. Trace level detection of NIAS in food packaging materials, through the application of scientifically recognized analytical strategies, is obligatory, and includes a description of the compounds involved.[55] A very low concentration of migrants and NIAS could be expected in this case. Sophisticated analytical techniques are therefore required.[11]

7.5.3 Oligomers

Usually, oligomers are considered as NIAS when they are detected as migrants from FCMs.[11] Nevertheless, oligomers are certainly created during the polymerization process and could be considered as polymer-specific accompanying substances. In practice, the oligomer patterns obtained are closely related to the polymerization process, hence they can give a clue about the polymer type. A good example of this are ε-caprolactam oligomers, which are extractable compounds associated with polyamide-type packaging materials.

ε-Caprolactam is the monomer used for the production of polyamide 6 (nylon), where a ring-opening polymerization takes place. Therefore, polyamide 6 is a mixture of linear and cyclic oligomers of ε-caprolactam.[56] Additionally, during the manufacture of polyurethane adhesives, polyesters are also used. As a result, some by-products can be formed, such as cyclic polyester oligomers coming from the reaction between constituents of multilayer material and adhesive. Table 7.2 provides structural information and possible MS/MS fragmentations of ε-caprolactam oligomers of food packaging origin.[56,57]

The reported presence of caprolactam and its higher molecular weight oligomers in packaging materials suggested that such oligomers may also be present in simulant solutions applied in migration tests and food products contacted by either nylon itself, nylon-containing materials or polyurethane adhesive in multilayer materials. The results of migration tests involving degradation compounds of cyclic oligomers in 3% acetic acid showed that the concentration of some cyclic oligomers derived from polyurethane adhesives can decrease over time.[57] It has also been demonstrated that the distribution of oligomers in the commercially available nylon-6 material is clearly different from that in the nylon layer of the multilayer material. This is probably due to the difference in the relative proportions of oligomers.[56]

Moreover, gamma-irradiated polyamide packaging has been studied and caprolactam has been detected and quantified during migration assays. It should be mentioned that the irradiation of packaged food is connected with its sterilization.[58] Although there is a maximum limit when gamma or beta irradiation is used (usually 10 kGy), the stability and safety of such packaging materials should be confirmed. It has also been concluded that caprolactam is a good irradiation marker owing to its presence in both types of analysed film, irradiated and non-irradiated, with a higher proportion in non-irradiated film samples.[59]

7.6 Active Packaging

Active packaging involves the deliberate incorporation of the active agent into the packaging system either in the packaging material or in the packaging headspace. On the one hand, this interactive technology allows the release of active compounds into the packaging atmosphere or packaged food. On the other, the absorption of undesirable substances from the surrounding

Table 7.2 Information on ε-caprolactam oligomers of food packaging origin.

Name	Exact mass	MS/MS fragments (m/z)	Structure
Caprolactam (n = 1)	113.0841	—	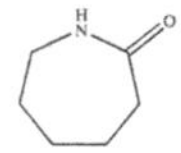
Caprolactam cyclic dimer (n = 2)	226.1682	227.1858, 209.1781, 114.0960, 96.0869	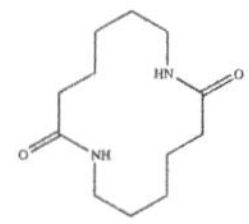
Caprolactam cyclic trimer (n = 3)	339.2523	340.2777, 322.2710, 227.1858, 209.1781, 114.0960	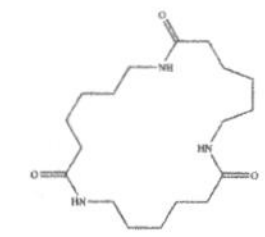
Caprolactam cyclic tetramer (n = 4)	452.3364	453.3656, 435.3436, 340.2812, 322.2737, 226.2067, 209.174, 114.1022	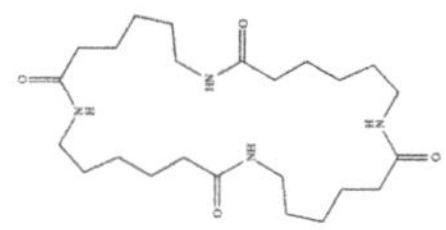
Caprolactam cyclic pentamer (n = 5)	565.4205	566.4750, 548.4605, 453.3907, 435.3682, 322.2737, 209.1917, 114.1148	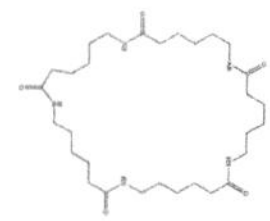
Caprolactam cyclic hexamer (n = 6)	678.5046	679.5062, 340.3403, 359.3439, 209.1732, 114.0914	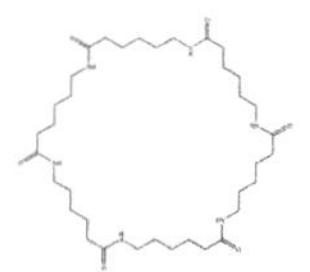
Hydrolysed caprolactam dimer (n = 2)	244.1787	245.1967, 228.1626, 227.1838, 132.1085, 114.0898	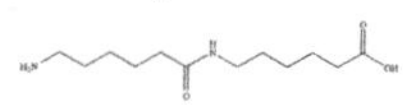
Hydrolysed caprolactam trimer (n = 3)	357.2628	358.2638, 341.2386, 340.2590, 322.2523, 245.1967, 228.1626, 227.1838, 209.1752, 132.1085, 114.0898	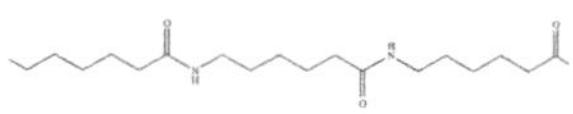

environment or packaged food is also possible. Active packaging can contain antioxidants or antimicrobials as the active agent.

Antioxidant performance can be achieved either by adding antioxidants to the packaging material that are released from it to the food or incorporating free radical scavengers in the packaging material. The latter approach has been demonstrated to be very efficient, because the antioxidant performance takes place without direct contact with the food and does not affect the organoleptic quality of packaged food.

Active packaging represents an innovative strategy for incorporating antioxidants in a polymer to prevent oxidative processes and extend the shelf life of food products.[60] There is a wide range of advantages of the application of active packaging: small amounts of active substances needed, localized activity, controlled release of antioxidants and elimination of processing steps

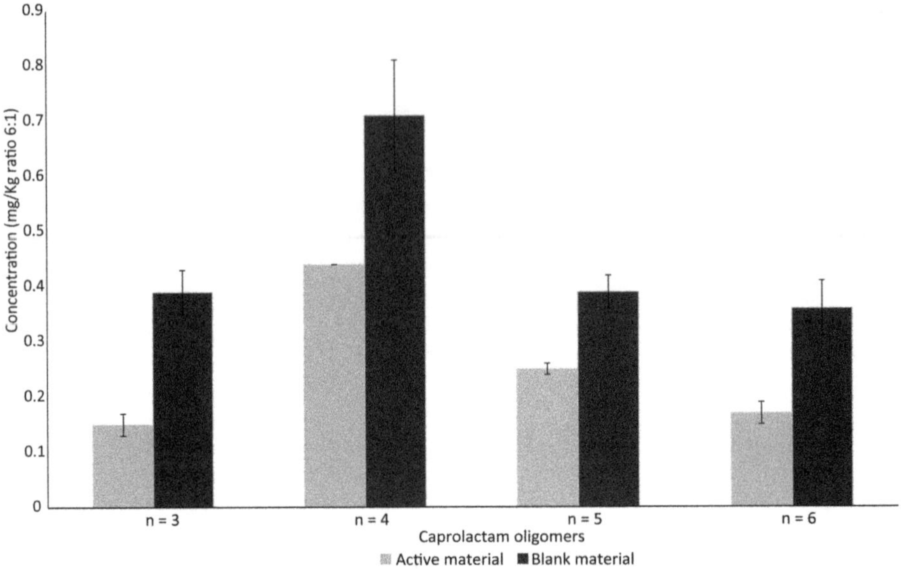

Figure 7.3 Migration of caprolactam oligomers (n = 3–6) from blank and active packaging into 10% ethanol analysed by UHPLC®-Q-TOF-MS^E.

such as mixing, immersion or spraying of food compared with direct addition of antioxidants to food products. Extensive research has been carried out looking at the effect of antioxidant packaging systems in controlling oxidative processes in fresh foods. However, only selected commercial applications can been found on the market.[61–65] Addition of antioxidants to the polymer matrix can influence the migration of polymer components. It can significantly reduce the concentration of some migrants such as non-volatile compounds and also prevent the migration of volatile compounds. An example is the migration of caprolactam oligomers from polyamide active packaging. Figure 7.3 compares the migration of caprolactam oligomers from blank polyamide *versus* an active polyamide based on green tea solution. It can be observed that the addition of active compounds such as catechins and caffeine to the polymer matrix significantly reduces the migration of caprolactam oligomers.

Microbial growth accelerates some changes in the aroma, colour and texture of food, resulting in a reduction in shelf life and a simultaneous increased risk of food-borne illnesses. The development of antimicrobial packaging represents a promising approach for the active control of the growth of microorganisms.[66,67]

Antimicrobial food packaging materials extend the lag phase and reduce the growth phase of microorganisms.[68] However, antimicrobial packaging materials need to be released from the FCM or have to be in direct contact with the food surface, as the antimicrobial substances need to reach the microorganism cells to kill them. These antimicrobials could either be immobilized on the surface of the material or migrate into the food. Therefore, antimicrobial packaging is effective on food products where the microbial contamination occurs at the surface.[69] Successful applications of antimicrobial packaging have been reported.[70–72] Tests have also been carried out successfully with foods and there are also some antimicrobial packaging materials on the market for processed foods.

The safety assessment should focus on the risks related to dietary exposure to chemicals due to the migration of active substances, their potential degradation and/or reaction products and their toxicological properties. Further, in the risk assessment of active packaging, target analysis of active compounds used as both antimicrobials and antioxidants should be included. For this purpose, high-resolution mass spectrometry (HRMS) techniques with high selectivity and sensitivity should be applied. Several approaches will be described later.

7.7 Packaging with Nanomaterials

Nanomaterials are engineered particles with at least one dimension shorter than 100 nm and characterized by very large surface-to-mass or surface-to-volume ratios. They may have different physical, chemical and biological properties compared with their larger bulk counterparts.

The applications of nanomaterials in numerous fields have expanded in recent years, including for food packaging, an area that has seen large increases over the last decade. Nanotechnology has the potential to offer many applications in the food industry, such as nutritional additives, stronger flavouring and colouring substances, antibacterial ingredients for food packaging and improvements in food structure and texture.

Nanotechnology-enabled food packaging can be divided into two different key aspects. The first is connected with improvements in packaging, where nanomaterials are mixed into the polymer matrix to give *e.g.* polymer–clay nanocomposites, to improve the gas barrier properties. The second is connected directly with the antioxidant/antimicrobial activity of packaging, where the nanoparticles (NPs) interact directly with the food or the environment to allow better protection of the food, examples being eugenol-loaded chitosan NPs for active packaging applications,[73] silver NPs as potent antimicrobial agents[74] and selenium NPs as free radical scavengers in food packaging.[75]

The main goal in the study of the migration of NPs is the development of new analytical techniques that are able to measure and quantify NPs in both the packaging and the food. Moreover, the major challenge in understanding the behaviour of migrated NPs in food simulants or food results from the difficulties connected with the presence of a complex matrix and, at the same time, the low concentrations of NPs. Another analytical challenge in the detection of NPs is the definition of whether the migrated compounds were present as NPs or ionic species. The migration of NPs from packaging materials can also be influenced by both the chemical interaction between the NPs and the polymer matrix, changes in the packaging film that might be induced by NPs and the composition of the food product. Other properties such as pH, migration time and temperature should also be taken into account.[76,77]

7.8 Legislation

To ensure food safety, the migration of detected substances should not exceed unacceptable amounts established by European legislation. The European Commission Regulation[78] on plastic materials intended to come into contact with food (Regulation EU 10/2011) presents specific rules for plastic materials and articles to be applied for their safe use. The mentioned legislation provides a positive list of authorized monomers, starting substances and additives that can be applied in the manufacture of FCMs. All amendments to Regulation EU 10/2011 should be taken into account (eight have been published so far). The listed substances are called intentionally added substances (IAS). Their specific migration limits (SMLs) and overall migration limits are provided in Annex I of the Regulation EU 10/2011. Non-intentionally added substances (NIAS) come from the degradation of additives, plastic materials and also impurities in the starting materials. Unfortunately, they are not listed in the Regulation. However, it was recognized by Regulation EU 10/2011 that NIAS are able to be formed during the manufacture and use of plastic materials for food contact. It is crucial to consider at least the main NIAS coming from the final packaging material because they are extremely relevant for risk assessment. Moreover, the already known NIAS should undoubtedly be included in the list of positive substances in the EU Regulation.

To ensure the overall quality of the plastic, the overall migration to food of all substances together may not exceed the overall migration limit (OML) of 60 mg kg^{-1} of food or 10 mg dm^{-2} of the contact material. Nevertheless, it should be emphasized that there are compounds such as adhesives components that are still not legislated for at the European level. The migration level of non-authorized substances should not exceed a level of 0.01 mg kg^{-1} of food or food simulant. For example, this is the case with some cyclic polyester oligomers and therefore the determination of their migration to food is very relevant.

Notwithstanding, these rules exclude CMR substances (substances classified as carcinogenic, mutagenic or toxic for reproduction). Generally, studies are focused on compounds with molecular weights up to 1000 Da. Substances with higher molecular weights are considered as physiologically not absorbable because of their size.

In summary, the launching onto the market of any new active material requires that migration tests be carried out to certify that these materials conform to the legislation and guarantee the health of consumers. Migration tests consist of the exposure of the plastic sample in contact with several established liquid simulants that mimic the behaviour of food. In the case of monolayer materials, total immersion is applied when performing migration tests, whereas in the case of multilayer materials, special migration cells should be used to expose to the simulant or food only one layer that has contact with the food product in the real packaging (one-face migration). Regulation EU 10/2011 lists 10% v/v ethanol, 3% w/v acetic acid, 20 and 50% v/v ethanol and vegetable oil as aqueous, acidic and fatty food simulants, respectively. Alternatively, isooctane and 95% v/v ethanol can be used instead of olive oil as fatty simulants. Regulation EU 10/2011 includes modified polyphenylene oxide

(Tenax®) as a simulant for solid dry foods. Experiments are performed by direct contact between small specimens of test material and Tenax® powder. As Tenax® is expensive compared with other simulants, it may be reused several times with intermediate clean-up processes based on Soxhlet extraction.

After exposure of the plastic samples to simulants for a determined time and temperature, the released substances, migrants, are analysed by appropriate analytical methods and their concentrations are compared with the established values. The plastic is considered safe and can be marketed if all the value found s are below their SMLs. The total concentration of all detected migrants is expressed as that corresponding to 6 dm^2 of material per kilogram of food simulant or food.[78]

7.9 Risk Assessment

Risk assessment of any plastic material for food applications is based on the analysis of migration assay results. When no toxicity data for any chemical substance that migrates are available, the threshold of toxicological concern (TTC) is applied. Originally, this theoretical approach was used in the assessment of indirect food additives (contact substances) and food flavourings; subsequently, it has been investigated and proposed for use in a wide range of regulatory areas. For the migrants that are not included in the list of authorized substance in Regulation EU 10/2011, the following procedure should be used to evaluate their health risk. First, the software Toxtree is applied to estimate the toxicity of migrants based on Cramer rules. The Cramer decision tree is an approach for classifying and ranking chemicals on the basis of their expected level of oral toxicity, which classifies the compounds into one of three toxicity classes and proposes a maximum daily intake. This theoretical approach is based on the chemical structure of any substance, but its application is valid only when experimental data are not available. Class I contains substances with simple chemical structures and for which efficient modes of metabolism exist, suggesting a low order of oral toxicity. The maximum daily intake for substances in class I is 1.8 mg per person per·day. Class II contains substances that possess structures that are less innocuous than class I substances, but do not contain structural features suggestive of toxicity such as those substances in class III. The maximum daily intake for substances in class II is 0.54 mg per person per·day. Class III contains substances with chemical structures that permit no strong initial presumption of safety or may even suggest significant toxicity or have reactive functional groups. The maximum daily intake for substances in class III is 0.09 mg per·person per·day.

Second, theoretical maximum migration amounts ($mg \cdot kg^{-1}$) of the migrants can be calculated from maximum daily intake based on the following equation:

$$EDI = \text{migration} \times \text{food intake} \times CF \tag{7.1}$$

where EDI is the estimated daily intake, which indicates the maximum daily intake for each substance per person; in Europe food intake is considered as

1 kg per person per day. *CF* is the consumption factor, which represents the fraction of the daily diet for a specific material. It is based on exposure and has been obtained and used by the US Food and Drug Administration (FDA). In European legislation no *CF* is applied. However, it could be applied only for migrants in the absence of other parameters; it cannot be used for compliance of any FCM. Finally, comparison of the theoretical maximum migration amounts with the migration results should be performed.[79,80]

7.10 Analytical Methods

The necessity for the qualitative and quantitative analysis of migrants has led to the development of a wide range of analytical methods. Basically, we can distinguish five different analytical approaches according to the type of sample analysed and type of analyte:

- methods for direct analysis of the polymer surface;
- methods for analysis of volatile compounds;
- methods for analysis of non-volatile compounds;
- methods containing a preconcentration step;
- methods for analysis of NPs.

A general scheme of the sample treatment procedures required for these studies is shown in Figure 7.4.

7.10.1 Methods for Direct Analysis of the Polymer Surface

Polymers before and after migration and also solid food simulants can be analysed directly with the following analytical techniques: atmospheric solids analysis probe (ASAP) MS, direct analysis in real-time (DART) MS and desorption electrospray ionization (DESI) MS. These methods are fast and do not require any sample handling. However, they should be used to determine only well-known NIAS owing to the complicated fragmentation patterns that are usually obtained. These methods are applied for target analysis. Additionally, polymer dissolution in solvents or extraction of polymers may be performed and analysis by GC-MS and LC-MS methods can be carried out.[81–83]

7.10.2 Methods Including a Preconcentration Step

Several sample preparation techniques can be used for the extraction of compounds from liquid samples such as food simulants.

Solid-phase microextraction (SPME) is a widely used solvent-free sampling technique for the analysis of volatile and semivolatile compounds. It is a variation of solid-phase extraction in which a thin fibre coated with extraction phase is placed inside the steel needle. The process of isolation of analytes by SPME consists of two stages: first, the sorption of analytes from the gas, liquid or solid sample into the fibre takes place, then desorption of the analytes into the injection port of the chromatograph is performed. SPME is

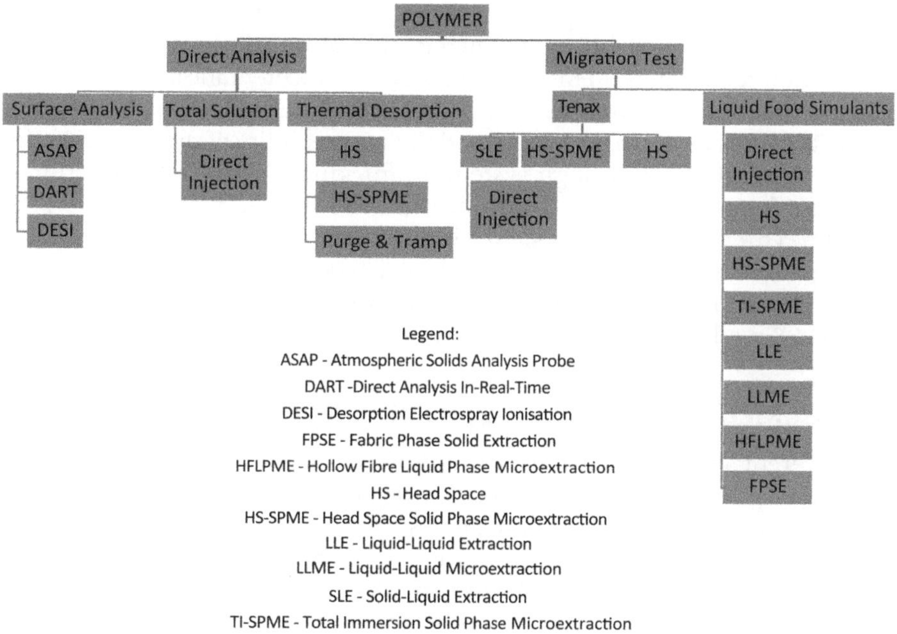

Figure 7.4 Sample treatment procedures for migration studies of FCMs.

often used in headspace mode (HS-SPME), where there is no direct contact between the fibre and the sample. In this mode, the analytes are taken from the headspace of the vial containing the sample. The great advantage is that it allows the directly analysis of aqueous samples and consequently food simulants with high contents of water, such as a 10% solution of ethanol or 3% acetic acid. Moreover, it can be used to evaluate interactions between compounds coming from the food and packaging used for its storage.[34,36,84]

A two-phase hollow-fibre liquid-phase microextraction (HFLPME) with a high degree automation is able to process up to six samples simultaneously by means of a multiple-channel syringe pump. The experimental setup allows dynamic extractions to be carried out with a considerable reduction of sample handling. Membrane extraction combines preconcentration, separation and cleanup steps in a single operation with high versatility, selectivity, cheapness and ease of both miniaturization and automation. The system has been applied for the determination in aqueous food simulants of migrants from prototypes of active packaging to assess their safety before marketing, showing detection limits in the ng g^{-1} range, relative standard deviations below 13% and concentration factors ranging from 83 to 338. The migration of 43 compounds, including terpenes, alkanes, plastic additives and degradation compounds, was reported.[85–87]

Another novel, simple, fast and sensitive analyte extraction method based on fabric-phase sorptive extraction (FPSE), followed by analysis using UPLC®-MS, can be applied to the analysis of common non-volatile plastic additives. FPSE, developed recently by Kabir and Furton,[88] is a new-generation sample preparation technique that effectively addresses the majority of the shortcomings associated with conventional sorptive sample preparation techniques. FPSE integrates the advanced material properties of sol–gel sorbents, the hydrophobic/hydrophilic properties of natural or synthetic fabrics as an active substrate and the permeability of the substrate that creates a pseudo flow-through system, mimicking solid-phase extraction to stimulate fast mass transfer from the aqueous sample matrix to the FPSE medium. Three FPSE media coated with different sol–gel sorbents characterized with different polarities, including sol–gel polydimethylsiloxane, sol–gel poly(ethylene glycol) and sol–gel polytetrahydrofuran, have been studied. All three FPSE media gave very satisfactory results.[89]

7.10.3 Methods for the Analysis of Volatile Compounds

Two different approaches can be distinguished depending on the presence or absence of sample treatment. Direct headspace coupled to GC-MS is a well-known, powerful technique for the analysis of volatile compounds. The main advantage resides in it being an automatic technique, able to analyse the most volatile migrants released from the FCM. However, in this technique only a fraction of the vapour in equilibrium with the sample is injected into the GC-MS system, which limits the sensitivity. Both volatile and semivolatile compounds can be analysed using SPME in headspace mode directly coupled to a GC-MS system. In this way, the sensitivity is considerably increased, although the SPME microfibre needs prior optimization. Then, quantitative analysis should be performed using exactly the same conditions as in SPME-GC-MS. An example of this technique is shown in Figure 7.5.

Most commercial GC-MS systems apply different types of ionization under vacuum conditions, such as chemical ionization (CI) and electron ionization (EI). CI is based on the formation of ions after reactions with reagent gas. It is considered to be a soft ionization in which only a few fragments are created and often the molecular form can be seen. On the other hand, EI is considered to be a hard ionization that produces extensive molecule fragmentation. Different structural information on molecules can be obtained by using different gas types as the fragmentation pattern depends on it.

The combination of GC-MS with an atmospheric pressure chemical ionization (APCI) source is an alternative technique for the analysis of volatile compounds and is called atmospheric pressure gas chromatography (APGC). Since this technique is based on APCI, it is a relatively soft ionization method compared with EI. Moreover, APGC operates at atmospheric pressure, which removes the restriction imposed by pumps, allowing a much wider range of flow rates for full GC separations.[13] An example is shown in Figure 7.6.

7.10.4 Methods for the Analysis of Non-volatile Compounds

LC-MS is a common technique for determining trace levels of compounds migrating from FCMs. The LC-MS system can be equipped with several mass analysers, each of which provides unique features capable of identifying,

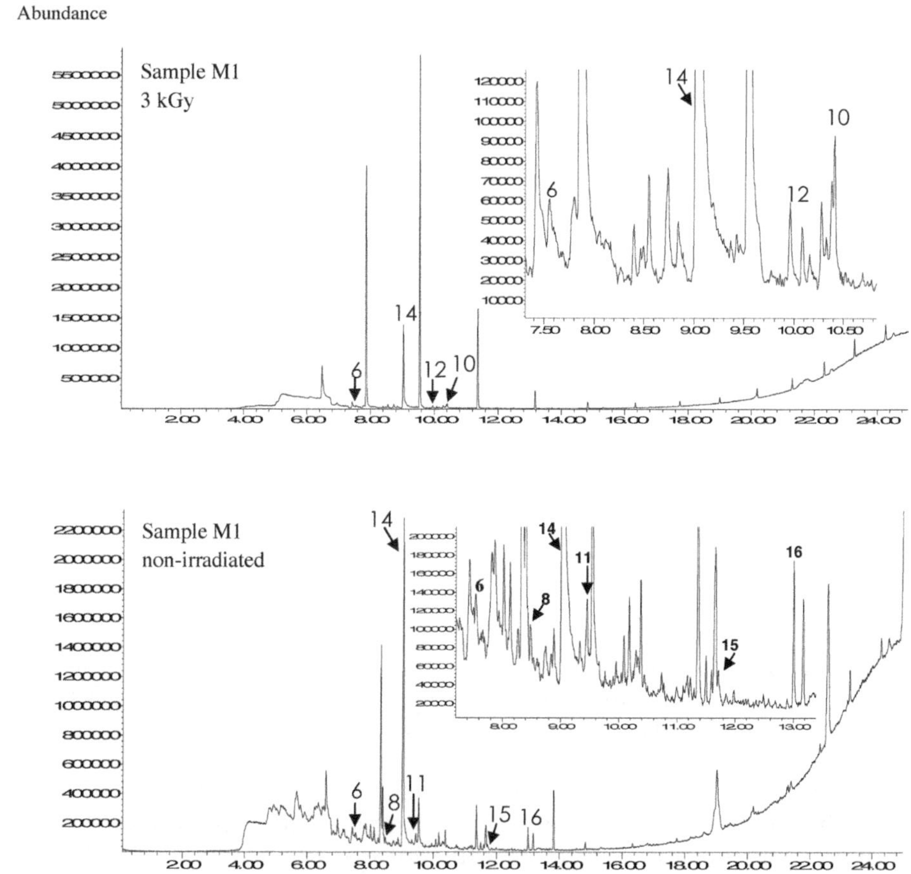

Figure 7.5 Chromatograms obtained by HS-SPME-GC-MS, using polydimethylsiloxane fibre, from multilayer polyamide 6 films (sample M1) after irradiation with 0 and 3 kGy. Reproduced from ref. 59 with permission from Springer Nature, Copyright 2008.

quantifying and resolving ambiguities by selecting appropriate ionization and acquisition parameters.

The use of hybrid instruments aims to unify several advantages in one instrument. A tandem quadrupole and time-of-flight (Q-TOF) instrument or an ion trap and time-of-flight (IT-TOF) instrument allows accurate mass determinations of the precursor ion and also of the product ions, providing information about fragmentation patterns. Detection of NIAS by ultra-performance liquid chromatography coupled to quadruple time-of-flight (UHPLC®-Q-TOF) with a novel technique called elevated energy mass spectrometry (MS^E) is the ultimate technology for comprehensive, reproducible profiling and characterization. It is one of the most powerful techniques used for the determination of non-volatile compounds. Simultaneous acquisition of low- and high-energy data with different energy ramps is possible. MS^E allows the generation of an exact mass precursor ion and the exact mass fragments ions

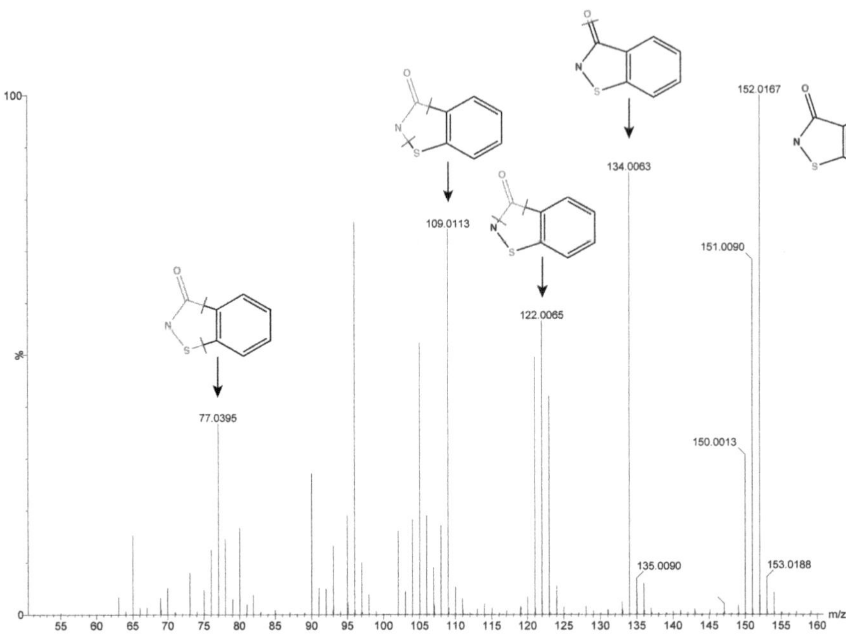

Figure 7.6 High-energy mass spectrum of the NIAS 1,2-benzothiazol-3(2*H*)-one from acrylic adhesive acquired by APGC-Q-TOF-MS^E. Reproduced from ref. 95 with permission from Elsevier, Copyright 2012.

in a single analytical run. Mass precursors and mass fragments are collected for each detectable compound in the sample. This represents excellent structural information and ensures the correct identification of unknown compounds. MS^E detection is faster than standard MS or tandem MS. An additional advantage is that the proposed method does not require time-consuming sample preparation and the solvent consumption is low.[17,90] Figure 7.7 shows an example of low- and high-energy ionization to elucidate the chemical structure of a NIAS.

The recent technique of ion mobility (IM) coupled to an HRMS detector is an additional tool for the separation and identification of compounds. It is based on the different mobilities of ions in the drift tube, where the movement depends on the shape, volume and weight of the ions. The measurement of the drift time of the ions, after a calibration process, is converted to a collision cross-section (CCS) value, which represents another parameter for identification, as it is characteristic of the molecule and does not depend on the chromatographic conditions. In this way, isomers of flat and spherical compounds behave very differently and provide different CCS values, which can be further used as an additional tool for identification.

Hybrid linear ion trap–high-resolution mass spectrometry (LTQ-Orbitrap®) has also allowed the identification of non-targeted and unknown compounds. It combines a high-resolution mass spectrometer, such as an Orbitrap® analyser, with an external accumulation device such as a linear ion trap, making possible multiple levels of fragmentation (MS^n) for the elucidation of analyte

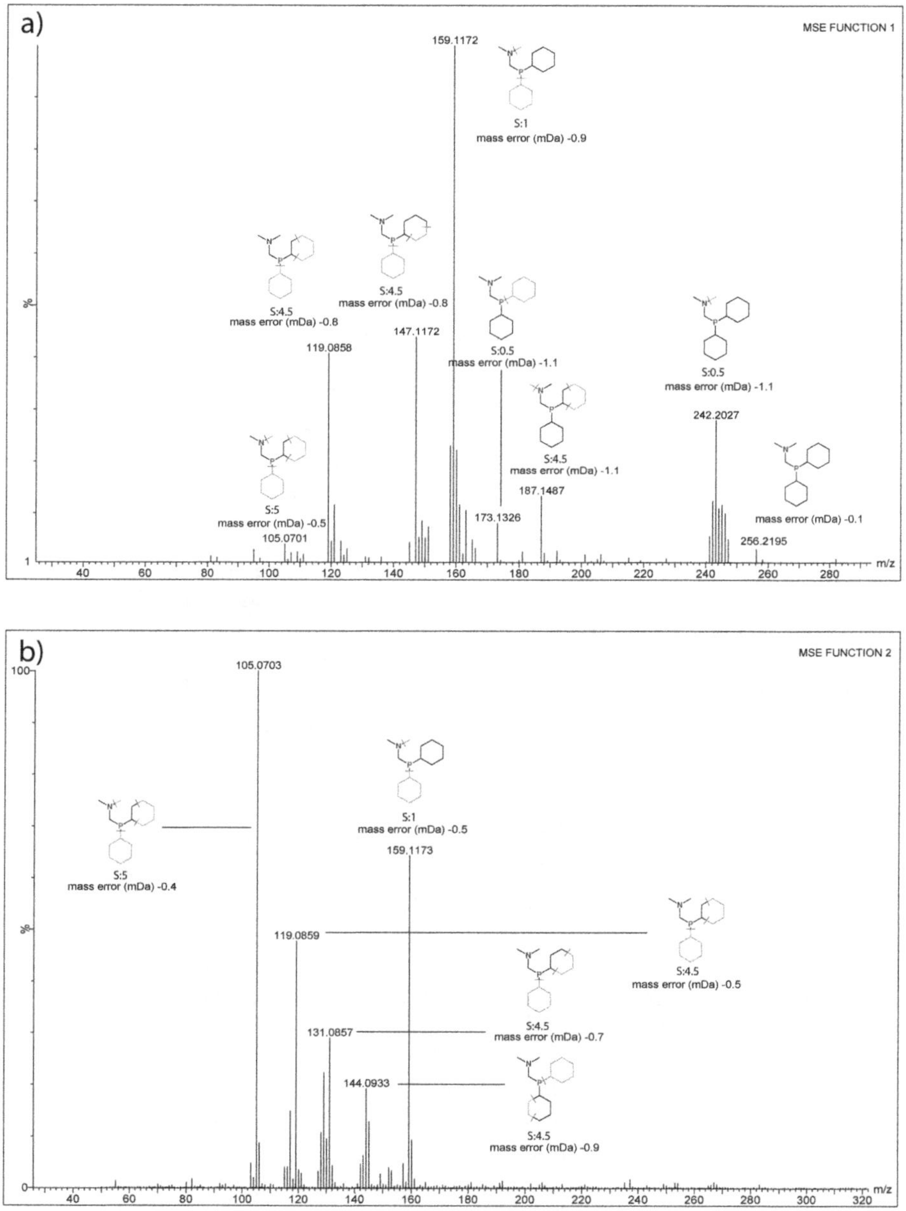

Figure 7.7 a) High-energy mass spectra (Function 1; cone voltage 30 V) and b) low-energy mass spectra (Function 2; cone voltage 30 V; collision ramp energy from 20 to 40 eV) to identify NIAS. Reproduced from ref. 13 with permission from John Wiley and Sons, Copyright © 2014 John Wiley & Sons, Ltd.

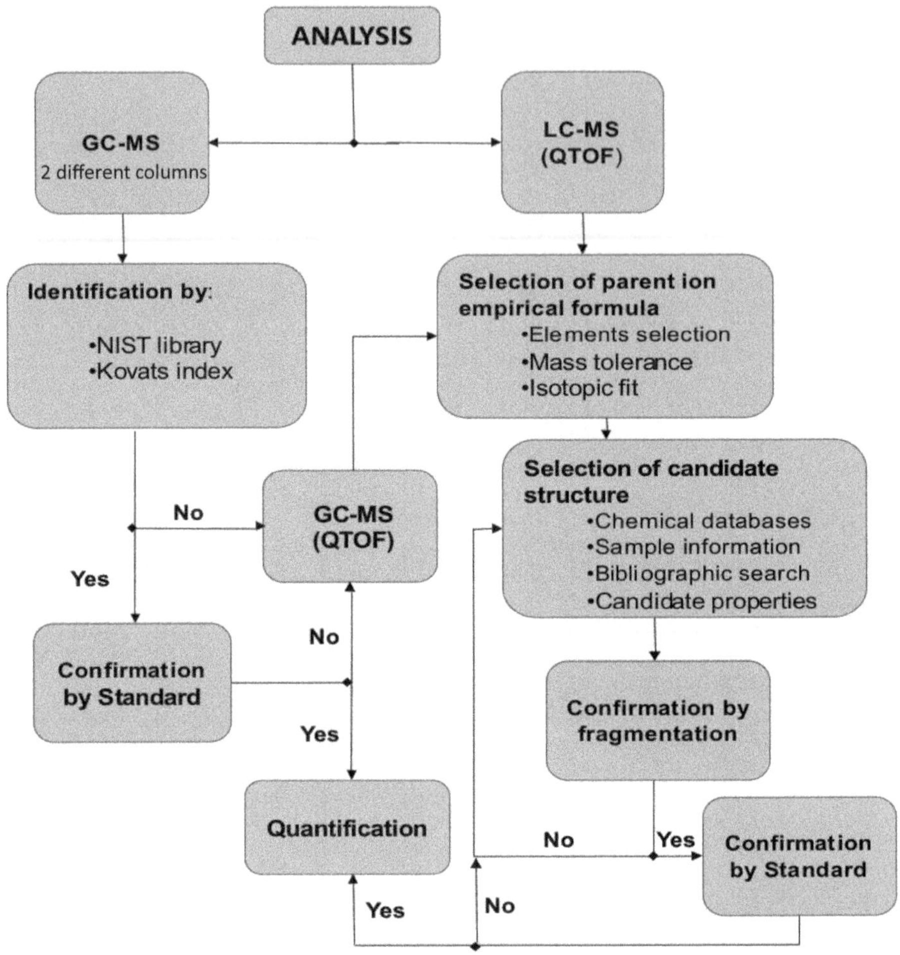

Figure 7.8 Decision tree for the analytical procedures in NIAS identification. Reproduced from ref. 11 with permission from Elsevier, Copyright 2013.

structures. The use of the LTQ-Orbitrap® instrument allows high-quality accurate mass determination and acquisition of MS^n spectra. Fourier transformation of the acquired transient allows wide mass range detection with high resolving power, mass accuracy and dynamic range. LTQ-Orbitrap® MS has two main advantages: high sensitivity in full-scan mode and the possibility of the determination of accurate masses of product ions.[91]

A scheme of the analytical procedures for the identification of volatile and non-volatile NIAS and unknowns is shown in Figure 7.8.

7.10.5 Methods for the Analysis of Nanoparticles

Spectrometric techniques such as single-particle inductively coupled plasma mass spectrometry (ICP-MS) have been applied to the detection of NPs and this technique permits the simultaneous determination of concentration and size.[92,93] However, ICP-MS is an expensive technique because of the use of large amounts of high-purity argon.

Techniques that are widely used to study NPs are transmission electron microscopy (TEM) and scanning electron microscopy (SEM), which are capable of detecting single NPs. However, it is difficult to perform such investigations in complex matrices such as food.[74]

The dynamic light scattering (DLS) technique, which measures the hydrodynamic diameters of particles, is another useful technique applied in the study of NPs in solutions. Limitations of this technique are poor sensitivity at dilute concentrations, non-selective material detection and failure to differentiate between NPs and other matrix components.[94]

Field flow fractionation (FFF) allows the separation of NPs. The separation process is based on physical forces and it is similar to chromatography. However, it requires connection to ICP-MS or ICP-AES (atomic emission spectrometry).[74,75]

References

1. V. Trinetta, Biodegradable Packaging. *Reference Module in Food Science*, Elsevier, 2016, DOI: 10.1016/B978-0-08-100596-5.03319-9.
2. K. Marsh and B. Bugusu, Food packaging–roles, materials, and environmental issues, *J. Food Sci.*, 2007, **72**, R39–R55.
3. C. Nerin, *Plastics and Polymers for Food Packaging Manufacturing. Reference Module in Food Science*, Elsevier, 2016, DOI: 10.1016/B978-0-08-100596-5.03195-4.
4. J. Muncke, *Chemical Migration from Food Packaging to Food. Reference Module in Food Science*, Elsevier, 2016, DOI: 10.1016/B978-0-08-100596-5.03311-4.
5. J. Muncke, *Hazards of Food Contact Material: Food Packaging Contaminants*, Elsevier, 2014, DOI: 10.1016/B978-0-12-378612-8.00218-3.
6. L. Tovar, J. Salafranca, C. Sánchez and C. Nerin, Migration studies to assess the safety in use of a new antioxidant active packaging, *J. Agric. Food Chem.*, 2005, **53**, 5270–5275.
7. J. A. Garde, R. Catalá and R. Gavara, Global and specific migration of antioxidants from polypropylene films into food simulants, *J. Food Prot.*, 1998, **61**, 1000–1006.
8. C. Nerín, K. Bentayeb and A. Rodriguez-Lafuente, Sampling techniques for the determination of migrants from packaging materials in food, in the curated reference collection, in *Chemistry, Molecular Sciences and Chemical Engineering*, Elsevier, 2012, pp. 357–379.
9. D. Till, Indirect food additive migration from polymeric food packaging materials, *Crit. Rev. Toxicol.*, 1987, **18**, 215–243.
10. K. Bhunia, S. S. Sablani and J. Tang, Migration of chemical compounds from packaging polymers during microwave, conventional heat treatment, and storage, *Compr. Rev. Food Sci. Food Saf.*, 2013, **12**, 523–545.
11. C. Nerín, P. Alfaro, M. Aznar and C. Domeño, The challenge of identifying non-intentionally added substances from food packaging materials: a review, *Anal. Chim. Acta*, 2013, **775**, 14–24.

12. E. Canellas, P. Vera and C. Nerín, Multiple headspace-solid phase microextraction for the determination of migrants coming from a self-stick label in fresh sausage, *Food Chem.*, 2016, **197**, 24–29.
13. E. Canellas, P. Vera and C. Nerin, Atmospheric pressure gas chromatography coupled to quadrupole-time of flight mass spectrometry as a tool for identification of volatile migrants from autoadhesive labels used for direct food contact, *J. Mass Spectrom.*, 2014, **49**, 1181–1190.
14. M. Aznar, P. Vera, E. Canellas, C. Nerín, P. Mercea and A. Störmer, Composition of the adhesives used in food packaging multilayer materials and migration studies from packaging to food, *J. Mater. Chem.*, 2011, **21**, 4358.
15. P. Vera, E. Canellas and C. Nerin, Migration of odorous compounds from adhesives used in market samples of food packaging materials by chromatography olfactometry and mass spectrometry (GC-O-MS), *Food Chem.*, 2014, **145**, 237–244.
16. E. Canellas, M. Aznar, C. Nerin and P. Mercea, Partition and diffusion of volatile compounds from acrylic adhesives used for food packaging multilayers manufacturing, *J. Mater. Chem.*, 2010, **20**, 5100–5109.
17. M. Aznar, A. Rodriguez-Lafuente, P. Alfaro and C. Nerín, UPLC-Q-TOF-MS analysis of non-volatile migrants from new active packaging materials, *Anal. Bioanal. Chem.*, 2012, **404**, 1945–1957.
18. E. Canellas, C. Nerín, R. Moore and P. Silcock, New UPLC coupled to mass spectrometry approaches for screening of non-volatile compounds as potential migrants from adhesives used in food packaging materials, *Anal. Chim. Acta*, 2010, **666**, 62–69.
19. C. Nerín, J. D. Gaspar, P. Vera, E. Canellas, M. Aznar and P. Mercea, Determination of partition and diffusion coefficients of components of two rubber adhesives in different multilayer materials, *Int. J. Adhes. Adhes.*, 2013, **40**, 56–63.
20. F. Isella, E. Canellas, O. Bosetti and C. Nerin, Migration of non intentionally added substances from adhesives by UPLC–Q-TOF/MS and the role of EVOH to avoid migration in multilayer packaging materials, *J. Mass Spectrom.*, 2013, **48**, 430–437.
21. P. Vera, E. Canellas and C. Nerin, Identification of non-volatile compounds and their migration from hot melt adhesives used in food packaging materials characterized by ultra-performance liquid chromatography coupled to quadrupole time-of-flight mass spectrometry, *Anal. Bioanal. Chem.*, 2013, **405**, 4747–4754.
22. E. Canellas, P. Vera and C. Nerín, Risk assessment derived from migrants identified in several adhesives commonly used in food contact materials, *Food Chem. Toxicol.*, 2015, **75**, 79–87.
23. D. Pezo, M. Fedeli, O. Bosetti and C. Nerin, *Anal. Chim. Acta*, 2012, **756**, 49–59.
24. P. Vera, B. Uliaque, E. Canellas, A. Escudero and C. Nerin, Identification and quantification of odorous compounds from adhesives used in food packaging materials by headspace solid phase extraction and headspace solid phase microextraction coupled to gas

chromatography-olfactometry-mass spectrometry, *Anal. Chim. Acta*, 2012, **745**, 53–63.
25. V. Souza and A. L. Fernando, Nanoparticles in food packaging: biodegradability and potential migration to food—a review, *Food Packag. Shelf Life*, 2016, **8**, 63–70.
26. A. L. Fernando, M. P. Duarte and A. Vatsanidou, Environmental aspects of fiber crops cultivation and use, *Ind. Crops Prod.*, 2015, **68**, 105–115.
27. K. Leja and G. Lewandowicz, Polymer biodegradation and biodegradable polymers—a review, *Pol. J. Environ. Stud.*, 2010, **19**(2), 255–266.
28. Y. Tokiwa, B. P. Calabia, C. U. Ugwu and S. Aiba, Biodegradability of plastics, *IJMS*, 2009, **10**, 3722–3742.
29. E. Chiellini, A. Corti, S. D'Antone and R. Baciu, Oxo-biodegradable carbon backbone polymers – oxidative degradation of polyethylene under accelerated test conditions, *Polym. Degrad. Stab.*, 2006, **91**, 2739–2747.
30. M. Wrona, J. Salafranca and C. Nerin, *J. Mater. Chem. C*, 2016, **5**, 463–469.
31. D. M. Wiles and G. Scott, Polyolefins with controlled environmental degradability, *Polym. Degrad. Stab.*, 2006, **91**, 1581–1592.
32. M. Wrona, P. Vera, D. Pezo and C. Nerin, Identification and quantification of odours from oxobiodegradable polyethylene oxidised under a free radical flow by headspace solid-phase microextraction followed by gas chromatography-olfactometry-mass spectrometry, *Talanta*, 2017, **172**, 37–44.
33. J. N. Hahladakis, C. A. Velis, R. Weber, E. Iacovidou and P. Purnell, An overview of chemical additives present in plastics: migration, release, fate and environmental impact during their use, disposal and recycling, *J. Hazard. Mater.*, 2018, **344**, 179–199.
34. C. Dutra, D. Pezo, M. T. de A. Freire, C. Nerin and F. G. R. Reyes, Determination of volatile organic compounds in recycled polyethylene terephthalate and high-density polyethylene by headspace solid phase microextraction gas chromatography mass spectrometry to evaluate the efficiency of recycling processes, *J. Chromatogr. A*, 2011, **1218**, 1319–1330.
35. J. S. Félix, P. Alfaro and C. Nerin, Pros and cons of analytical methods to quantify surrogate contaminants from the challenge test in recycled polyethylene terephthalate, *Anal. Chim. Acta*, 2011, **687**, 67–74.
36. É. C. Oliveira, Y. Echegoyen, S. A. Cruz and C. Nerin, Comparison between solid phase microextraction (SPME) and hollow fiber liquid phase microextraction (HFLPME) for determination of extractables from post-consumer recycled PET into food simulants, *Talanta*, 2014, **127**, 59–67.
37. B. Geueke, K. Groh and J. Muncke, Food packaging in the circular economy: overview of chemical safety aspects for commonly used materials, *J. Cleaner Prod.*, 2018, **193**, 491–505.
38. M. Wrona, D. Pezo and C. Nerin, Rapid analytical procedure for determination of mineral oils in edible oil by GC-FID, *Food Chem.*, 2013, **141**, 3993–3999.
39. D. Pezo, M. Wrona, A. Rodriguez-Lafuente and C. Nerin, *Food Chem.*, 2012, **134**, 405–411.

40. K. Fiselier and K. Grob, Determination of mineral oil paraffins in foods by on-line HPLC–GC–FID: lowered detection limit; contamination of sunflower seeds and oils, *Eur. Food Res. Technol.*, 2009, **229**, 679–688.
41. M. Aznar, P. Alfaro, C. Nerin, E. Jones and E. Riches, Progress in mass spectrometry for the analysis of set-off phenomena in plastic food packaging materials, *J. Chromatogr. A*, 2016, **1453**, 124–133.
42. M. Aznar, C. Domeño, C. Nerin and O. Bosetti, Set-off of non volatile compounds from printing inks in food packaging materials and the role of lacquers to avoid migration, *Dyes Pigm.*, 2015, **114**, 85–92.
43. M. Aznar, J. Gómez-Estaca, D. Vélez, V. Devesa and C. Nerín, Migrants determination and bioaccessibility study of ethyl lauroyl arginate (LAE) from a LAE based antimicrobial food packaging material, *Food Chem. Toxicol.*, 2013, **56**, 363–370.
44. P. Vera, M. Aznar, P. Mercea and C. Nerin, Study of hotmelt adhesives used in food packaging multilayer laminates. Evaluation of the main factors affecting migration to food, *J. Mater. Chem.*, 2011, **21**, 420–431.
45. N. Haider and S. Karlsson, Loss of Chimassorb 944 from LDPE and identification of additive degradation products after exposure to water, air and compost, *Polym. Degrad. Stab.*, 2001, **74**, 103–112.
46. T. P. McNeal J. E. Biles, T. H. Begley, J. C. Craun, M. L. Hopper and C. A. Sack, Determination of suspected endocrine disruptors in foods and food packaging, in *Analysis of Environmental Endocrine Disruptors.* ACS Symposium Series, 1999, Vol. 747, ch. 4, pp. 33-52, ISBN13: 9780841236509, eISBN: 9780841217539, DOI: 10.1021/bk-2000-0747.ch004.
47. A. Dąrowska, A. Borcz and J. Nawrocki, Aldehyde contamination of mineral water stored in PET bottles, *Food Addit. Contam.*, 2010, **20**, 1170–1177.
48. J. Alin and M. Hakkarainen, Microwave heating causes rapid degradation of antioxidants in polypropylene packaging, leading to greatly increased specific migration to food simulants as shown by ESI-MS and GC-MS, *J. Agric. Food Chem.*, 2011, **59**, 5418–5427.
49. L. Burman, A.-C. Albertsson and A. Höglund, Solid-phase microextraction for qualitative and quantitative determination of migrated degradation products of antioxidants in an organic aqueous solution, *J. Chromatogr. A*, 2005, **1080**, 107–116.
50. K. Vinković, M. Rožić and N. Galić, Development and validation of an HPLC method for the determination of endocrine disruptors bisphenol A and benzophenone in thermochromic printing inks, *J. Liq. Chromatogr. Relat. Technol.*, 2017, **40**, 959–966.
51. M. Suman, E. De Dominicis and I. Commissati, Trace detection of the chlorohydrins of epoxidized soybean oil in foodstuffs by UPLC-ESI-MS/MS, *J. Mass Spectrom.*, 2010, **45**, 996–1002.
52. L. Coulier, H. G. M. Orbons and R. Rijk, Analytical protocol to study the food safety of (multiple-) recycled high-density polyethylene (HDPE) and polypropylene (PP) crates: influence of recycling on the migration and formation of degradation products, *Polym. Degrad. Stab.*, 2007, **92**, 2016–2025.

53. S. Ebnesajjad, *Plastic Films in Food Packaging*, Elsevier, 2013, pp. 345–388.
54. J. Murphy, in *Additives for Plastics Handbook*, Elsevier, 2001, pp. 1–3, DOI: 10.1016/B978-185617370-4/50003-8.
55. B. Brüschweiler, Assessment of non-intentionally added substances from food contact materials in food: Which way to go?, *Toxicol. Lett.*, 2014, **229**, S34.
56. D. Jenke, Identification of caprolactam oligomers and related compounds in aqueous extracts of nylon-6, *J. Appl. Polym. Sci.*, 2005, **95**, 1262–1274.
57. S. Úbeda, M. Aznar, P. Vera, C. Nerín, L. Henríquez, L. Taborda and C. Restrepo, Overall and specific migration from multilayer high barrier food contact materials – kinetic study of cyclic polyester oligomers migration, *Food Addit. Contam.: Part A*, 2017, **34**, 1784–1794.
58. J. Salafranca, I. Clemente, F. Isella, C. Nerin and O. Bosetti, Influence of oxygen and long term storage on the profile of volatile compounds released from polymeric multilayer food contact materials sterilized by gamma irradiation, *Anal. Chim. Acta*, 2015, **878**, 118–130.
59. J. S. Félix, M. Monteiro, J. M. Manzoli, M. Padula, D. Pezo, J. Romero and C. Nerin, Identification and migration of degradation compounds from irradiation of multilayer polyamide 6 films for meat foodstuffs and cheese, *Anal. Bioanal. Chem.*, 2008, **391**, 847–857.
60. T. Bolumar, M. L. Andersen and V. Orlien, Antioxidant active packaging for chicken meat processed by high pressure treatment, *Food Chem.*, 2011, **129**, 1406–1412.
61. P. Montero-Prado, A. Rodriguez-Lafuente and C. Nerin, Active label-based packaging to extend the shelf-life of 'Calanda' peach fruit: changes in fruit quality and enzymatic activity, *Postharvest Biol. Technol.*, 2011, **60**, 211–219.
62. M. Wrona, K. Bentayeb and C. Nerin, *Food Control*, 2015, **54**, 200–207.
63. M. Wrona, C. Nerin, M. J. Alfonso and M. Á. Caballero, Antioxidant packaging with encapsulated green tea for fresh minced meat, *Innovative Food Sci. Emerging Technol.*, 2017, **41**, 307–313.
64. R. Gherardi, R. Becerril, C. Nerin and O. Bosetti, Development of a multilayer antimicrobial packaging material for tomato puree using an innovative technology, *LWT–Food Sci. Technol.*, 2016, **72**, 361–367.
65. D. Carrizo, G. Gullo, O. Bosetti and C. Nerin, Development of an active food packaging system with antioxidant properties based on green tea extract, *Food Addit. Contam., Part A Chem. Anal. Control Exposure Risk Assess.*, 2014, **31**, 364–373.
66. J. P. Kerry, New packaging technologies, materials and formats for fast-moving consumer products, in *Innovations in Food Packaging*, ed. J. H. Han, Academic, 2nd edn, 2014.
67. J. P. Kerry, M. N. O'Grady and S. A. Hogan, Past, current and potential utilisation of active and intelligent packaging systems for meat and muscle-based products: a review, *Meat Sci.*, 2006, **74**, 113–130.
68. J. H. Han, Antimicrobial Food Packaging, in *Novel Food Packaging Techniques*, Elsevier, 1st edn, 2003.

69. N. Nguyen Van Long, C. Joly and P. Dantigny, Active packaging with antifungal activities, *Int. J. Food Microbiol.*, 2016, **220**, 73–90.
70. F. Akrami, A. Rodríguez-Lafuente, K. Bentayeb, D. Pezo, S. Ghalebi and C. Nerín, Antioxidant and antimicrobial active paper based on Zataria (*Zataria multiflora*) and two cumin cultivars (*Cuminum cyminum*), *LWT–Food Sci. Technol.*, 2015, **60**, 929–933.
71. S. Manso, D. Pezo, R. Gómez-Lus and C. Nerín, Diminution of aflatoxin B1 production caused by an active packaging containing cinnamon essential oil, *Food Control*, 2014, **45**, 101–108.
72. L. Gutiérrez, C. Sánchez, R. Batlle and C. Nerín, New antimicrobial active package for bakery products, *Trends Food Sci. Technol.*, 2009, **20**(2), 92–99.
73. S. Woranuch and R. Yoksan, Eugenol-loaded chitosan nanoparticles: I. Thermal stability improvement of eugenol through encapsulation, *Carbohydr. Polym.*, 2013, **96**(2), 578–585.
74. M. Carbone, D. Tommasa Donia, G. Sabbatella and R. Antiochia, Silver nanoparticles in polymeric matrices for fresh food packaging, *J. King Saud Univ., Sci.*, 2016, **28**(4), 273–279.
75. P. Vera, Y. Echegoyen, E. Canellas, C. Nerin, M. Palomo-Siguero, Y. Madrid-Albarrán and C. Camara, Nano selenium as antioxidant agent in a multilayer food packaging material, *Anal. Bioanal. Chem.*, 2016, **408**(24), 6659–6670.
76. K. K. Kuorwel, M. J. Cran and J. D. Orbell, Review of mechanical properties, migration, and potential applications in active food packaging systems containing nanoclays and nanosilver, *Compr. Rev. Food Sci. Food Saf.*, 2015, **14**, 411.
77. M. Correia and K. Loeschner, Detection of nanoplastics in food by asymmetric flow field-flow fractionation coupled to multi-angle light scattering: possibilities, challenges and analytical limitations, *Anal. Bioanal. Chem.*, 2018, **14**, 4501–4513.
78. *Commission Regulation (EU) No. 10/2011 of 14 January 2011 on Plastic Materials and Articles Intended to Come into Contact with Food*, 2011.
79. R. Kroes and G. Kozianowski, Threshold of toxicological concern (TTC) in food safety assessment, *Toxicol. Lett.*, 2002, **127**, 43–46.
80. I. Dewhurst and A. G. Renwick, Evaluation of the threshold of toxicological concern (TTC) – challenges and approaches, *Regul. Toxicol. Pharmacol.*, 2013, **65**, 168–177.
81. M. Haunschmidt, C. W. Klampfl, W. Buchberger and R. Hertsens, Rapid identification of stabilisers in polypropylene using time-of-flight mass spectrometry and DART as ion source, *Analyst*, 2009, **135**, 80.
82. R. J. Fussell, D. Chan and M. Sharman, An assessment of atmospheric-pressure solids-analysis probes for the detection of chemicals in food, *TrAC Trends Anal. Chem.*, 2010, **29**, 1326–1335.
83. M. R. L. Paine, P. J. Barker and S. J. Blanksby, Ambient ionisation mass spectrometry for the characterisation of polymers and polymer additives: a review, *Anal. Chim. Acta*, 2014, **808**, 70–82.

84. N. P. Brunton, D. A. Cronin, F. J. Monahan and R. Durcan, A comparison of solid-phase microextraction (SPME) fibres for measurement of hexanal and pentanal in cooked turkey, *Food Chem.*, 2000, **68**, 339–345.
85. D. Pezo, J. Salafranca and C. Nerin, Development of an automatic multiple dynamic hollow fibre liquid-phase microextraction procedure for specific migration analysis of new active food packagings containing essential oils, *J. Chromatogr. A*, 2007, **1174**, 85–94.
86. J. Salafranca, D. Pezo and C. Nerin, Assessment of specific migration to aqueous simulants of a new active food packaging containing essential oils by means of an automatic multiple dynamic hollow fibre liquid phase microextraction system, *J. Chromatogr. A*, 2009, **1216**, 3731–3739.
87. M. M. Rosero, M. Aguirre, D. Pezo, G. Taborda, C. Dussán and C. Nerin, Solventless microextraction techniques for determination of trihalomethanes by gas chromatography in drinking water, *Water Air Soil Pollut.*, 2011, **223**, 667–678.
88. A. Kabir and K. G. Furton, *Fabric Phase Sorptive Extractors, US patent* US15069430, 2016.
89. M. Aznar, P. Alfaro, C. Nerín, A. Kabir and K. G. Furton, Fabric phase sorptive extraction: an innovative sample preparation approach applied to the analysis of specific migration from food packaging, *Anal. Chim. Acta*, 2016, **936**, 97–107.
90. M. Wrona, D. Pezo, E. Canellas and C. Nerin, Ultra high performance liquid chromatography coupled to quadruple time-of-flight with MS(E) technology used for qualitative analysis of non-volatile oxidation markers in sliced packed mushrooms (*Agaricus bisporus*), *J. Chromatogr. A*, 2016, **1432**, 73–83.
91. J. Regueiro, C. Sánchez-González, A. Vallverdú-Queralt, J. Simal-Gándara, R. Lamuela-Raventós and M. Izquierdo-Pulido, *Food Chem.*, 2014, **152**, 340–348.
92. Y. Echegoyen and C. Nerin, Nanoparticle release from nano-silver antimicrobial food containers, *Food Chem. Toxicol.*, 2013, **62**, 16–22.
93. Y. Echegoyen, S. Rodríguez and C. Nerin, Nanoclay migration from food packaging materials, *Food Addit. Contam., Part A Chem. Anal. Control Exposure Risk Assess.*, 2016, **33**, 530–539.
94. M. Wrona, M. J. Cran, C. Nerin and S. W. Bigger, Development and characterisation of HPMC films containing PLA nanoparticles loaded with green tea extract for food packaging applications, *Carbohydr. Polym.*, 2017, **156**, 108–117.
95. E. Canellas, P. Vera, C. Domeño, P. Alfaro and C. Nerín, Atmospheric pressure gas chromatography coupled to quadrupole-time of flight mass spectrometry as a powerful tool for identification of non intentionally added substances in acrylic adhesives used in food packaging materials, *J. Chromatogr. A*, 2012, **1235**, 141–148.

CHAPTER 8

Non-target/High-resolution Mass Spectrometric Strategies Combined with Proper Data Handling and Software Elaboration Applied to Food Contact Materials Issues

MARÍA JOSÉ GÓMEZ RAMOS, MARÍA JESÚS MARTÍNEZ BUENO, CARMEN FERRER, MARÍA MURCIA-MORALES AND AMADEO RODRÍGUEZ FERNÁNDEZ-ALBA*

Department of Physics and Chemistry, Almería University, Agrifood Campus of International Excellence (ceiA3), Ctra. Sacramento s/n, La Cañada de San Urbano, 04120 Almería, Spain
*E-mail: amadeo@ual.es

8.1 Introduction

The production and use of food contact materials (FCMs) have increased in recent years. More than one-third of packaging materials are made of plastic.[1,2] The polymers most frequently used in food packaging are polypropylene (PP) and different types of polyethylene (PE), including low-density PE

Food Chemistry, Function and Analysis No.10
Food Contact Materials Analysis: Mass Spectrometry Techniques
Edited by Michele Suman
© The Royal Society of Chemistry 2019
Published by the Royal Society of Chemistry, www.rsc.org

(LDPE), linear low-density PE (LLDPE) and high-density PE (HDPE).[3] Modern FCMs made of plastics are designed to fulfil multiple purposes, such as the protection of food products from external sources of contamination. They are also intended for the prevention of food deterioration by protecting the products from light, humidity, oxygen and foreign aroma compounds and by controlling/inhibiting pathogen growth, in order to improve the shelf life and stability of the products during storage. Nevertheless, chemical contamination of the food *via* migration from the packaging can also occur. Food packaging materials are often multilayer materials and are manufactured by the combination of different polymers joined by adhesive layers that in most cases are made of polyurethane. When multilayer materials are used in food packaging, migration can take place not only from the internal side of the packaging (food contact surface) but also from internal layers due to diffusion and partition processes.[3–5] Consequently, several components can migrate to food products placed in contact with plastic materials. Many different classes of substances have been reported in food migration studies, such as colorants, surfactants, printing ink components, adhesives, plastic additives (antioxidants, UV absorbers and stabilizers), monomers and oligomers.[6–12]

The European Commission Regulation on plastic materials and articles intended to come into contact with food, Regulation EU 10/2011,[13] has specific rules for plastic materials to be applied for their safe use. This Regulation establishes a positive list of those compounds authorized for use in plastic formulations and manufacturing and provides specific migration limits for a large number of molecules that can pose a threat to human health. The list contains more than 150 compounds, including monomers or other starting substances, additives (excluding colorants), polymer production aids (excluding solvents) and macromolecules obtained from microbial fermentation.[13] Such substances are known as intentionally added substances (IASs). However, it is important to take into account the possible release of substances that are not used intentionally in food packaging materials. These substances are commonly referred to as non-intentionally added substances (NIASs), and do not appear in lists of permitted ingredients. NIASs may be present as a result of the interactions between different ingredients in the packaging materials, from degradation of the material components and mainly from the impurities in the raw materials used to make the plastic.[14] For non-authorized substances, their migration should not exceed a level of 0.01 mg kg^{-1} of food/simulant. The identification and control of these substances are a relevant issue that deserves close attention to ensure food quality and safety.

8.2 Analysis and Identification

For the analysis of substances that migrate from FCMs, there are standardized tests, established by Regulation EU 10/2011,[13] that are applicable to all materials that come in direct contact with food. However, the identification of unknown migrants from FCMs is very complex and challenging, owing to the many existing chemical classes of substances, low concentration levels, incomplete information about the ingredients used to manufacture the FCMs, the lack of analytical standards for many of them and the fact that many of these substances are not included in chemical or spectral databases.

Nowadays, liquid chromatography (LC) coupled to mass spectrometry (MS) is the most frequently used technique for the identification of thermally unstable and non-volatile migrant compounds. Sanchis *et al.*[15] reviewed the ionization sources used to analyse food packaging contaminants by LC-MS. They concluded that electrospray ionization (ESI) is a more commonly used technique than atmospheric pressure chemical ionization (APCI), which can be explained by the fact that ESI provides a higher degree of sensitivity for most polar and semipolar compounds.[14,16] Different mass analysers have been used in LC-MS analysis related to FCM issues, including low-resolution (single and triple quadrupole or ion trap)[17–19] and high-resolution MS instruments [time-of-flight (TOF) or Orbitrap®].[8,10,20,21] Low-resolution MS strategies have been applied for the identification of NIASs using commercial or in-house libraries. However, it limitations because the degree of confidence is based on unit mass resolution, which can be the origin of some false positives and negatives, as noted previously by Gallart-Ayala *et al.*[16] By contrast, high-resolution accurate mass spectrometry (HRAMS) techniques provide accurate mass measurement in full-scan mode with high sensitivity, high resolution power (>30 000 FWHM) and accurate mass measurement (1–5 ppm). This methodology allows the identification of the chemical structures of unknown compounds based on the information about fragmentation patterns provided by the mass spectra of the precursor ion and also of the product ions, by selecting different data acquisition modes (Figure 8.1). Hence the use of highly sensitive advanced analytical techniques in combination with software tools for data acquisition and data treatment is a good option for gaining increased confidence in peak assignments. To date, hybrid instruments such as quadrupole time-of-flight (Q-TOF) and Q-Orbitrap® have been the most extensively used to analyse migrants in FCM applications based on HRAMS methods.[4,7,8,10,20–30]

8.2.1 Data Acquisition Mode

A targeted analysis is an MS approach that uses multiple stages of MS (MS^n, $n = 2$ or 3) for ions of specific mass (m/z) at specific retention times (t_R). The values of m/z and t_R are defined in an inclusion list that is derived from a previous analysis. By contrast, in suspected–targeted analysis, a specific list of compounds is screened, but no reference compounds are available for confirmation, whereas non-targeted analysis is performed without the use of any inclusion list (the aim is to analyse everything). Up to now, few studies

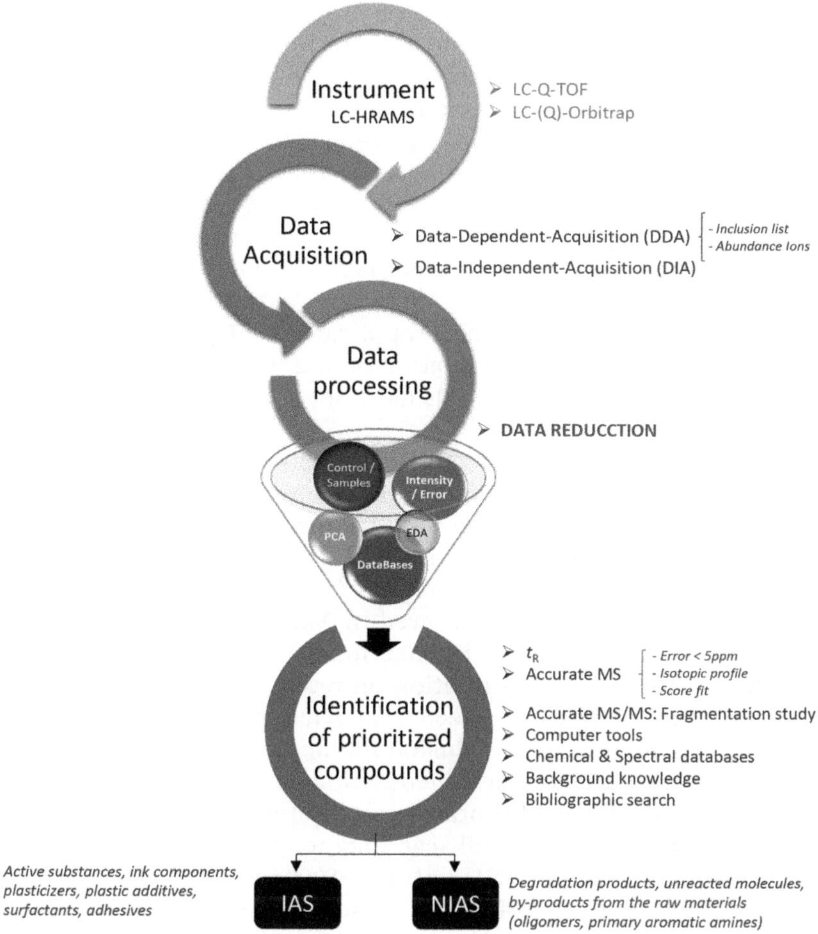

Figure 8.1 General workflow for the analytical procedure used in the targeted and non-targeted analysis of FCMs using high-resolution accurate mass spectrometry (HRAMS).

have been carried out on the application of LC-HRAMS methods to identify migrants from FCMs by targeted analysis, with or without the use of reference standards, such as primary aromatic amines (PAAs),[6,31] bisphenol,[32] colouring agents and plastic[11,33] in different plastic materials. Most of the studies published on the application of LC-HRAMS are based on non-targeted screening.[4,7,20–30,34]

Two different data acquisition modes can be used to perform targeted and non-targeted screening analysis: data-dependent and data-independent acquisition (DDA and DIA, respectively). In the DDA mode, the instrument performs the data collection using predetermined rules, switching between a full-spectrum MS survey and a full-spectrum product ion (MS/MS) analysis-dependent mode, in one chromatographic run. The switching is controlled automatically by the intensity of precursor ions observed, but can be also performed using other additional criteria such as isotope pattern, charge state or specific *m*/*z* values on an inclusion list or an exclusion list. In the DIA mode, all ions within a selected *m*/*z* range are fragmented and analysed in a second stage of tandem MS in order to detect as many unknown compounds as possible. The switching between MS and MS/MS is performed automatically to obtain fragments for all precursor ions present. MS and MS/MS data for unknown compounds are also acquired simultaneously in one chromatographic run. As summarized in Table 8.1, most of the studies published on the application of LC-HRAMS methods to identify migrants from FCMs are based on the use of DIA experiments. In contrast, the utility of DDA methods has not yet been very exploited in this field.

Regarding DDA experiments, most of the studies published combined targeted analysis to determine the presence of common additives (such as antioxidants and UV absorbers in food contact plastics) and non-targeted screening to characterize oligomers derived from polycarbonate degradation and traces of colouring agents.[9,11,12,33] For targeted data-dependent experiments, a full-scan event followed by MS/MS scan events of the Top *N* precursor ions (the *N* most intense ions) included in a list was performed. The list included information on retention times, *m*/*z* values and optimal collision energies of selected analytes in each study. In the case of non-targeted analysis, the inclusion list option was disabled. In recent study, Paseiro-Cerrato *et al.*[9] identified 29 non-volatile oligomers from polyester cans intended for food contact uses, by application of a non-targeted data-dependent experiment. Full-scan data were acquired from a selected mass range (*m*/*z* 100–800 for ESI+; *m*/*z* 100–1000 for ESI−), whereas data-dependent MS/MS data were obtained by setting several criteria such as a maximum injection time, a loop count, an intensity threshold, an isolation window and a dynamic inclusion. In this study, two internal databases were used in order to identify as many compounds as possible. One database contained all substances approved through the food contact notification system in 2013 and the other was prepared with all monomers and additives listed in the EU positive list.

On the other hand, in the case of DIA experiments, only three studies have been reported e on the application of targeted analysis for the identification of migrants from FCMs.[6,31,32] Mattarozi *et al.*[6] and Sanchis *et al.*[31] used an LC-Orbitrap® mass analyser for the determination of migration levels from food packaging of PAAs, which are toxic compounds and suspected to be human carcinogens. Pérez-Palacios *et al.*[32] determined bisphenol-type endocrine-disrupting compounds (EDCs) in food contact recycle-paper materials using a Q-TOF instrument. In all of these studies, targeted extracted fragment ions from high-resolution, accurate-mass MS/MS data were collected. Any detected precursor ions were quantified using extracted ion chromatograms (XICs) at narrow mass tolerance (<5 ppm) in combination with the

Table 8.1 Overview of studies on the analysis of FCMs, including NIASs, based on targeted and untargeted analysis using high-resolution accurate mass spectrometry (HRAMS).[a]

Food packaging material/food simulant or extraction	Instrument (brand)	Analysis/data acquisition	Data processing/ data reduction strategy	Identification/databases	IASs/NIASs identified	Ref.
Six novel active film substances with antimicrobial and/ or antioxidant agents – combination of PP, EVOH and PET/food simulant A (aqueous) and D (fat simulant)	UPLC-Q-TOF-MS (Xevo G2, Waters)	Non-target analysis, elucidation of unknown migrants/MS^E mode – low and high CE in the same run	Visual comparison between the chromatograms from the control and sample solutions PCA (MarkerLynx)	Accurate masses of precursor and fragment ions Background knowledge MassFragment tool – two fragments with a low score (<3)/ChemSpider, SciFinder	Active substances used in the packaging (*e.g.* caffeine, carvacrol, citral); unknown compounds used as additives in varnishes or polymers (*e.g.* glycerols, adipate)/citral reaction compounds or impurities	22
Multilayer packaging material containing ink and PE/simulant D and simulant E (extraction with MeOH)	UPLC-Q-TOF-MS (Xevo G2, Waters)	Non-target analysis, elucidation of NIASs/MS^E mode (low and high CE in the same run)	Visual comparison between data for blank and sample materials PCA (MarkerLynx software)	Accurate mass of precursor and fragmentation ions (MassLynx) Background knowledge Bibliographic search ChemSpider, SciFinder	17 compounds coming from inks: IASs (sebacates, glycol ethers); NIASs (cyclic polyesters)	4

(*continued*)

Table 8.1 (*continued*)

Food packaging material/food simulant or extraction	Instrument (brand)	Analysis/data acquisition	Data processing/data reduction strategy	Identification/databases	IASs/NIASs identified	Ref.
Materials used for food packaging (OPP, PET, PP, aluminium) containing printing inks/simulant E (extraction with acetone), simulant D	UPLC-Q-TOF-MS (Xevo G2, Waters)	Non-target analysis/ MS^E mode (low and high CE in the same run)	Comparison between data for materials with and without ink (samples–blanks)	Accurate masses of precursor and fragmentation ions (MassLynx)	17 IASs: ink components such as colorants (Solvent Red 49), plasticizers (citric acid triethyl ester, TBoAC), surfactants (Schercodine L), printing ink solvents (tripropylene glycol monobutyl ether)	10
				Background knowledge Bibliographic search Confirmation with pure standards (when possible) Confirmation by IM-MS: determination of the collision cross-section (CCS) of the ions detected (related to their key physicochemical properties)/ChemSpider, SciFinder	1 NIA (triphenylphosphine oxide, oxidation product of TPP)	

Paper and board (P&B) FCM (*e.g.* pizza box)/ethanol extraction	UPLC-Q-TOF-MS (6550 iFunnel, Agilent)	Non-target analysis (ESI+/−)/full scan in data-independent all ions MS/MS mode	Automated integration and deconvolution	Locate molecular ion and adducts	75 tentatively identified substances (GC and LC), 15 selected for further testing *in vitro*, only 7 were commercially available	8
			Fractionation–effect-directed analysis (EDA) – fractions exhibiting *in vitro* activity (AhR)	Database hits – score >85 (mass accuracy, isotope distribution and isotope spacing), error <10 ppm	7 IASs: dyes, printing ink components, adhesives, additives	
			Most prominent spectral peaks	MS/MS spectra	8 NIASs	
			Home-made database	Home-made database with ~2100 entries, containing chemicals suspected (or known) to be present in food packing of P&B (IASs, NIASs described in scientific articles, from legislative lists and inventory lists) ChemSpider, PubChem		

(*continued*)

Table 8.1 (*continued*)

Food packaging material/food simulant or extraction	Instrument (brand)	Analysis/data acquisition	Data processing/data reduction strategy	Identification/databases	IASs/NIASs identified	Ref.
Polycarbonated food-contact plastics/extraction by magnetic stirring and hand shaking with chloroform, methanol and acetone	UPLC-Q-Orbitrap® (Q-Exactive, ThermoFisher Scientific)	Target and non-target analysis (in both positive and negative modes), elucidation of NIAS/full-scan MS-ddMSMS (Top *N*)[b] – for targeted analysis using an inclusion list and for untargeted analysis without an inclusion list	Intensity threshold 5×10^4	Accurate mass of precursor and interpretation of fragmentation spectrum	IASs: colouring agents, plastic additives (antioxidants, UV absorbers, stabilizers, whitening agent)	33
			Home-made database	Comparison of isotopic patterns with a simulator Bibliographic search/target compounds Home-made database with a selection of the most used additives based on a literature overview Available chemical databases	NIASs: BPA–polycarbonate degradation products	
Plastic reusable containers – polycarbonate tableware/food simulant D and isooctane (food simulant D2)	UPLC-Q-Orbitrap® (Q-Exactive, ThermoFisher Scientific)	Non-target analysis, elucidation of NIASs (ESI +/−)/full-scan MS–data dependent MS/MS experiments with inclusion list	Main detected signals	Identification was carried out by reconstructing the mass fragmentation spectrum of the most representative molecular ion visible in the full-scan chromatograms	IASs: BPA, colouring agents (Solvent Red 179, Solvent Yellow 184, Solvent Yellow 232)	40
			Inclusion list	Visualization of the isotopic pattern and the matching between the exact mass and the accurate mass	NIASs: PC degradation products and unreacted molecules	
			Mass error ≤2 ppm	MS/MS spectra obtained in the same run were used for confirmatory purpose		

Plastic reusable containers – polycarbonate tableware/dark chocolate as a model for food with a high fat content – diethyl ether extraction followed by liquid–liquid extraction with acetonitrile and C_{18} SPE cleanup	UPLC-Q-Orbitrap® (Q-Exactive, ThermoFisher Scientific)	Target and non-target analysis, elucidation of unknown chemicals/full-scan MS-ddMSMS and tSIMddMSMS, switching in positive and negative modes. For targeted analysis using an inclusion list and for untargeted analysis without an inclusion list	Inclusion list Mass error ≤2 ppm	Accurate mass and interpretation of fragmentation spectrum Isotopic ratios matching (comparison of isotopic patterns by a simulator)	IASs: colouring agents, plastic additives (Tinuvin 234 and 327, Uvitex OB, Cyasorb 5411, Yellow 184) NIASs: BPA–PC degradation products and unreacted oligomers	11
Multilayer packaging material (PLA, Ecovio) containing a water-based biodegradable adhesive/simulant E (extraction with MeOH)	UPLC-Q-TOF-MS (Xevo G2, Waters)	Non-target analysis, elucidation of NIASs/MS^E mode (low and high CE in the same run)	Migration studies with packaging materials conducted after identification of the adhesive composition	Accurate masses of precursor and fragmentation ions (mass tolerance 3 mDa, MassLynx) Background knowledge Bibliographic search Confirmation with pure standards (when possible) Fragmentation patterns of combination products ChemSpider, SciFinder	3 IASs (adipic acid, 2-methyl-2*H*-isothiazol-3-one, 1,2-benzisothiazol-3(2*H*)-one), 4 NIASs (neoformed lactones from the reaction between other components)	7

(*continued*)

Table 8.1 (*continued*)

Food packaging material/food simulant or extraction	Instrument (brand)	Analysis/data acquisition	Data processing/data reduction strategy	Identification/databases	IASs/NIASs identified	Ref.
Multilayer materials prepared by combinations of paper, PE, PET, BOPP, PLA and Ecovio – all containing a water-based acrylic adhesive/simulant E (extraction with MeOH)	UPLC-Q-TOF-MS (Xevo G2, Waters)	Non-target analysis/ MS^E mode (low and high CE in the same run)	Mass range 10–120 Da	Accurate masses of precursor and fragmentation ions (mass tolerance 3 mDa, MassLynx)	4 IASs: sufactants (TMDD and TMDD ethoxylates)	23
			Study of the adhesive chromatogram (identification of components)	Background knowledge	1 NIAS: malonic acid derivative from methyl methacrylate	
			Comparison between data of blanks and migration assays	Bibliographic search ChemSpider, SciFinder		
Two multilayer plastic packaging – combinations of PE and nylon materials/food simulants B and C	LC-Q-TOF-MS (X500R, Sciex)	Non-target analysis, elucidation of unknown migrants/full-scan MS and MS/MS modes in the same run, SWATH	Case–control filter (ratio ×10)	Accurate masses of precursor and fragment ions (mass tolerance <2 ppm, score fit ≥80%)	3 IASs: caprolactam and 2 unregulated compounds used as additives in the polymers	21
			Selection of most abundant compounds (signal intensity $\geq 1 \times 10^6$ counts)	*In silico* fragmentation study MS/MS spectra deconvolution Bibliographic search ChemSpider, METLIN	23 NIASs: caprolactam oligomers and by-products from the raw materials (mainly cyclic oligomers coming from polyurethane adhesives)	

Multilayer packaging material containing polyurethane adhesive PU1. Materials: aluminium, metallized PE and PE/EVOH/PE/simulant E, extraction with MeOH	UPLC-Q-TOF-MS (Xevo G2, Waters)	Non-target analysis/ MS^E mode (low and high CE in the same run)	Comparison between data for blank and sample materials	Accurate masses of precursor and fragmentation ions (MassLynx) Background knowledge Bibliographic search ChemSpider	19 NIASs: potential migrants, most of them cyclic compounds (lactones) or adipic based	24
New active nanomaterials with antimicrobial agents – combination of PP-based nanocomposites functionalized with ZnO NPs/food simulants A and B	UPLC-Q-TOF-MS (6550, Agilent)	Non-target analysis/ full-scan and MS/MS modes	Relevant intensity signals Home-made database	Accurate MS, error <5 ppm Isotopic profile Score (MS) ≥75% Elucidation of mass fragmentation, error (MS/MS) <15 ppm Home-made database (60 chemicals) including additives, monomers, starting substances used in polymer manufacture and phthalates	3 NIASs: 2,4,6-triamino-1,3,5-triazine, azepan-2-one (caprolactam) and (2*E*)-3-phenylprop-2-enal	25

(*continued*)

Table 8.1 (*continued*)

Food packaging material/food simulant or extraction	Instrument (brand)	Analysis/data acquisition	Data processing/ data reduction strategy	Identification/databases	IASs/NIASs identified	Ref.
A developed polymer – PLA, PL and ZnO NPs/food simulants A and B	UPLC-Q-Orbitrap® (Q-Exactive focus, ThermoFisher Scientific)	Non-target analysis, elucidation of NIAS/full-scan, all ion fragmentation MS/MS (AIF)	Most intense signals: ≥1 × 10^5 S/N threshold ≥10 Intensity tolerance ≥30% Max. shift: 0.5 min PCA, volcano plot	Mass Frontier 7.0 software was used to match each fragment obtained with a fragment of the proposed candidate with compound Discoverer 2.0 software in combination with databases/*mz*Cloud (mass spectral library), ChemSpider	4 NIASs: *N*,*N*-diethyldodecanamide, *N*-[(9*Z*)-9-octadecen-1-yl]acetamide, 1-palmitoylglycerol and glycerol stearate	20
Plastic laminated materials/food simulant B	LTQ Orbitrap® XL hybrid FTMS (ThermoFisher Scientific)	Target strategy – analysis of 22 migrating primary aromatic amines/full-scan HRMS	Signal intensity >10^4	Identification with pure standards of the amines selected Accurate masses of the protonated ions (mass tolerance <5 ppm, t_R ±2.5%)	22 NIASs: primary aromatic amines arising from isocyanates (polyurethane adhesives) and azo dyes or pigments	6

Two polyester–polyurethane lacquers – mixture of diacids and diol monomers (NPG, HD, DEG, EG)/extraction with acetonitrile	UPLC-Q-Orbitrap® (Q-Exactive, ThermoFisher Scientific)	Non-target strategy – analysis of predicted migrating substances and elucidation of unpredicted substances/full-scan mode, a selection of 22 precursor ions were fragmented in targeted-MS/MS mode	Open-source R-environment (scms and CAMERA packages), for noise filtering and deconvolution	Chromatographic considerations	17 NIASs: caprolactam oligomers; cyclic oligomers	12
			Most intense migrating m/z 155–1064	Investigation of fragmentation patterns		
			Home-made database	Home-made database with predicted polyester oligomer combinations from a selection of diols and diacids		
Plastic baby bottles – PP and PA/food simulant D1	UPLC-Q-TOF-MS (Xevo G2, Waters)	Non-target analysis, elucidation of unknown migrants/MS^E mode – low and high CE in the same run	Home-made database, mass error (±0.002 Da)	Accurate mass of the (de)protonated molecule	IASs: laurolactam monomer (plastic monomer), bis(3,4-dimethylbenzylidene) sorbitol (clarifier for PP), Irganox 1010, dibutyl phthalate, diisobutyl phthalate	26
				Product ion fragments Characteristic isotopic ions Bibliographic search Home-made database containing common migrating compounds and plastic additives – 50 standards, more than 800 compounds authorized for plastic FCMs by EU (exact mass of the empirical formula)	NIASs: laurolactam dimer and trimer, 2,5-bis(5′-*tert*-butyl-2-benzoxazolyl) thiophene, ethyl 4-ethoxybenzoate	

(*continued*)

Table 8.1 (*continued*)

Food packaging material/food simulant or extraction	Instrument (brand)	Analysis/data acquisition	Data processing/data reduction strategy	Identification/databases	IASs/NIASs identified	Ref.
Polyester can coatings/extraction with acetonitrile	UPLC-Q-Orbitrap® (Q-Exactive, ThermoFisher Scientific)	Non-target analysis (ESI +/−/full-scan MS-ddMSMS with inclusion list	High-intensity peaks	Home-made database containing most probable monomers employed in the coating and all of the possible combinations of the potential monomers (≤1000 Da) including linear and cyclic structures	IASs: monomers and additives (IPA, TA, NA, COH and CRM)	9
			Good peak shape and appropriate t_R	Two other internal databases were used – one contained all substances approved through the food contact notification system through 2013 and the other list was prepared with all monomers and additives listed in the EU positive list	NIASs: 29 oligomers, all corresponding to the reaction of carboxylic acids and diols, except for *t*BuPh (4-*tert*-butylphenol) which is a mono-alcohol	
			Home-made databases containing monomers, oligomers and additives	MS/MS spectra were used for confirmatory purposes		

Recycled-paper materials/FUSLE with methanol	UPLC-Q-TOF-MS (Micro-tof-Q, Bruker Daltonik)	Target analysis/full-scan MS	Analysis of the target compounds – BPA, BPF, BADGE, BFDGE	Quantification – with analytical standard – by ion extraction with a ±20 mDa *m*/*z* window	IAS: bisphenol-type EDCs – BPA and BPF	32
18 multilayer films for food packaging [PET, OPA, PE(EVOH), PE] with polyurethane adhesives/simulant B	UPLC-Q-TOF-MS (Xevo G2, Waters)	Non-target analysis/ MS^E mode (low and high CE in the same run)	Comparison between data for blank and sample materials using PCA (MarkerLynx)	Accurate masses of precursor and fragmentation ions (MassLynx) Background knowledge Bibliographic search Confirmation by MS/MS (t_R ±0.1 min, mass tolerance 0.02 Da) ChemSpider, SciFinder	40 cationic NIASs including primary aromatic amines derived from residual isocyanates	27
Nylon kitchen utensils/simulant B	Orbitrap® MS Exactive (ThermoFisher Scientific)	Target strategy – analysis of 8 primary aromatic amines and another 77 migrants/full-scan MS, full-scan MS all ions fragmentation	Signal intensity $>10^4$	Home-made database for migrants including bisphenols, phthalates and amines	8 primary aromatic amines, 77 migrating substances (phthalates, bisphenols, perfluorinated compounds, *etc.*)	31
			S/N threshold ≥5	Identification with pure standards of the amines selected: similar ion ratios (<30% difference), t_R ±0.2 min		
			Home-made database	Accurate masses of precursor and fragmentation ions (<5 ppm)		

(*continued*)

Table 8.1 (*continued*)

Food packaging material/food simulant or extraction	Instrument (brand)	Analysis/data acquisition	Data processing/data reduction strategy	Identification/databases	IASs/NIASs identified	Ref.
Two flexible multilayer bags (PET, aluminium, PA, CPP) containing polyurethane adhesive/simulants A, B, D	UPLC-Q-TOF-MS (Xevo G2, Waters)	Non-target analysis, elucidation of NIAS/MS^E mode (low and high CE in the same run)	Migration studies: analysis of compounds that migrate from the multilayer material to the food simulant	Accurate masses of precursor and fragmentation ions (MassLynx)	17 IASs: caprolactam and its oligomers, plasticizers (tributyl acetyl citrate), antioxidants (Irganox 1010), lubricants (glycerol monotridecanoate and other derivatives)	29
				Background knowledge Bibliographic search Confirmation with pure standards (when possible) In some cases, NIAS identification was based on the exact mass and fragmentation patterns of the possible combination products of compounds in the samples ChemSpider, SciFinder	16 NIASs: oligomer esters coming from polyurethane adhesive	

20 PET pellets, some of them recycled/total dissolution (HFIP), solvent extraction (DCM), simulant B or D	UPLC-Q-TOF-MS (Xevo G2, Waters)	Non-target analysis/MS^E mode (low and high CE in the same run)	Identification of potential oligomers in the PET sample chromatogram Principal component analysis (PCA)	Accurate masses of precursor and fragmentation ions (MassLynx) Background knowledge Bibliographic search	14 NIASs: oligomers composed of TPA, DEG, EG	28
Cardboard with hot-melt adhesives based on EVA and APAO enriched in propene/simulant E, extraction with MeOH	UPLC-Q-TOF-MS (Xevo G2, Waters)	Non-target analysis/MS^E mode (low and high CE in the same run)	Most intense signals $\geq 10^4$ Comparison between data for substrates and sample materials	Accurate masses of precursor and fragmentation ions (MassLynx) Confirmation with pure standards (when possible) Background knowledge Bibliographic search ChemSpider, SciFinder	7 compounds from adhesives: 5 IASs (4 amides, 1 aromatic aldehyde) and 2 unidentified structures	30

[a] PP, polypropylene; PA, polyamide; EVOH, ethylene–vinyl alcohol copolymer; PET, poly(ethylene terephthalate); PE, polyethylene; PC, polycarbonates; food simulant A, 10% v/v ethanol; food simulant B, 3% w/v acetic acid; food simulant C, 20% v/v ethanol; food simulant D, 95% v/v ethanol; food simulant D1, 50% v/v ethanol, simulant for milk; food simulant D2, vegetable oil or 95% v/v ethanol and isooctane; food simulant E, Tenax® (for dry food); SIR, selected ion recording; DDA, data-dependent acquisition; DIA, data-independent acquisition; SWATH, sequential windowed acquisition of all theoretical MS; NPG, neopentyl glycol; HD, 1,6-hexanediol; DEG, diethylene glycol; EG, ethylene glycol; PLA, poly(lactic acid); PL, polylimonene; ZnO PNs, zinc oxide nanoparticles; S/N, signal-to-noise ratio; ASE, accelerated solvent extraction; FCM, food contact material; AhR, aryl hydrocarbon receptor activity; SPE, solid-phase extraction; FUSLE, focused ultrasonic solid–liquid extraction; BPA, bisphenol A; PBF, bisphenol F; BADBE, bisphenol A diglycidyl ether; BFDGE, bisphenol F diglycidyl ether; EDCs, endocrine-disrupting compounds; IPA, isophthalic acid; TA, terephthalic acid; NA, *cis*-5-norbornene-*endo*-2,3-dicarboxylic acid; CHO, 1,4-cyclohexanedimethanol; CRM, 2,2,5,7,8-pentamethyl-6-chromanol; EVA, ethylene–vinyl acetate; APAO, amorphous polyolefin; OPA, oriented polyamide; TBoAC, tributyl *O*-acetylcitrate; TMDD, 2,4,7,9-tetramethyl-5-decyne-4,7-diol; OPP, oriented polypropylene; BOPP, bioriented propylene; TPP, triphenylphosphine; TPA, terephthalic acid; HFIP, hexafluoro-2-propanol; DCM, dichloromethane.

[b] A full-scan MS-ddMSMS (Top *N*) experiment consists of a full-scan event followed by MS/MS scan events of Top *N* precursors in the inclusion list or the *N* most abundant ions at that time.

retention times and simultaneously confirmed using MS/MS by applying a post-acquisition targeted data extraction approach.

To date, most studies have been based on the use of non-targeted data-independent experiments.[4,7,8,10,20–30] This data acquisition mode has significant advantages in identifying unknown compounds compared with other acquisition methods, because it allows the comprehensive detection of all peaks detectable in a sample and the acquisition of the corresponding tandem mass spectrum without establishing any mass filters or intensity threshold prior to injection. It is worth emphasizing that they all used a Q-TOF mass analyser for the identification of non-targeted migrants, except one, which employed a Q-Orbitrap® instrument.[20]

Gómez *et al.*[21] applied, for the first time, a "Sequential Windowed Acquisition of All THeoretical" (SWATH) MS analysis for the identification of unexpected migrants from two commercially multiplastic materials used for purée and juice from fruit and vegetable packaging. It was based on a DIA method, which is an emerging technology in HRAMS for FCM issues. So far, SWATH data acquisition has been used effectively in proteomics and metabolite identification,[35] in clinical and forensic toxicology[36,37] and in the environmental field.[38] The approach allowed the identification of 26 potential migrants (mainly cyclic oligomers coming from polyurethane adhesives), five of which were reported for the first time in this study.[21]

In another recent study, a Q-Orbitrap® mass spectrometer operating at a mass-resolving power of 70 000 FWHM was used for the identification of NIASs from two monolayer films prepared with poly(lactic acid) (PLA) and polylimonene (PL).[20] A DIA experiment was performed using "full-scan/all ion fragmentation MS2" (AIF) analysis, in order to obtain unequivocal identification of the potential NIAS by obtaining its product ion-scan spectrum with HRAMS. In the AIF mode, all precursor ions are fragmented without a preselection in the quadrupole. The fragmentation is obtained with the HCD (higher energy collision-induced dissociation) cell. During the filling of the HCD collision cell, the energy can be set to step between specified percentage values around the chosen middle energy regardless of the ion's characteristics. This approach allowed the identification of four LC-amenable NIASs, mainly amide derivatives coming from slipping and antiblocking additives.

Two studies were found on the application of an "All Ion MS/MS" method for the identification of relevant substances used in paper and board food contact materials[8] and for the structural characterization of NIASs released from two different films prepared with PP and PL.[25] In the "All Ion MS/MS" mode, MS/MS data can be acquired using two different conditions: (1) with a low value of the fragmenter voltage and (2) with one or multiple high-energy values. The low-energy spectra predominantly show the molecular (or precursor) ions for the compounds whereas the high-energy spectra provide the precursors plus their fragment ions. Bengtström *et al.*[8] applied two energy values (110 and 120 V), whereas Martínez-Bueno *et al.*[25] used only one high-energy value (360 V).

Finally, acquisition in the MS^E (elevated energy mass spectrometry) mode has been the most reported approach to carry out non-targeted analysis by DIA experiments. MS^E records data without any discrimination or preselection. The main chemicals reported in these studies were active substances used in food packaging, such as adhesives, colorants, plasticizers, surfactants, printing ink solvents, oligomers and cyclic polyesters.[4,7,10,21–24,26,28–30] Similarly to the "All Ion MS/MS" mode, the data acquisition in MS^E can be performed at low, high or multiple collision energies (CEs), in order to obtain as much information as possible about the tandem mass spectra. The low-energy spectra provide information about the precursor ion whereas the high-energy spectra provide information about their fragment ions. Onghena *et al.*[26] combined low and high collision energy in the same run for the identification of unknown plastic migrants from baby bottles. They observed that a limited fragmentation took place when the low-energy function was used. By contrast, for the high-energy function the fragmentation was enhanced. Despite this, the applied strategy allowed the tentative identification of seven relevant compounds, laurolactam monomer (plastic monomer), bis(3,4-dimethylbenzylidene)sorbitol (clarifier for PP), 2,5-bis(5′-*tert*-butyl-2-benzoxazolyl) thiophene, Irganox 1010 (antioxidant), dibutyl and diisobutyl phthalate (plasticizers) and ethyl 4-ethoxybenzoate.

In all cases, the workflow for carrying out non-targeted screening by DIA experiments comprised retrospective analysis and data processing using mass spectral libraries of accurate masses and commercial or home-made databases, including different types of chemicals suspected to be present in FCMs, such as additives, monomers, phthalates or pigments, to identify compounds.

8.2.2 Data Processing in Non-target Screening

Currently, there is no standard protocol for how to proceed with the data processing and identification of FCMs, in particular after non-targeted screening approaches, in which samples are analysed without *a priori* knowledge of the contaminants of interest. HRAMS data are highly complex with a large number of molecular features and require a series of steps for interpretation that can be challenging. Typically, the general workflow for the analysis of FCMs involves extraction of molecular features, advanced data processing tools, including data reduction strategies, and finally structure identification with mass spectral libraries or structure elucidation of the prioritized chemicals and confirmation if analytical standards are available or can be synthesized. An in-depth study of the sample, the material and the manufacturing process would be of significant help for the subsequent identification of the migrants.

The extraction of useful information from large full-scan HRAMS-acquired data sets requires automatic data processing, which depends on the vendor software, but all procedures follow similar approaches. Following data acquisition, the data generated are processed using software that employs algorithms to identify automatically (ideally all) the detectable chemicals or molecular features in accurate mass data from the raw chromatogram. This generates a peak list of molecular features with accurate masses, retention times and ion abundances. Then, possible molecular formulae are ranked according to a weighted score based on the mass accuracy and the isotope distribution and isotope spacing information. Some of the studies for the identification of FCMs (Table 8.1) use specific data preprocessing treatment, such as noise filtering, automated integration and deconvolution of the peaks from the chromatograms, to resolve coeluting ions, enabling pairs of signals to be obtained potentially originating from the same compound, such as isotopologues, adducts or fragments.[8,12] Omer *et al.*[12] used specific software, the open-source CAMERA R-package, where a feature corresponds to a chromatographic peak from a given ion.

8.2.2.1 Data Reduction Strategies for HRAMS-acquired Data

Among the principal challenges of a non-targeted screening approach for the identification of FCMs is the number of peaks generated by the HRAMS instrument, which can be on the order of several thousand per sample. Therefore, processing of non-targeted datasets is time intensive, making it advantageous to reduce the number of relevant peaks prior to attempting compound identification. Filtering or data reduction strategies are critical for prioritizing the most relevant and interesting masses in the samples for their subsequent accurate identification or assignment of the most likely structure to the measured accurate mass.

Different data reduction strategies have been used for the identification of FCMs (Table 8.1). A simple strategy and one of the most commonly used approaches for the prioritization and identification of NIASs is a case–control comparison or comparison between the chromatograms from a control (or blanks) and sample material solutions.[4,10,21–24,27,30] This strategy provides very important information about the compounds directly related to each sample and has proven to be extremely valuable in these FCM studies and in other fields, such as clinical chemistry and environmental chemistry.[38,39] In the case of FCM screening, the case–control comparison has been made visually,[4,10,22] using software programs[21,23] or statistical comparisons.[4,20,22,27,28] There are several computer tools for helping with the comparison. New software packages such as MarkerLynx™ (Waters), Mass Frontier™ (Thermo Scientific) and Sciex OS (Sciex) have been used in FCM studies to help with in the prioritization of the chemicals related to the materials.

Other strategies for data reduction used in FCM screening studies include intensity threshold,[6,8,9,12,20,21,25,30,31,33,40] mass error threshold,[11,26,40] signal-to-noise ratio threshold,[9] mass ranges[12,23] and home-made databases or inclusion lists containing common migrating chemicals suspected (or known) to be present in food packing, such as plastic additives, starting substances used in polymer manufacture, compounds authorized for plastic FCMs by the EU and monomers and possible combinations of the potential monomers.[8,9,25,26,33] The selection of the chemicals included in the home-made databases was based in most cases on a literature overview and substances from legislative and inventory lists.

In many cases, the number of peaks left after using the data reduction approach is still too large to be identified. Hence the peak lists need to be further reduced and more than one approach could be necessary to reduce the size of the search space to manageable levels. As can be observed in the example in Figure 8.2, in a study of the non-targeted analysis of migrating chemicals coming from multilayer plastic packaging materials, Gomez *et al.*[21] combined two different filtering approaches to decrease the number of ions to be elucidated from 3149 to 21 spectra. First, applying a sample–control comparison filter using an area ratio of 10, they reduced the number of masses to only 57 spectra, which were characteristic of the studied packaging material. To reduce the number of spectra further to investigate and focus attention on the most abundant compounds, they set the signal intensity to be greater than 1×10^6 counts, decreasing the number of chemicals of interest to only 21 ions to be reviewed.

Other studies have combined different approaches, such as intensity threshold, mass error threshold and a home-made database;[25,26,31,33,40] case–control comparison and intensity threshold or mass range selection;[23,30] intensity threshold, signal-to-noise ratio threshold and multivariate analysis;[20] and intensity threshold, mass error threshold, good peak shape, appropriate retention time and a home-made database.[9] Effect-directed analysis (EDA) was also used in one study on the non-targeted screening of contaminants in paper and board FCMs.[8] In this study, Bengtström *et al.*[8] applied a step-by-step strategy, including high-performance liquid chromatography (HPLC) fractionation and *in vitro* testing of each subsample, combined with intensity threshold and a home-made database to prioritize chemicals with potentially adverse health effects in paper and board FCMs. The fractionation reduced the number of substances to be identified to the fractions exhibiting *in vitro* activity. They also reduced the number of peaks to be identified, using a cut-off response based on the threshold of toxicological concern (TTC). In spite of this exhaustive data reduction strategy, it turned out to be a complex task to identify the causative agents of toxicity in real samples. It was suggested that an improvement of the developed method could be to fine-tune the fractionation further in order to narrow down the number of substances present in each fraction even more.

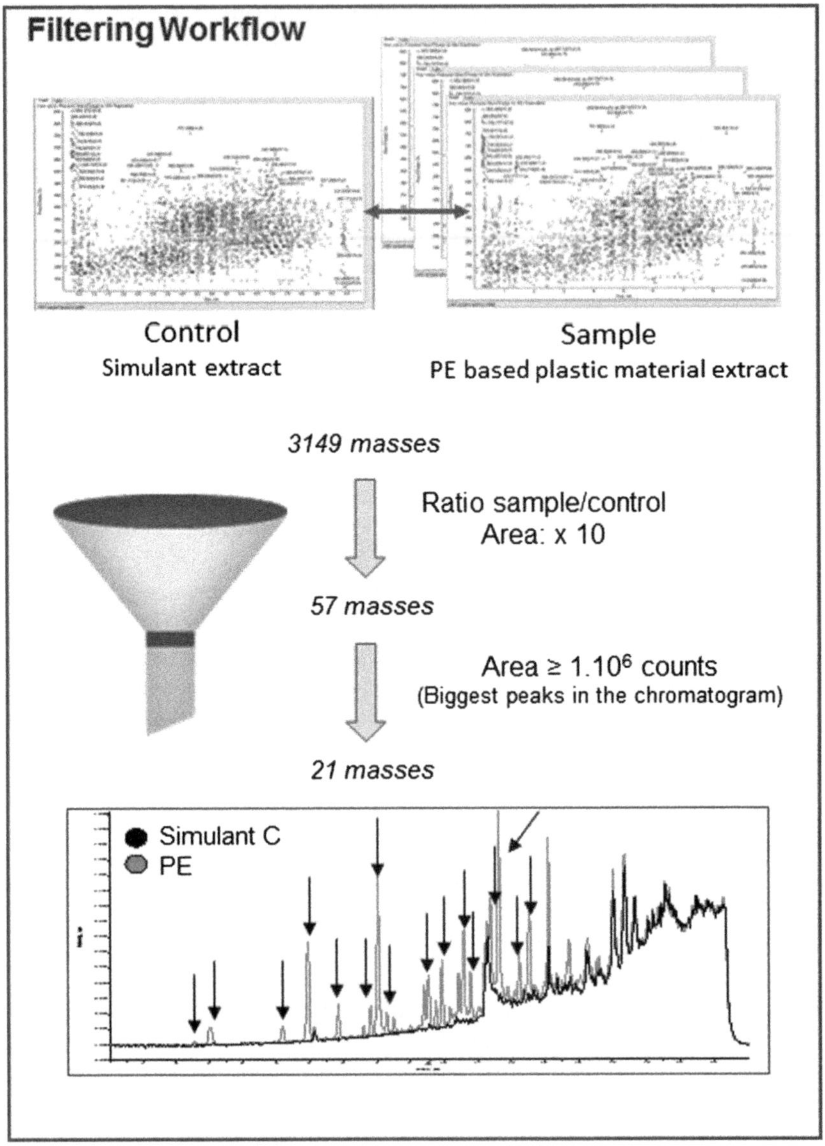

Figure 8.2 Example of a data reduction workflow for the prioritization of migrants in non-targeted analysis using high-resolution accurate mass spectrometry (HRAMS).

8.2.2.2 Identification of Prioritized Chemicals

The identification of prioritized chemicals in FCMs involves all information available from the spectra obtained by MS and MS/MS (molecular formula, isotope pattern and fragments), compound databases and literature and background information on the sample.

The first step in the identification of a non-target chemical is to determine the elemental composition of the precursor ion by means of empirical formula generator software. The empirical formula generator considers a combination of elements and identifies theoretical empirical formulae that correspond to the experimental masses. The elemental composition is then linked to a chemical structure using chemical compound database websites such as ChemSpider, SciFinder and PubChem, containing structures and properties of millions of natural and synthetic organic chemicals, or online mass spectral libraries such us MassBank or *mz*Cloud. In this step one could also make a comparison of isotopic patterns provided by a simulator.[11,33,40] The next step is to evaluate whether the product ions detected in the tandem mass spectrum could be linked to the fragments generated from the chemical structures of the candidates proposed. The similarity between predicted and experimental spectra is used to rank the candidate structures. This task can be carried out with the help of analytical software that uses chemically intelligent algorithms to see if the fragments obtained experimentally match the fragments of the candidate proposed, such as MassFragment (Waters),[4,7,10,22–24,27–30] Mass Frontier (ThermoFisher Scientific)[20] or Analytics (Sciex).[21] For the final selection of compounds from the proposed candidates, background knowledge about the materials under study (material composition and the manufacturing process) and a literature review are essential.

Depending on the study objectives, the screening of non-target compounds falls into (i) suspect screening of migrants, with *a priori* information from available chemical characterization in databases and the literature, but no reference standards, and (ii) analysis of unknown migrants, with neither *a priori* information nor chemical characterization in databases or reference standards, hence the structure elucidation of chemicals from HRAMS data is necessary. This is the most complex and challenging analysis and it may apply to new chemical formulations, by-products or transformation products formed from parent species, as is the case with most NIASs, for example, and is an important new area of research, as in some instances these secondary products are more toxic than the parent. Some studies on the identification of FCMs using HRAMS (Table 8.1) have used suspect screening of chemicals with the help of home-made or chemical databases.[8,9,24,25,30,31] However, there are also many studies where structure elucidation of unknown migrants and NIASs from HRAMS data or a combination of suspect analysis and analysis of unknowns have been carried out.[4,7,10–12,20–22,26,33,40] Aznar and co-workers[4,10,22] identified unknown compounds used as additives in varnishes or polymers (*e.g.* glycerols, adipate), cyclic polyesters in multilayer packaging material containing ink and PE and triphenylphosphine oxide, an oxidation product of triphenylphosphine. Other unknown FCMs and NIASs identified using these methods include bisphenol A–polycarbonate degradation products and unreacted oligomers detected in polycarbonate materials;[11,33,40] neoformed lactones and malonic acid in multilayer packaging material containing biodegradable adhesives;[23] caprolactam oligomers and by-products from raw materials, mainly cyclic oligomers coming from polyurethane adhesives in multilayer packaging material;[12,21,24,28,29] 2,4,6-triamino-1,3,5-triazine, azepan-2-one (caprolactam) and (2*E*)-3-phenylprop-2-enal in new active nanomaterials with antimicrobial agents;[25] *N*,*N*-diethyldodecanamide, *N*-[(9*Z*)-9-octadecen-1-yl] acetamide, 1-palmitoylglycerol and glycerol stearate in a developed polymer;[20] primary aromatic amines arising from isocyanates (polyurethane adhesives) and azo dyes in plastic laminated materials;[6,27] laurolactam dimer and trimer; 2,5-bis(5′-*tert*-butyl-2-benzoxazolyl)thiophene and ethyl 4-ethoxybenzoate in plastic baby bottles;[26] and oligomers corresponding to the reaction of carboxylic acids and diols and the mono-alcohol *t*BuPh (4-*tert*-butylphenol) in polyester can coatings.[9]

Figure 8.3 shows an example of the tentative identification of a NIAS in a developed food packaging nano film, the amide *N*,*N*-diethyldodecanamide,[20] illustrating the resulting full-scan spectrum in MS and MS/MS modes for this compound obtained by LC-Q-Orbitrap®-MS, including the structure of the fragment ions proposed by the Mass Frontier software.

Figure 8.4 presents an example of the tentative identification of another unknown migrant found in a multilayer plastic packaging material, in this case a cyclic ester made up of phthalic acid and diethylene glycol (1 : 2) (PA–DEG–DEG).[21] Cyclic ester oligomers were previously reported to be undesirable by-products generated during polyester synthesis and have been found in previous migration studies from multilayer high barrier FCMs;[29] however, the presence of this oligomer had not been previously described in the literature. Figure 8.4 display the results obtained after non-target screening, XIC, MS and MS/MS information for this oligomer, Formula Finder results and ChemSpider results. In Formula Finder. The results were automatically ranked by mass accuracy (MS and MS/MS) and the matching of the isotope pattern. For m/z 325.1280 the formula $C_{16}H_{20}O_7$ was ranked first, with a score of 94.9%. As can be observed, the ChemSpider search found 176 known structures for the highest priority elemental composition ($C_{16}H_{20}O_7$). The knowledge of the sample and the automatic elucidation of the tandem mass spectrum led to the tentative identification of PA–DEG–DEG. The main fragments successfully matched the proposed structure, with a plausible structure assigned to all six dominant fragments.

8.3 Risk Assessment of NIASs

8.3.1 The Threshold of Toxicological Concern (TTC) Approach. Cramer Rules

There are no toxicity data for all existing chemical substances and for most of the NIASs. When a new NIAS is identified, an assessment of risk is necessary. The problem is how to evaluate the safety of these materials if no

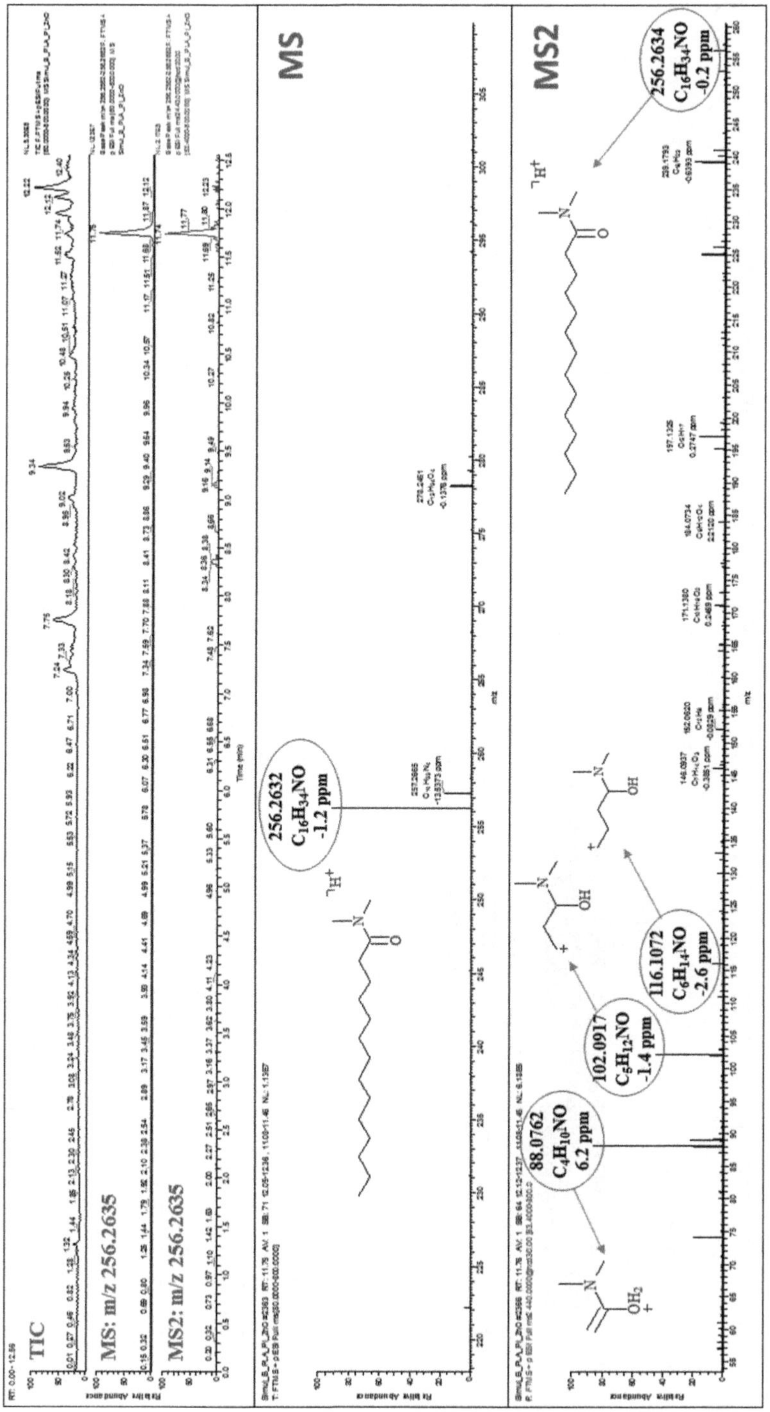

Figure 8.3 Example of NIAS identification in food packaging nano films. Full-scan spectrum in MS and MS/MS spectrum obtained by LC-Q-Orbitrap®-MS for *N*,*N*-diethyldodecanamide.

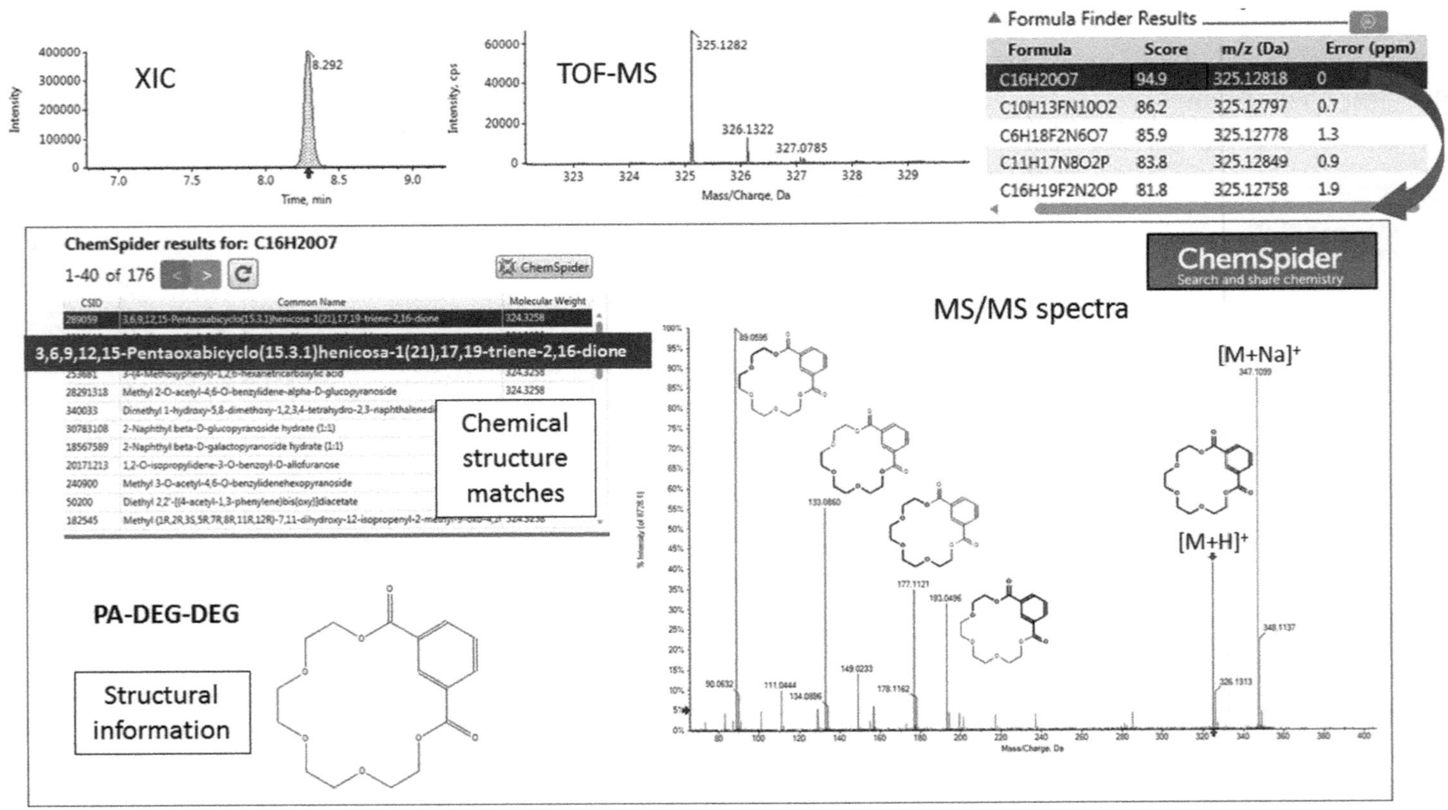

Figure 8.4 Example of the tentative identification of an unknown migrant found in a multilayer plastic packaging material, a cyclic ester made up of phthalic acid and diethylene (PA–DEG–DEG). Visual display of the results after non-target screening using LC-ESI-Q-TOF-MS. XIC, TOF-MS and MS/MS spectra, Formula Finder results, ChemSpider search and automatic elucidation of the tandem mass spectrum.

analytical standards are available, as only a theoretical evaluation could be made. An interesting option is to use the threshold of toxicological concern (TTC) approach, which may be applied to evaluate materials for their potential toxicity when exposure is very low. In the absence of data, this practical tool sets up levels of human exposure that are considered to have no appreciable risk to human health.[41]

The TTC approach is based on the Cramer rules, which estimate the theoretical toxicity of compounds according to their molecular structure. There are three classes of toxicity according to these rules: low (class I), moderate (class II) and high (class III). A maximum intake value is recommended by the Cramer rules for each toxicity group: values for class I, II and III are 1.8, 0.54 and 0.09 mg per person per day, respectively. The software Toxtree v2.6.13, developed by the JRC Commission, is an open-source application that is able to estimate toxic hazards by applying a decision tree approach following Cramer rules (http://toxtree.sourceforge.net/). This software can be used to estimate the theoretical toxicity of the identified compounds.

The TTC approach has a relatively long history of use in the evaluation of FCMs. The TTC strategy has been applied in some studies for estimating the toxicity of identified NIASs or unknown FCMs by non-targeted analysis. Canellas *et al.*[23] applied the TTC approach to the safe design of an acrylic adhesive for food contact laminates, since there were neither specific migration limits according to the Regulation EU 10/2011[13] nor other toxicity data for the migrants identified. The migrants were classified in Cramer class II and III with self-derived specific migration limits (SMLs) of 0.54 and 0.09 mg kg^{-1}, respectively. For one of the chemicals, a surfactant polymer, the migration values obtained were higher than the self-derived SML. As a result, the authors decided to remove this migrant from the formulation. The same group[7] also used the TTC approach to evaluate whether neoformed compounds found in the adhesive used in biodegradable food packaging could endanger human health. These neoformed compounds were classified in the lowest toxicity class and they concluded that taking into account these theoretical toxicity data, the adhesive used in the multilayer food packing was safe for food contact applications. In another study on the identification of unknown migrant compounds from adhesives used in FCMs,[30] the TTC approach was applied in order to check the possible human risk of the identified migrants, five amides and one compound classified in Cramer toxicity class II. None of the migration values exceeded the recommended Cramer exposure values. The TTC approach was also used in a very recent study evaluating the toxicity of 23 NIASs identified in multilayer plastic packaging materials[21] and comparing two plastic packaging materials. According to the Cramer rules, most of the identified NIASs were classified in the highest toxicity class and the comparison of the studied materials showed that both materials were very similar.

8.4 Conclusion

Over the last decade, a wide range of functional materials employed in packaging technology in order to maintain the quality of foodstuffs have been developed and introduced onto the market. Hence interest in the identification of potential contaminants migrating into food has increased considerably. However, the identification of unknown migrants from FCMs is a challenge since neither chemical nor spectral databases nor analytical standards are available.

In this chapter, we have addressed some of the most relevant data acquisition modes and the data handling used for the identification/characterization of migrants from FCMs. LC-MS has been the most frequently used technique for the identification of thermally unstable and non-volatile migrant compounds. To date, ESI has been the most commonly used ionization source rather than APCI in the analysis of food packaging contaminants.

We have also discussed the most recent approaches in the LC-HRAMS analysis of food packaging contaminants. Although low-resolution MS strategies have been applied for the analysis of food packaging contaminants, the most extensively used approaches to the analysis of migrants in FCM applications are based on HRAMS methods. The use of different MS acquisition strategies and HRAMS is one of the best ways to avoid false positives or even false negatives and we presented some relevant examples concerning the analysis of food packaging contaminants.

In summary, there are few applications in the literature concerning the targeted analysis of food packaging contaminants. Most of the studies published on the application of LC-HRAMS are based on non-targeted screening. To date, DIA experiments have been a more commonly used data acquisition mode than DDA to identify migrants from FCMs.

Regarding data processing, the main challenge in a non-targeted screening approach for FCM identification is the number of peaks generated by the HRAMS instrument, making it necessary to reduce the number of relevant peaks prior to attempting compound identification. In this chapter, we have discussed the different filtering and data reduction strategies used for the prioritization and identification of FCMs, including case–control comparison, intensity threshold, signal-to-noise ratio threshold, mass range restrictions, home-made databases and even effect-directed analysis (EDA) in one of the studies. In most cases, a combination of more than one approach was necessary to reduce the size of the search space to manageable levels. The identification of prioritized chemicals in FCMs involved all information available from MS and MS/MS, chemical and spectral databases and literature and background information on the samples. Some studies on the identification of FCMs used suspect screening of chemicals with the help of home-made or chemical databases, however, there are also many studies where structure elucidation of unknown migrants and NIASs from HRAMS data has been carried out. Some of the migrants identified using these acquisition and data treatment strategies were unknown IASs used as additives in varnishes or polymers or NIASs, such as polycarbonate degradation products and unreacted oligomers detected in polycarbonate materials, cyclic oligomers coming from polyurethane adhesives or primary aromatic amines arising from isocyanates (polyurethane adhesives) and azo dyes in plastic laminated materials.

Finally, we briefly discussed food packaging migration risk assessment studies. There are no toxicity data for all existing chemical substances and for most of the NIASs. Therefore, in the absence of data and analytical standards, a theoretical evaluation based on the TTC approach and Cramer rules has been applied in some studies to evaluate the safety of these materials. The TTC approach was used for the safe design of an acrylic adhesive for food contact laminates, to evaluate if neoformed compounds found in the adhesive used in biodegradable food packaging could endanger human health, in order to check the possible human risk of the identified migrants in adhesives used in FCMs and to evaluate the potential toxicity of NIASs identified in multilayer packaging materials.

Acknowledgements

The authors wish to thank the European Union's Seventh Programme for Research and Technological Development FP7-ERANET-SUSFOOD "CEREAL Project" for financial support along with the National Institute for Agricultural and Food Research and Technology, ERA33-CEREAL-UAL. Dr María José Gómez Ramos acknowledges funding obtained from the Spanish Government for a Ramón y Cajal Research Fellowship (RYC-2015-17959). Dr María Jesús Martínez Bueno thanks the Marie Skłodowska-Curie Actions (MSCA) for individual fellowships (H2020-MSCA-IF-2015#707816).

References

1. B. Athenstädt, M. Fünfrocken and T. C. Schmidt, Migrating components in a polyurethane laminating adhesive identified using gas chromatography/mass spectrometry, *Rapid Commun. Mass Spectrom.*, 2012, **26**, 1810–1816.
2. J. Muncke, Exposure to endocrine disrupting compounds *via* the food chain: Is packaging a relevant source?, *Sci. Total Environ.*, 2009, **407**, 4549–4559.
3. M. Hoppe, P. de Voogt and R. Franz, Identification and quantification of oligomers as potential migrants in plastics food contact materials with a focus in polycondensates - A review, *Trends Food Sci. Technol.*, 2016, **50**, 118–130.

4. M. Aznar, C. Domeño, C. Nerín and O. Bosetti, Set-off of non volatile compounds from printing inks in food packaging materials and the role of lacquers to avoid migration, *Dyes Pigm.*, 2015, **114**, 85–92.
5. J. Veyrand, *et al.*, Identification of unknown compounds from polyester cans coatings that may potentially migrate into food or food simulants, *Anal. Bioanal. Chem.*, 2017, **34**, 49–59.
6. M. Mattarozzi, F. Lambertini, M. Suman and M. Careri, Liquid chromatography-full scan-high resolution mass spectrometry-based method towards the comprehensive analysis of migration of primary aromatic amines from food packaging, *J. Chromatogr. A*, 2013, **1320**, 96–102.
7. E. Canellas, P. Vera and C. Nerín, UPLC-ESI-Q-TOF-MS(E) and GC-MS identification and quantification of non-intentionally added substances coming from biodegradable food packaging, *Anal. Bioanal. Chem.*, 2015, **407**, 6781–6790.
8. L. Bengtström, *et al.*, Non-targeted screening for contaminants in paper and board food-contact materials using effect-directed analysis and accurate mass spectrometry, *Food Addit. Contam., Part A: Chem., Anal., Control, Exposure Risk Assess.*, 2016, **33**, 1080–1093.
9. R. Paseiro-Cerrato, S. MacMahon, C. D. Ridge, G. O. Noonan and T. H. Begley, Identification of unknown compounds from polyester cans coatings that may potentially migrate into food or food simulants, *J. Chromatogr. A*, 2016, **1444**, 106–113.
10. M. Aznar, P. Alfaro, C. Nerín, E. Jones and E. Riches, Progress in mass spectrometry for the analysis of set-off phenomena in plastic food packaging materials, *J. Chromatogr. A*, 2016, **1453**, 124–133.
11. C. Bignardi, A. Cavazza, C. Laganà, P. Salvadeo and C. Corradini, Optimization of mass spectrometry acquisition parameters for determination of polycarbonate additives, degradation products and colorants migrating from food contact materials to chocolate, *J. Mass Spectrom.*, 2018, **53**, 83–90.
12. E. Omer, *et al.*, Elucidation of non-intentionally added substances migrating from polyester-polyurethane lacquers using automated LC-HRAMS data processing, *Anal. Bioanal. Chem.*, 2018, 1–13, DOI: 10.1007/s00216-018-0968-z.
13. European Commission, Union Guidance on Regulation (EU) No 10 / 2011 on plastic materials and articles intended to come into contact with food as regards information in the supply chain , *Off. J. Eur. Union,* 2013, 1–32.
14. C. Nerin, P. Alfaro, M. Aznar and C. Domeño, The challenge of identifying non-intentionally added substances from food packaging materials: A review, *Anal. Chim. Acta*, 2013, **775**, 14–24.
15. Y. Sanchis, V. Yusà and C. Coscollà, Analytical strategies for organic food packaging contaminants, *J. Chromatogr. A*, 2017, **1490**, 22–46.
16. H. Gallart-Ayala, O. Núñez and P. Lucci, Recent advances in LC-MS analysis of food-packaging contaminants, *TrAC, Trends Anal. Chem.*, 2013, **42**, 186–204.

17. D. J. Kim and K. T. Lee, Determination of monomers and oligomers in polyethylene terephthalate trays and bottles for food use by using high performance liquid chromatography- electrospray ionization-mass spectrometry, *Polym. Test.*, 2012, **31**, 490–499.
18. S. K. Mortensen, X. T. Trier, A. Foverskov and J. H. Petersen, Specific determination of 20 primary aromatic amines in aqueous food simulants by liquid chromatography-electrospray ionization-tandem mass spectrometry, *J. Chromatogr. A*, 2005, **1091**, 40–50.
19. F. Kühne, *et al.*, N-nitrosamines migrating from food contact materials into food simulants: analysis and quantification by means of HPLC-APCI-MS/MS, *Food Addit. Contam., Part A: Chem., Anal., Control, Exposure Risk Assess.*, 2018, **35**, 792–805.
20. M. J. Martínez-Bueno, *et al.*, Identification of non-intentionally added substances in food packaging nano films by gas and liquid chromatography coupled to orbitrap mass spectrometry, *Talanta*, 2017, **172**, 68–77.
21. M. J. Gómez, A. Lozano and R. A. Férnandez-Alba, High-resolution mass spectrometry with data independent acquisition for the comprehensive non-targeted analysis of migrating chemicals coming from multilayer plastic packaging materials used for fruit purée and juice, *Talanta*, 2019, **191**, 180–192.
22. M. Aznar, A. Rodriguez-Lafuente, P. Alfaro and C. Nerin, UPLC-Q-TOF-MS analysis of non-volatile migrants from new active packaging materials, *Anal. Bioanal. Chem.*, 2012, **404**, 1945–1957.
23. E. Canellas, P. Vera and C. Nerín, Migration assessment and the 'threshold of toxicological concern' applied to the safe design of an acrylic adhesive for food-contact laminates, *Food Addit. Contam., Part A: Chem., Anal., Control, Exposure Risk Assess.*, 2017, **34**, 1721–1729.
24. F. Isella, E. Canellas, O. Bosetti and C. Nerin, Migration of non intentionally added substances from adhesives by UPLC-Q-TOF/MS and the role of EVOH to avoid migration in multilayer packaging materials, *J. Mass Spectrom.*, 2013, **48**, 430–437.
25. M. J. Martínez-Bueno, *et al.*, Characterization of non-intentionally added substances (NIAS) and zinc oxide nanoparticle release from evaluation of new antimicrobial food contact materials by both LC-Q-TOF-MS, GC-Q-TOF-MS and ICP-MS, *Anal. Methods*, 2016, **8**, 7209–7216.
26. M. Onghena, *et al.*, Identification of substances migrating from plastic baby bottles using a combination of low-resolution and high-resolution mass spectrometric analysers coupled to gas and liquid chromatography, *J. Mass Spectrom.*, 2015, **50**, 1234–1244.
27. D. Pezo, M. Fedeli, O. Bosetti and C. Nerín, Aromatic amines from polyurethane adhesives in food packaging: The challenge of identification and pattern recognition using Quadrupole-Time of Flight-Mass SpectrometryE, *Anal. Chim. Acta*, 2012, **756**, 49–59.
28. S. Ubeda, M. Aznar and C. Nerín, Determination of oligomers in virgin and recycled polyethylene terephthalate (PET) samples by UPLC-MS-Q-TOF, *Anal. Bioanal. Chem.*, 2018, **410**, 2377–2384.

29. S. Úbeda, *et al.*, Overall and specific migration from multilayer high barrier food contact materials–kinetic study of cyclic polyester oligomers migration, *Food Addit. Contam., Part A: Chem., Anal., Control, Exposure Risk Assess.*, 2017, **34**, 1784–1794.
30. P. Vera, E. Canellas and C. Nerin, Identification of non-volatile compounds and their migration from hot melt adhesives used in food packaging materials characterized by ultra-performance liquid chromatography coupled to quadrupole time-of-flight mass spectrometry, *Anal. Bioanal. Chem.*, 2013, **405**, 4747–4754.
31. Y. Sanchis, C. Coscollà, M. Roca and V. Yusà, Target analysis of primary aromatic amines combined with a comprehensive screening of migrating substances in kitchen utensils by liquid chromatography-high resolution mass spectrometry, *Talanta*, 2015, **138**, 290–297.
32. D. Pérez-Palacios, M. Á. Fernández-Recio, C. Moreta and M. T. Tena, Determination of bisphenol-type endocrine disrupting compounds in food-contact recycled-paper materials by focused ultrasonic solid-liquid extraction and ultra performance liquid chromatography-high resolution mass spectrometry, *Talanta*, 2012, **99**, 167–174.
33. C. Bignardi, A. Cavazza, C. Corradini and P. Salvadeo, Targeted and untargeted data-dependent experiments for characterization of polycarbonate food-contact plastics by ultra high performance chromatography coupled to quadrupole orbitrap tandem mass spectrometry, *J. Chromatogr. A*, 2014, **1372**, 133–144.
34. C. Nerin, *et al.*, Characterisation and migration properties of silicone materials during typical long-term commercial and household use applications: A combined case study, *J. Chromatogr. A*, 2016, **48**, 133–144.
35. A. E. Lewandowska, *et al.*, Qualitative and Quantitative Analysis of Proteome and Peptidome of Human Follicular Fluid Using Multiple Samples from Single Donor with LC-MS and SWATH Methodology, *J. Proteome Res.*, 2017, **16**, 3053–3067.
36. A. T. Roemmelt, A. E. Steuer, M. Poetzsch and T. Kraemer, Liquid chromatography, in combination with a quadrupole time-of-flight instrument (LC Q-TOF), with sequential window acquisition of all theoretical fragment-ion spectra (SWATH) acquisition: Systematic studies on its use for screenings in clinical and foren, *Anal. Chem.*, 2014, **86**, 11742–11749.
37. M. P. Elmiger, M. Poetzsch, A. E. Steuer and T. Kraemer, Assessment of simpler calibration models in the development and validation of a fast postmortem multi-analyte LC-Q-TOF quantitation method in whole blood with simultaneous screening capabilities using SWATH acquisition, *Anal. Bioanal. Chem.*, 2017, **409**, 6495–6508.
38. S. L. Kaserzon, A. L. Heffernan, K. Thompson, J. F. Mueller and M. J. Gomez Ramos, Rapid screening and identification of chemical hazards in surface and drinking water using high resolution mass spectrometry and a case-control filter, *Chemosphere*, 2017, **182**, 656–664.

39. A. Rotander, *et al.*, Novel fluorinated surfactants tentatively identified in firefighters using liquid chromatography quadrupole time-of-flight tandem mass spectrometry and a case-control approach, *Environ. Sci. Technol.*, 2015, **49**, 2434–2442.
40. C. Bignardi, A. Cavazza, C. Laganà, P. Salvadeo and C. Corradini, Release of non-intentionally added substances (NIAS) from food contact polycarbonate: Effect of ageing, *Food Control*, 2017, **71**, 329–335.
41. S. Bhatia, *et al.*, Comparison of Cramer classification between Toxtree, the OECD QSAR Toolbox and expert judgment, *Regul. Toxicol. Pharmacol.*, 2015, **71**, 52–62.

CHAPTER 9

Migration of Nanomaterials from Food Contact Materials

RUUD J. B. PETERS*[a] AND EDDO J. HOEKSTRA*[b]

[a]RIKILT Wageningen University and Research, Akkermaalsbos 2, 6708-WB Wageningen, The Netherlands; [b]DG Joint Research Centre of the European Commission, Via E. Fermi 2749, 21027 Ispra (VA), Italy
*E-mail: ruudj.peters@wur.nl, eddo.hoekstra@ec.europa.eu

9.1 Introduction

The use of engineered nanomaterials (ENMs) in food contact applications has created enormous interest in recent years.[1–3] However, nanomaterials can be used in plastic food contact materials (FCMs) only when they are explicitly authorized.[4] At present there are 12 nanomaterials that are authorized to be used in plastic FCMs, with the following restrictions: (1) no migration, (2) only to be used in specific plastics and up to a certain concentration and (3) the agglomerates in the plastic shall have diameters >100 nm, which may consist of primary nanoparticles that have diameters of <100 nm. The first FCM is kaolin (FCM 410), which shall consist of particles smaller than 100 nm and shall be incorporated at a concentration of less than 12% w/w in an ethylene–vinyl alcohol copolymer (EVOH), in

Food Chemistry, Function and Analysis No.10
Food Contact Materials Analysis: Mass Spectrometry Techniques
Edited by Michele Suman
© The Royal Society of Chemistry 2019
Published by the Royal Society of Chemistry, www.rsc.org

which the layer in direct contact with the food provides a functional barrier preventing migration of particles into the food. The second is carbon black (FCM 411), which shall consist of aggregated particles in the size range 100–1200 nm and originating from primary particles of size 10–300 nm. The aggregates may form agglomerates within the size distribution from 0.3 μm to the millimeter size range. Similarly, synthetic amorphous silicon dioxide (FCM 504) and silanated silicon dioxide (FCM 87) shall consist of aggregated particles of size range 0.1–1 μm originating from primary particles of size 1–100 nm. The aggregates may form agglomerates within the size distribution from 0.3 μm to the millimeter size range. Titanium nitride nanoparticles (FCM 807) are authorized for the use in poly(ethylene terephthalate) (PET) in concentrations up to 20 mg kg^{-1}. The agglomerates in PET shall have a diameter of 100–500 nm, consisting of primary titanium nitride nanoparticles that have diameters of approximately 20 nm. Butadiene–ethyl acrylate–methyl methacrylate–styrene copolymer crosslinked with divinylbenzene (FCM 859) is only to be used as particles in non-plasticized poly(vinyl chloride) (PVC) at up to 10% w/w in contact with all food types at room temperature or below, including long-term storage. When used together with FCM 998 and/or FCM 1043, the restriction of 10% w/w applies to the sum of those substances. The diameter of particles shall be >20 nm and for at least 95% by number it shall be >40 nm. The same restrictions are applied to the use of a non-crosslinked butadiene–ethyl acrylate–methyl methacrylate–styrene copolymer (FCM 998), a methacrylic acid–ethyl acrylate–*n*-butyl acrylate–methyl methacrylate–butadiene copolymer (FCM 1016) and a butadiene–ethyl acrylate–methyl methacrylate–styrene copolymer crosslinked with 1,3-butanediol dimethacrylate (FCM 1043). Another authorized FCM is montmorillonite clay modified with dimethyldialkyl(C_{16}–C_{18}) ammonium chloride (FCM 1030), which is only to be used at levels up to 12% w/w in polyolefins in contact with dry foods. The sum of the specific migration of 1-chlorohexadecane and 1-chlorooctadecane shall not exceed 0.05 mg kg^{-1} food. The material can contain platelets in nanoform that are only in one dimension <100 nm. Finally, zinc oxide nanoparticles coated with [3-(methacryloxy)propyl]trimethoxysilane (FCM 1046) and uncoated zinc oxide nanoparticles (FCM 1050) are authorized for the use in unplasticized polymers.

In recent years, numerous studies have reported on the release of nanomaterials from polymer-based FCMs. However, there is some debate about the possible migration of nanomaterials from packaging since the diffusion rates of nanoparticles in a polymer material are so low that only very small nanoparticles (around 1 nm) in a polyolefin-based material are expected to migrate.[5] Migration studies involving nano-enabled FCMs often face challenges associated with detecting, characterizing and quantifying ENMs and the ability to distinguish them from ionic species. Based on the literature, this chapter will give an insight into the reported nanoparticle migrations, consider their possible mechanisms, and which analytical techniques are used and available for migration testing.

9.2 Release of Nanomaterials from Food Contact Materials

Incorporation of nanomaterials or nanotechnological devices in packaging materials or storage containers in order to lengthen the storage time while keeping the products fresh is considered the most important type of application of nanotechnology in the food area for the near future.[1,3] Nanocomposites can improve mechanical strength, reduce weight, increase heat resistance and improve the barriers against oxygen, carbon dioxide, ultraviolet radiation, moisture and volatiles of food package materials. Nanoscale inclusions consist mostly of nanoclays, carbon nanoparticles and nanoscale metals and oxides.[2,6–8] Table 9.1 gives an overview of migration studies that were found in the literature. In general, three groups of nanomaterials were involved in the identified studies: nanoclay (6 out of 34), nanosilver (22 out of 34) and nano-sized metal oxides, TiN and carbon black (6 out of 34).

Nanoclay (bentonite, montmorillonite) is a naturally occurring substance with platelets whose thicknesses are in the nanoscale size range. Considering the probable extent of its use, there have been surprisingly few studies investigating the migration of nanoclay constituents into food simulants or foods. In some of the studies, migration of elemental components from the nanoclay (particularly Si) into food and acidic food simulants has been detected from food packaging material, although the overall migration in all cases (0–9.5 mg kg^{-1}) was significantly lower than the overall migration limit of 60 mg kg^{-1} foodstuff in Europe (note: according to legislation, the official maximum overall migration is 10 mg dm^{-2}. By using a value of 6 dm^2 kg^{-1} food, this translates into 60 mg kg^{-1} food). Migration of Al from nanoclay was minimal (0–1 mg kg^{-1} food) and even lower than the concentrations typically found in foods. Only two studies examined the migration of nanoclay particles *per se* and in both instances their presence in food simulants was not detected, indicating no nanoclay particle migration.

At present, silver is the most commonly investigated antibacterial nanoparticle. Silver nanoparticles are added to food packaging materials in several ways, for instance, as finely dispersed nanosilver particles embedded in containers and coatings,[9–12] as a silver-based zeolite in poly(lactic acid) biocomposites,[13,14] and as gelatin–silver nanocomposites.[15] The antimicrobial properties of the food packaging materials were improved by incorporating any of these materials but only the finely dispersed nanosilver particles embedded in the containers are currently used in food packaging, the other methods being still in the research stage.[16] Unlike nanoclay, the antimicrobial function of nanosilver in food packaging materials means that it is intended that silver ions will be released to prevent food spoilage. Hence a balance between what is considered too little to be effective and too much from a safety perspective needs to be achieved. A considerable number of migration studies were found for nanosilver-containing polymer composites

Table 9.1 Overview of migration studies of nanoparticle-based FCMs.[a,b]

Polymer matrix	Commercially available	NP content[c]/mg kg^{-1}	Food/simulant	Analytical technique	Migration[c]/mg kg^{-1}	NPs detected	Reference
Clay (montmorillonite)-based FCMs							
Starch–polyester films	No	—	Food (lettuce and spinach)	AAS	None (Fe, Mg) 3–6 (Si)	Not investigated	Avella *et al.*, 2005[20]
PET bottles	Yes	—	Simulant (3% acetic acid, 10% ethanol, isooctane)	SEM-EDX ICP-MS	None (Sb + Al)	No	Chaudhry *et al.*, 2008[21]
PLA films	No	50 000	Simulant (95% ethanol)	AF4-MALS-ICP-MS	None (Mg)	No	Schmidt *et al.*, 2009[22]
Wheat gluten/nanoclay films	No	50 000	Simulant (distilled water, 3% acetic acid, 15% ethanol, olive oil, Tenax®, agar gel)	ICP-MS	0–1 (Al) 0–4.5 (Si)	Not investigated	Mauricio-Iglesias *et al.*, 2010[23]
PET bottles	No	30 000	Simulant (3% acetic acid)	ICP-MS	0.18–0.34 (Al) 6–9.5 (Si)	Not investigated	Farhoodi *et al.*, 2014[24]
PLA composites	No	40 000	Simulant (distilled water)	ICP-MS	None (Al, Fe, Mg, Si, Ca)	Not investigated	Maisanaba *et al.*, 2014[25]
Ag NP-based FCMs							
PE films	No	—	Simulant (distilled water) Food (malt broth extract, apple juice)	AAS	1–1.9 0.26–0.38 (malt broth) 0.2–0.25 (apple juice)	Not investigated	Nobile *et al.*, 2004[26]
PP container	Yes	—	Simulant (3% acetic acid) Food (apple sauce, pizza, lasagna)	ICP-MS	None (Ag, Sn, Ba) <0.002 (Ag)	No	Chaudhry *et al.*, 2008[21]

(*continued*)

Table 9.1 (*continued*)

Polymer matrix	Commercially available	NP content[c]/mg kg^{-1}	Food/simulant	Analytical technique	Migration[c]/mg kg^{-1}	NPs detected	Reference
PLA–clay	No	340–3400	0.03% HNO_3	Voltammetry	0.036	Not investigated	Busolo *et al.*, 2010[27]
PE bags	Yes	100	Simulant (pure water, 4% acetic acid, 95% ethanol, hexane)	SEM-EDX AAS	0.0045–0.024	Yes, 300 nm (SEM) spherical as in product	Huang *et al.*, 2011[28]
PE film	Yes	234	Simulant (3% acetic acid, 95% ethanol)	ICP-MS	3.9–13 (acetic acid) 0.5–0.6 (ethanol)	Not investigated	Song *et al.*, 2011[29]
PVC films	No	5000–50 000	Food (raw chicken breast)	ICP-MS	0.06–2.76	Not investigated	Cushen *et al.*, 2013[30]
LDPE bag and PP containers	Yes	3300	Simulant (3% acetic acid, 50% ethanol)	SEM-EDX spICP-MS	0.001–0.019	Yes (SEM, 10–200 nm)	Echegoyen and Nerin, 2013[31]
PLA composites	No	—	Simulant (10% ethanol, isooctane)	ICP-MS	0.0018–0.004 (ethanol) 0.00028–0.0012 (isooctane)	Not investigated	Fortunati *et al.*, 2013[32]
LDPE films	No	1–22	Simulant (distilled water, 3% acetic acid, 10% ethanol) Food (apple juice)	AAS	0.4–1 0.6–1.4	Not investigated	Jokar *et al.*, 2014[33]
Nano-Ag–TiO_2 polymer composite	Yes	—	Food (various)	ICP-MS	0.002–0.0057	Not investigated	Metak *et al.*, 2013[34]

PP containers PE bags	Yes	9–37	Simulant (water, 3% acetic acid, 10% ethanol, olive oil)	TEM-EDX spICP-MS	PP containers: 0.0057 (acetic acid) 0.0029 (water and ethanol) None (olive oil) PE bags: <0.0003	Yes (12% of migrated Ag, SEM 20–100 nm, spICP-MS 100–350 nm)	von Goetz *et al.*, 2013 [35]
LDPE films	No	50–250	Simulant (3% acetic acid, 10% ethanol, 95% ethanol, isooctane)	ICP-MS	0.014–6	Not investigated	Bott *et al.*, 2014 [36]
PE films	No	—	Simulant (distilled water, 3% acetic acid)	TEM ICP-MS	0.05–0.5 (Ag NPs) 0.0003–0.05 (Ag ion)	Yes (TEM)	Cushen *et al.*, 2014 [37]
PE composite	No	5000	Food (raw chicken breast)	ICP-MS	0.018–0.03	Not investigated	Cushen *et al.*, 2014 [38]
LDPE film (Ag/ZnO NP)	No	—	Simulant (aqueous, no further details provided)	ICP-MS	<0.001 (Ag) 2.44 (Zn)	Not investigated	Panea *et al.*, 2014 [39]
HDPE bag	Yes	28	Simulant (distilled water, 3% acetic acid, 10% ethanol, 95% ethanol)	AF4-ICP-MS SEM ICP-MS	0.017 (30% as Ag NPs)	Yes (AF4-ICP-MS, 40–60 nm)	Artiaga *et al.*, 2015 [40]
LDPE	No	5000–50 000	Simulant (3% acetic acid, 10% ethanol)	SEM ICP-AES	1.3	Yes (SEM, 215–945 nm)	Hannon *et al.*, 2015 [41]

(continued)

Table 9.1 (*continued*)

Polymer matrix	Commercially available	NP content[c]/mg kg^{-1}	Food/simulant	Analytical technique	Migration[c]/mg kg^{-1}	NPs detected	Reference
LDPE PP PES	Yes	7–36	Simulant (distilled water, 3% acetic acid)	TEM ICP-MS	0.0038	No	Addo Ntim *et al.*, 2015[42]
PE	Yes	29–34	Food (chicken meat)	ICP-MS	None	Not investigated	Gallocchio *et al.*, 2016[43]
PE HDPE	Yes	1–31	Simulant (distilled water, 3% acetic acid, 10% ethanol)	TEM-EDX spICP-MS	0.0019	Yes (spICP-MS, 23–89 nm)	Mackevica *et al.*, 2016[44]
PC	Yes	28–62	Simulant (distilled water, 3% acetic acid, 10% ethanol, 95% ethanol)	SEM-EDX spICP-MS	0.011	Yes (spICP-MS, 18–30 nm)	Ramos *et al.*, 2016[45]
PP	Yes	30	Simulant (distilled water, 4% acetic acid, hexane)	SEM-EDX ICP-MS	0.4–1.3	Yes (SEM, 40–400 nm)	Liu *et al.*, 2016[46]
TiN NP-based FCMs							
LDPE	No	100–1000	Simulant (3% acetic acid, 95% ethanol, isooctane)	TEM ICP-MS	<0.00024 (acetic acid) <0.0001 ethanol, isooctane)	No	Bott *et al.*, 2014[47]
Carbon black NP-based FCMs							
LDPE PS	No	25 000–50 000	Simulant (3% acetic acid, 95% ethanol, isooctane)	AF4-MALS	<0.0006	No	Bott *et al.*, 2014[48]

SiO_2 NP-based FCMs							
LDPE films	No	5000–30 000	Simulant (3% acetic acid, 95% ethanol, isooctane)	AF4-MALS	<0.0003	No	EFSA, 2014[49]
ZnO NP-based FCMs							
LDPE films	No	20 000–30 000	Simulant (3% acetic acid, 95% ethanol, isooctane)	ICP-AES ICP-MS	7.6–17.3 (acetic acid) 0.08 (ethanol) <0.1 (isooctane)	Not investigated	EFSA, 2015[50]
PP	Yes	200	Simulant (distilled water, 4% acetic acid, hexane)	SEM-EDX ICP-MS	0.001–0.01	Yes (SEM, 50–300 nm)	Liu *et al.*, 2016[51]
TiO_2 NP-based FCMs							
PE	Yes	250	Simulant (3% acetic acid, 50% ethanol)	ICP-MS LSPA	0.0014–0.012 (acetic acid) 0.0005–0.001 (ethanol)	Yes (LSPA in acetic acid, 50–150 nm)	Lin *et al.*, 2014[52]

[a]Abbreviations of materials: HDPE, high-density polyethylene; LDPE, low-density polyethylene; PC, polycarbonate; PE, polyethylene; PES, polyether sulfone; PET, poly(ethylene terephthalate); PP, polypropylene; PLA, poly(lactic acid); PS, polystyrene; NP nanoparticle.
[b]Abbreviations of analytical techniques: AAS, atomic absorption spectrometry; AES, atomic emission spectrometry; AF4, asymmetric flow-field flow fractionation; ICP, inductively coupled plasma; MS, mass spectrometry; spICP, single-particle ICP; LSPA, laser particle size analysis; SEM, scanning electron microscopy; TEM, transmission electron microscopy; EDX, energy-dispersive X-ray spectrometry; MALS, multi-angle light scattering.
[c]Conversion from units of mg dm^{-2} into mg kg^{-1} food assumed a contact area of 6 dm^2 kg^{-1} food. This assumption is used by European authorities when assessing results from migration testing into foods or food simulants.[53].

or coatings. Overall, the results from these studies suggest that the production method of the nanocomposites (*e.g.* incorporation or coating, surfactant modification), starting silver concentration, temperature, time and choice of contact medium are all factors that may have an effect on the extent of silver ion migration into food simulants. In general, the rate of migration increases when nanosilver is coated onto the food packaging material or when surfactants are added, when the storage temperature and length of storage increase and when the acidity of the contact medium increases. There appears to be a specific length of storage after which a steady-state release of silver is achieved. This is supported also by a repeat contact migration experiment, which found that silver migration decreased considerably (by an order of magnitude) after first contact.[16]

The specific migration of a substance into food depends on several parameters: the polymer, the food intended to come into contact with that polymer, the surface area-to-volume ratio between the polymer and the food and the contact time and temperature. If the food is already packed, the food is analyzed to check for compliance of the specific migration of the substance. If not, the specific migration can be tested by selecting the relevant food simulants representing the food with which the material or article is intended to come into contact. Worst-case foreseeable use test conditions need to be selected such as contact time and temperature. Several of the studies listed in Table 9.1 attempted to investigate whether nanoparticles *per se* migrate from nanoparticle-containing polymer composites into food simulant solutions. The data for nanosilver in Table 9.1 show that mixed results were obtained. While nanoparticles have been detected in a number of studies, a theoretical model by Simon *et al.* predicts that migration of nanoparticles from packaging to food by simple diffusion is possible only for very small nanoparticles, *i.e.* 1 nm, that are embedded in a polymer matrix that has a low dynamic viscosity, *e.g.* low-density polyethylene (LDPE).[17] Silver nanoparticles detected in the studies listed in Table 9.1 range in size from 10 to 950 nm, clearly larger than the possible size migration limit of around 1 nm. The fraction of nanoparticulate Ag relative to total Ag ranged from below 1% to 70%.[31,35,45] It is also noteworthy that all but one of the studies using techniques that are able to identify nanoparticulate Ag reported the presence of Ag-nanoparticles in food simulants. If so many studies identify nanoparticles in the migration matrix, then other release mechanisms apart from diffusion should be at work. A review by Noonan *et al.* presented a number of release mechanisms that may result in the migration of nanoparticles from FCMs.[18]

Desorption is a release mechanism that is restricted to nanoparticles located on the material or substrate surface. In that case, the release of particles is controlled by electrostatic interactions between the nanoparticle and the polymer substrate. It is important to point out that in desorption, unlike in the diffusion mechanism, mobility of the nanoparticles within the matrix is not a limiting factor for release and therefore particle size and shape are not major considerations. Of more importance here are the characteristics of the migration liquid (pH, ionic strength), the temperature, the fluid velocity and vibrations. Diffusion is the mechanism that has traditionally been used to describe contaminant transport from a material to the contact medium. In the case of nanoparticles in an FCM, the nanoparticles first have to migrate inside the FCM to the interfacial region before they are released into the external medium or migrating matrix (food or food simulant). Classical contaminants are far smaller than most nanoparticles and, since the diffusion coefficient decreases exponentially with size, this results in very small diffusion coefficients for nanoparticles with sizes larger than a few nanometers.[17,47] Dissolution involves the transformation of a nanoparticle from its particulate form into its ionic constituents. According to Noonan *et al.*,[18] there is some debate about whether nanoparticles migrate to the surface and are then dissolved or whether the ions desorb from the nanoparticle surface while it is still dispersed in the FCM matrix. Von Goetz *et al.* observed a significant decrease in migration in repeated migration tests already after the first use.[35] This finding suggests that release of silver occurred mainly from the uppermost layer of the FCM where the food simulant can enter the porous FCM surface. A secondary formation of nanoparticles from migrating ions, as in the case of chemical reduction of Ag ions to Ag nanoparticles, could not be excluded in the literature, but is seen as less likely.[35,44] Degradation of the FCM matrix by chemical or mechanical decomposition may lead to the release of embedded nanoparticles. Decomposition can be caused by physical abrasion, heating, exposure to UV radiation and hydrolysis, or by photoactive nanoparticles (such as TiO_2) that can generate reactive oxygen species in response to exposure to UV radiation.[19] None of the studies in Table 9.1 attempted to explain the migration mechanisms of nanoparticles and these therefore remain hypothetical. To understand better the mechanisms and make safety assessments of the migration of nanoparticles from FCMs to food, there is a need for high-quality analytical methods that can distinguish between the different release mechanisms.

9.3 Analytical Challenges of Migration Testing

Migration studies involving nano-enabled FCMs often face challenges associated with detecting, characterizing and quantifying ENMs and the ability to distinguish them from ionic species. In conventional migration testing of molecular substances, an analytical method is needed to determine whether a certain substance is present in the migration matrix and what its concentration is. In the case of migrating nanoparticles the situation is more complex. The analytical method not only needs to determine the presence and concentration of the nanoparticles, but should also be able to discriminate between nanoparticles and ionic material of the same chemical composition and to characterize the properties of the particles such as size, shape and agglomeration status. Currently, no single analytical technique can fulfill all of these requirements. From Table 9.1, it is clear that, to assess the release of nanoparticles from FCMs into aqueous matrices, a combination of particle size separation techniques, total element quantitation methods and microscopic identification techniques have been applied.

SEM and TEM are the most widely used techniques to determine the size of particles and visualize their morphology. To obtain particle size distributions with a statistical assurance, a large number of particles need to be counted, which can be a problem since the particle concentrations in migration liquids (simulants) are generally low. The particle concentrations required for detecting a sufficient number of particles with electron microscopic techniques are typically in the mg L^{-1} range. Concentration techniques such as ultracentrifugation and evaporation can be used to overcome this problem; however, these sample preparation methods can lead to artifacts, such as agglomeration of nanoparticles and precipitation of ions. Energy-dispersive X-ray spectrometry (EDX) is used in combination with SEM, and sometimes TEM, to determine the elemental composition of detected particles. Since the EDX bundle is often significantly larger than the nanoparticle itself, the EDX spectrum also contains a lot of background. Samples containing water or other liquids can be observed by environmental SEM (ESEM) or environmental TEM (ETEM) systems, which allow a higher vapor pressure around the sample to be maintained at the expense of maximum resolution.[54,55] Drying samples for classical SEM/TEM is another option; however, this step can also lead to artifacts such as induced agglomeration and precipitation. Sample preparation methods that were used in the studies listed in Table 9.1 include preconcentration by evaporation followed by air drying on the SEM grid[45] and drop casting of migration solutions on the SEM grid followed by air drying.[31,41] The major challenge with electron microscopy is not so much the sample preparation but rather the low nanoparticle concentration in the migration liquids.

Laser light scattering has also been used and, in the form of dynamic light scattering, it is one of the most commonly used tools for determining the hydrodynamic diameter of particles. However, it is limited by its inability to characterize polydispersed particles in solution reliably. A size separation technique such as asymmetric flow-field flow fractionation (AF4) presents a solution to the polydispersity problems faced by dynamic light scattering analysis by allowing the separation of particles into monodispersed fractions prior to light scattering detection. AF4 is mostly combined with multi-angle light scattering (MALS) to size nanoparticles in an aqueous matrix. Bott *et al.* and EFSA used this technique to determine carbon black nanoparticles in LDPE and PS[48] and SiO_2-based nanoparticles in LDPE, respectively.[49] The separation process in AF4 combines the effects of a laminar flow profile with an exponential concentration profile of analyte components caused by their interactions with a cross-flow applied perpendicular to the flow of the carrier liquid as shown in Figure 9.1.

AF4 (and other field-flow fractionation techniques) have been interfaced with various analytical techniques such as ICP-MS and ICP atomic

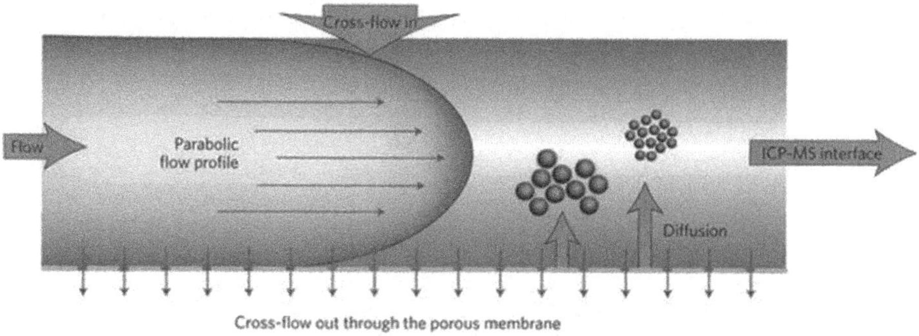

Figure 9.1 In AF4, the liquid in the separation channel has a parabolic flow profile. The cross-flow will push particles to the membrane that forms the porous channel bottom. The particles diffuse back from the membrane, with smaller particles diffusing more than larger particles. As a consequence, smaller particles will be more prevalent in the high-flow region of the parabolic flow profile and elute earlier than larger particles.

emission spectrometry (ICP-AES) for effective detection, characterization and quantification of metal and metal oxide ENMs. AF4 coupled with ICP-MS was used by Artiaga *et al.* to study Ag nanoparticles in migration matrices from commercial food containers.[40] Before analysis, the migration solution was dried by evaporation and resuspended in 0.5 mL of the AF4 carrier liquid, sonicated and filtered. Ag nanoparticles with diameters in the range 40–60 nm were determined in the migration liquids. For AF4-ICP-MS, the sample mass concentrations need to be in the $\mu g\ L^{-1}$ to $mg\ L^{-1}$ range. Particle sizes that can be determined with AF4-ICP-MS are in the range 5–500 nm. In general, obtaining quantitative size information by AF4 is challenging and therefore the technique is often combined with MALS in a AF4-MALS-ICP-MS configuration. In that case, AF4 will separate particles according to size, MALS will determine the hydrodynamic particle size and ICP-MS will quantify the particles. Schmidt *et al.* used this technique to determine the presence of nanoclay particles from a PLA material containing a nanoclay.[22]

Traditional ICP-MS has been extensively used for the quantitative determination of ionic species in migration studies because of its high sensitivity. The main limitation of traditional ICP-MS is that it can only determine total element concentrations. Nowadays, single-particle ICP-MS (spICP-MS) is a powerful characterization tool, particularly in aqueous media, as a result of its ability to determine simultaneously the concentration and size distribution of nanoparticles. In this technique, the sample, an aqueous suspension, is introduced continuously into an ICP-MS system that acquires data with high time resolution. Before introduction, the sample is diluted to such an extent that particles enter the plasma one at a time. Following nebulization, a fraction of the nanoparticles enter the plasma, where they are vaporized and the individual atoms are ionized, resulting in a cloud of ions as indicated

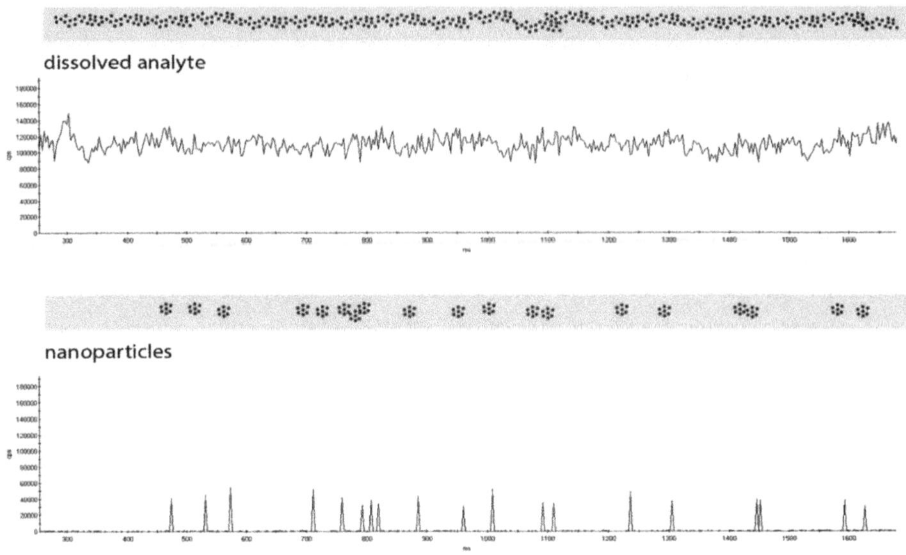

Figure 9.2 The upper trace shows the situation in traditional ICP-MS: a dissolved analyte enters continuously, producing a constant signal. The lower trace shows the situation for spICP-MS: clouds of ions enter the mass spectrometer, resulting in a discontinuous signal where each peak is formed by one particle.

in Figure 9.2. This cloud of ions is sampled by the mass spectrometer and detected as a signal pulse in the detector. A typical run time is 60 s and produces a time scan.

Under certain conditions (sufficiently dilute samples, known and fixed composition of the particles, *etc.*), the number of pulses detected per second is directly proportional to the particle number concentration in the sample while the intensity of the signal pulse is directly proportional to the mass of the detected nanoparticle. Assuming a certain particle shape (*e.g.* spherical) and composition, one can calculate the diameter of the particle. The method has the advantages of being fast, number based and, by virtue of the mass spectrometric detection, very selective. The technique is also very sensitive and typical particle concentrations for obtaining a particle size distribution are in the range 10^6–10^8 particles L^{-1}, which corresponds to particle mass concentrations in the ng L^{-1} range. The technique can not only detect particles at very low concentrations, but can also differentiate between nanoparticulate and ionic forms of the analyte. Sample preparation for spICP-MS involves, in the case of migration liquids, direct analysis without the need for sample preparation[35,44] or simple dilution with ultrapure water.[45] Although this technique holds great promise towards becoming the technique of choice for the complete detection, characterization and quantification of nanoparticles in solution, it is limited by the inability to detect particles of size <20 nm. Figure 9.3 shows the spICP-MS time

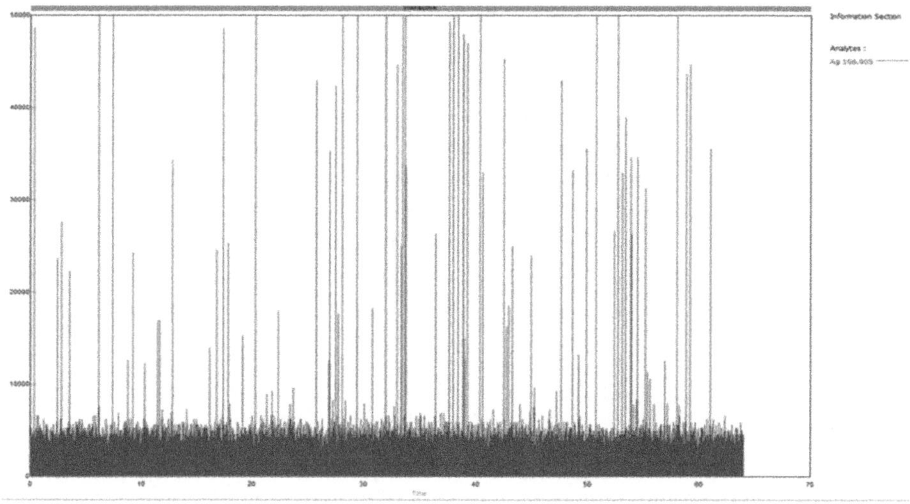

Figure 9.3 spICP-MS time scan of the migration matrix (distilled water) of plastic wire that was enabled with silver nanoparticles and intended for 3D printers. The spikes are individual silver nanoparticles and the broad band at the bottom is ionic silver.

scan of the migration matrix (distilled water) of plastic wire enabled with Ag nanoparticles with sizes in the range 60 nm and intended for the use in 3D printers. The "spikes" in the time scan are the detected Ag nanoparticles and the broad noise band at the bottom reflects the presence of ionic Ag. The ratio between ionic and nanoparticulate Ag can be determined directly from the analysis data.

9.4 Conclusion and Outlook

This chapter has presented some of the analytical possibilities and difficulties in determining the release of nanoparticles from FCMs, with an overview of published studies on migration testing. The determination of the nanoparticle release of FCMs is of importance for making assessments of the likely exposure of consumers to nanoparticles from dietary sources, especially since food packaging is the most likely food-related application of nanomaterials. The evaluation of the literature revealed that most studies have dealt with the release of silver nanoparticles and a much smaller number with the release of nanoclays and other nanomaterials. Methods that have been used to detect nanoparticle releases include SEM/TEM techniques to detect particles, AAS, ICP-AES and especially ICP-MS to determine total element concentrations and combined techniques such as AF4-MALS-ICP-MS to detect nanoparticle size and concentration. In recent years, spICP-MS has been introduced as a technique that can determine nanoparticulate and ionic material in one analysis. spICP-MS may also produce information on whether nanoparticles migrate as whole particles or as dissolved ions. Even in cases in which whole particles are observed, there is still the question of whether such particles are released in a particular manner or whether they originate from a secondary formation from released ions. So far, the migration studies found in the literature ignored release mechanism questions. Hence improved analytical techniques are required to answer such questions. In addition, a general challenge when analyzing nanoparticles released from FCMs is the lack of validated methods and of suitable reference materials to ensure accuracy of the results with respect to determined nanoparticle sizes and concentrations.

References

1. Q. Chaudhry, M. Scotter, J. Blackburn, B. Ross, A. Boxall, L. Castle, E. Aitken and R. Watkins, Applications and implications of nanotechnologies for the food sector, *Food Addit. Contam., Part A*, 2008, **25**, 241–258.
2. N. Bumbudsanpharoke and S. Ko, Nano-food packaging: an overview of market, migration research and safety regulations, *J. Food Sci.*, 2015, **80**, R910–R923.
3. Q. Chaudry, L. Castle and R. Watkins, *Nanoscience & Nanotechnology Series No 42: Nanotechnologies in Food*, The Royal Society of Chemistry, London, UK, 2017.
4. EC, European Commission Regulation (EU) No 10/2011 of 14 January 2011 on plastic materials and articles intended to come into contact with food, *Off. J. Eur. Communities: Legis.*, 2011, **12**, 1–89.
5. M. Jokar, G. A. Pedersen and K. Loeschner, Six open questions about the migration of engineered nano-objects from polymer-based food-contact materials: a review, *Food Addit. Contam., Part A*, 2017, **34**, 434–450.
6. C. Silvestre and S. Cimmino. *Ecosustainable Polymer Nanomaterials for Food Packaging*, CRC Press, New York, US, 2013.
7. R. Peters, P. Brandhoff, S. Weigel, H. Marvin, H. Bouwmeester. H. Aschberger, H. Rauscher, V. Amenta, M. Arena, F. Botelho Moniz, S. Gottardo and A. Mech, *Inventory of Nanotechnology Applications in the Agricultural, Feed and Food Sector*, EFSA supporting publication, 2014:EN-621.
8. N. Pradhan, S. Singh, N. Ojha, A. Shrivastava, A. Barla, V. Rai and S. Bose, Facets of Nanotechnology as Seen in Food Processing, Packaging and Preservation Industry, *BioMed Res. Int*, 2015, DOI: 10.1155/2015/365672.
9. M. Benelmekki, M. Torrell, E. Xuriguera, F. Vaz and V. Teixeira, Structure and properties of silver clusters implanted in PET by PVD sputtering for active packaging applications, *J. Nanopart. Res.*, 2012, **18**, 105–116.
10. A. Martínez-Abad, G. Sánchez, J. M. Lagaron and M. J. Ocio, Development and characterization of silver-based antimicrobial ethylene-vinyl alcohol copolymer (EVOH) films for food-packaging applications, *J. Agric. Food Chem.*, 2012, **60**, 5350–5359.
11. E. Llorens, A. Llorens, E. Lloret, P. A. Picouet, R. Trbojevich and A. Fernandez, Metallic-based micro and nanocomposites in food contact

materials and active food packaging, *Trends Food Sci. Technol.*, 2012, **24**, 19–29.

12. M. Carbone, D. Tommasa Donia, G. Sabbatella and R. Antiochia, Silver nanoparticles in polymeric matrices for fresh food packaging, *J. King Saud Univ., Sci.*, 2016, **28**, 273–279.
13. M. A. Busolo, P. Fernandez, M. J. Ocio and J. M. Lagaron, Novel silver-based nanoclay as an antimicrobial in polylactic acid food packaging coatings, *Food Addit. Contam., Part A*, 2010, **27**, 1617–1626.
14. A. Fernandez, E. Soriano, P. Hernandez-Munoz and R. Gavara, Migration of Antimicrobial Silver from Composites of Polylactide with Silver Zeolites, *J. Food Sci.*, 2010, **75**, E186–E193.
15. D. Halder, A. Mitra, S. Bag, U. Raychaudhuri and R. Chakraborty, Study on Gelatin-Silver Nanoparticle Composite Towards the Development of Bio-Based Antimicrobial Film, *J. Nanosci. Nanotechnol.*, 2010, **11**, 10374–10378.
16. A. Mackevica, M. E. Olsson and S. F. Hansen, Silver nanoparticle release from commercially available plastic food containers into food simulants, *J. Nanopart. Res.*, 2016, **18**, 1–11.
17. P. Simon, Q. Chaudhry and D. Bakos, Migration of engineered nanoparticles from polymer packaging to food – a physicochemical view, *J. Food Nutr. Res.*, 2008, **47**, 105–113.
18. G. O. Noonan, A. J. Whelton, D. Carlander and T. V. Duncan, Measurement methods to evaluate engineered nanomaterial release from food contact materials, *Compr. Rev. Food Sci. Food Saf.*, 2014, **13**, 679–692.
19. D. L. Wang, S. S. Wason, L. P. Sung, I. H. Tseng, C. J. Bouis and R. Fernando, Effect of TiO_2 pigment type on the UV degradation of filled coatings, *J. Coat. Technol. Res.*, 2011, **8**, 19–33.
20. M. Avella, J. J. De Vlieger, M. E. Errico, S. Fischer, P. Vacca and M. G. Volpe, Biodegradable starch/clay nanocomposite films for food packaging applications, *Food Chem.*, 2005, **93**, 467–474.
21. Q. Chaudhry, L. Castle, E. Bradley, J. Blackburn, R. Aitken, A. Boxall. *Assessment of Current and Projected Applications of Nanotechnology for Food Contact Materials in Relation to Consumer Safety and Regulatory Implications*, Final report Project A03063, Food Standards Agency, 2008, http://www.food.gov.uk/sites/default/files/a03063.pdf.
22. B. Schmidt, J. H. Petersen, C. Bender Koch, D. Plackett, N. R. Johansen, V. Katiyar and E. H. Larsen, Combining asymmetrical flow field-flow fractionation with light-scattering and inductively coupled plasma mass spectrometry for characterization of nanoclay used in biopolymer nanocomposites, *Food Addit. Contam., Part A*, 2009, **26**, 1619–1627.
23. M. Mauricio-Iglesias, S. Peyron, V. Guillard and N. Gontard, Wheat gluten nanocomposite films as food-contact materials: Migration tests and impact of a novel food stabilization technology (high pressure), *J. Appl. Polym. Sci.*, 2010, **116**, 2526–2535.
24. M. Farhoodi, S. M. Mousavi, R. Sotudeh-Gharebagh, Z. Emam-Djomeh and A. Oromiehie, Migration of Aluminum and Silicon from PET/Clay

Nanocomposite Bottles into Acidic Food Simulant, *Packag. Technol. Sci.*, 2014, **27**, 161–168.
25. S. Maisanaba, D. Gutiérrez-Praena, M. Puerto, M. Llana-Ruiz-Cabello, S. Pichardo, R. Moyano, A. Blanco, M. Jordá-Beneyto and Á. Jos, In vivo toxicity evaluation of the migration extract of an organomodified clay-poly(lactic) acid nanocomposite, *J. Toxicol. Environ. Health, Part A*, 2014, **77**, 731–746.
26. M. A. Nobile, M. Cannarsi, C. Altieri, M. Sinigaglia, P. Favia, G. Iacoviello and R. D'Agostino, Effect of Ag-containing Nano-composite Active Packaging System on Survival of Alicyclobacillus acidoterrestris, *J. Food Sci.*, 2004, **69**, E379–E383.
27. M. A. Busolo, P. Fernandez, M. J. Ocio and J. M. Lagaron, Novel silver-based nanoclay as an antimicrobial in polylactic acid food packaging coatings, *Food Addit. Contam., Part A*, 2010, **27**, 1617–1626.
28. Y. Huang, S. Chen, X. Bing, C. Gao, T. Wang and B. Yuan, Nanosilver migrated into food-simulating solutions from commercially available food fresh containers, *Packag. Technol. Sci.*, 2011, **24**, 291–297.
29. H. Song, B. Li, Q. B. Lin, H. J. Wu and J. Chen, Migration of silver from nanosilver-polyethylene composite packaging into food simulants, *Food Addit. Contam., Part A*, 2011, **49**, 1–5.
30. M. Cushen, J. Kerry, M. Morris, M. Cruz-Romero and E. Cummins, Migration and exposure assessment of silver from a PVC nanocomposite, *Food Chem.*, 2013, **139**, 389–397.
31. Y. Echegoyen and C. Nerín, Nanoparticle release from nano-silver antimicrobial food containers, *Food Chem. Toxicol.*, 2013, **62**, 16–22.
32. E. Fortunati, M. Peltzer, I. Armentano, A. Jiménez and J. M. Kenny, Combined effects of cellulose nanocrystals and silver nanoparticles on the barrier and migration properties of PLA nano-biocomposites, *J. Food Eng.*, 2013, **118**, 117–124.
33. M. Jokar and R. Abdul Rahman, Study of silver ion migration from melt-blended and layered-deposited silver polyethylene nanocomposite into food simulants and apple juice, *Food Addit. Contam., Part A*, 2014, **31**, 734–742.
34. A. Metak and T. Ajaal, Investigation on polymer based nano-silver as food packaging materials, *Int. J. Biol. Vet. Agric. Food Eng.*, 2013, **7**, 772–778.
35. N. von Goetz, L. Fabricius, R. Glaus, V. Weitbrecht, D. Günther and K. Hungerbühler, Migration of silver from commercial plastic food containers and implications for consumer exposure assessment, *Food Addit. Contam., Part A*, 2013, **30**, 612–620.
36. J. Bott, A. Störmer, R. Franz. A comprehensive study into the migration potential of nano silver particles from food contact polyolefins, in *Chemistry of Food, Food Supplements and Food Contact Materials: From Production to Plate*, American Chemical Society, 2014, pp. 51–70.
37. M. Cushen, J. Kerry, M. Morris, M. Cruz-Romero and E. Cummins, Silver migration from nanosilver and a commercially available zeolite filler

polyethylene composites to food simulants, *Food Addit. Contam., Part A*, 2014, **31**, 1132–1140.

38. M. Cushen, J. Kerry, M. Morris, M. Cruz-Romero and E. Cummins, Evaluation and simulation of silver and copper nanoparticle migration from polyethylene nanocomposites to food and an associated exposure assessment, *J. Agric. Food Chem.*, 2014, **62**, 1403–1411.
39. B. Panea, G. Ripoll, J. González, Á. Fernández-Cuello and P. Albertí, Effect of nanocomposite packaging containing different proportions of ZnO and Ag on chicken breast meat quality, *J. Food Eng.*, 2014, **123**, 104–112.
40. G. Artiaga, K. Ramos, L. Ramos, C. Cámara and M. Gómez-Gómez, Migration and characterisation of nanosilver from food containers by AF4-ICP-MS, *Food Chem.*, 2015, **166**, 76–85.
41. J. C. Hannon, J. P. Kerry, M. Cruz-Romero, S. Azlin-Hasim, M. Morris and E. Cummins, Assessment of the migration potentialof nanosilver from nanoparticle coated low density polyethylene food packaging into food simulants, *Food Addit. Contam., Part A*, 2015, **33**, 167–178.
42. S. Addo Ntim, T. A. Thomas, T. H. Begley and G. O. Noonan, Characterisation and potential migration of silver nanoparticles from commercially available polymeric food contact materials, *Food Addit. Contam., Part A*, 2015, **32**, 1003–1011.
43. F. Gallocchio, V. Cibin, G. Biancotto, A. Roccato, O. Muzzulon, L. Carmen, B. Simone, L. Manodori, A. Fabrizi and I. Patuzzi, *et al.* Testing nano-silver food packaging to evaluate silver migration and food spoilage bacteria on chicken meat, *Food Addit. Contam., Part A*, 2016, **33**, 1063–1071.
44. A. Mackevica, M. E. Olsson and S. F. Hansen, Silver nanoparticle release from commercially available plastic food containers into food simulants, *J. Nanoparticle Res.*, 2016, **18**, 1–11.
45. K. Ramos, M. M. Gómez-Gómez, C. Cámara and L. Ramos, Silver speciation and characterization of nanoparticles released from plastic food containers by single particle ICPMS, *Talanta*, 2016, **151**, 83–90.
46. J. Liu, R. Guan, F. Lyu, M. Liu, J. Gao and G. Cao, Migration of Nanosilver into Food-simulating Solutions from Polypropylene Disposable Spoons, *Adv. J. Food Sci. Technol.*, 2016, **11**, 110–116.
47. J. Bott, A. Störmer and R. Franz, A model study into the migration potential of nanoparticles from plastics nanocomposites for food contact, *Food Pack. Shelf Life*, 2014, **2**, 73–80.
48. J. Bott, A. Störmer and R. Franz, Migration of nanoparticles from plastic packaging materials containing carbon black into foodstuffs, *Food Addit. Contam., Part A*, 2014, **31**, 1769–1782.
49. EFSA, Statement on the safety assessment of the substance silicon dioxide, silanated, FCM Substance No 87 for use in food contact materials, *EFSA J.*, 2014, **12**(6), 3712.
50. EFSA, Scientific opinion on the safety evaluation of the substance zinc oxide, nanoparticles, uncoated and coated with [3-(methacryloxy)propyl] trimethoxysilane, for use in food contact materials, *EFSA J.*, 2015, **13**, 4063, http://www.efsa.europa.eu/en/efsajournal/doc/4063.pdf.

51. J. Liu, J. Hu, M. Liu, G. Cao, J. Gao and Y. Luo, Migration and characterization of nano-zinc oxide from polypropylene food containers, *Am. J. Food Technol.*, 2016, **11**, 159–164.
52. Q. B. Lin, H. Li, H. N. Zhong, Q. Zhao, D. H. Xiao and Z. W. Wang, Migration of Ti from nano-TiO2-polyethylene composite packaging into food simulants, *Food Addit. Contam., Part A*, 2014, **31**, 1284–1290.
53. EFSA, Opinion of the Scientific Panel on Food Additives, Flavourings, Processing Aids and Materials in Contact with Food (AFC) on a request from the Commission related to a 7th list of substances for food contact materials (Question No EFSA-Q-2003-076, EFSA-Q-2004-144, EFSA-Q-2004-166,EFSA-Q-2004-082,EFSA-Q-2003-204,EFSA-Q-2003-205, EFSA-Q-2003-206) adopted on 29 March 2005 by written procedure, *The EFSA J.*, 2005, **201**, 1–28. http://www.efsa.europa.eu/en/efsajournal/doc/201a.pdf.
54. A. Dudkiewicz, K. Tiede, K. Loeschner, S. Jensen, E. Jensen and Wierzbicki, *et al.*, Characterization of nanomaterials in food by electron microscopy, *Trends Anal. Chem.*, 2011, **30**, 28–43.
55. G. Singh, C. Stephan, P. Westerhoff, D. Carlander and T. V. Duncan, Measurement methods to detect, characterize and quantify engineered nanomaterials in food, *Compr. Rev. Food Sci. Food Saf.*, 2014, **13**, 693–704.

CHAPTER 10

Solutions Commonly Applied in Industry and Outsourced to Expert Laboratories

THOMAS GUDE*

Grünaustrasse 23, 8953 Dietikon, Switzerland
*E-mail: thomas.gude@sqts.ch

10.1 Introduction

Food contact materials (FCMs) have been an increasing part of the work of third-party (expert) laboratories for several years. The handling of FCM issues needs a good understanding of two different "worlds" – the materials world and the food world. Typically, laboratories are specialized in either materials or food. In the past, this differentiation has been responsible for the fact that analytically, only minimum types of tests have been performed. This can be seen in the limited number of methods that are fully validated and standardized for the analysis of food simulants and/or food itself – see EN Series EN 13130.[1] Most of these methods are not based on mass spectrometry, either gas chromatography-mass spectrometry (GC-MS) or liquid chromatography-mass spectrometry (LC-MS), but mainly use classical detectors, *e.g.* UV detectors. As FCMs came increasingly into the focus of food safety discussions, MS detection also became more popular. Nowadays, with few exceptions, the analysis of FCMs in third-party (expert) laboratories is based on MS with a clear tendency towards high-resolution MS.

Food Chemistry, Function and Analysis No.10
Food Contact Materials Analysis: Mass Spectrometry Techniques
Edited by Michele Suman
© The Royal Society of Chemistry 2019
Published by the Royal Society of Chemistry, www.rsc.org

In addition to its methodological work, significant tasks of a third-party (expert) laboratory include consultancy regarding test conditions, starting with clear instructions on how to handle samples for chemical analysis for the purpose of determining trace level concentrations, the construction of the experimental setup (using migration cells *versus* total immersion, *etc.*) and finally the assessment of the results. These tasks are increasing more and more in the light of the discussion of non-intentionally added substances (NIASs), which is strongly linked to good manufacturing practice (GMP). The origin of samples (extracts, migration solutions or even food itself) and possible contamination during manufacture and transport or in the laboratory itself play a major role and have to be highlighted by laboratories. Modern assessment also increasingly includes structure elucidation if "unknown" signals are detectable. In the case when substances can be identified *via* MS, the results have to be assessed against several different legal requirements as at present there is no world-wide harmonization of FCMs.

The general workflow when handling FCMs can be divided into the following steps:

- Consultancy
    - defining analytical strategy: extraction, migration or analysis from food
    - legal background
    - defining simulants
    - defining time/temperature conditions
- Sample preparation/sample handling
    - avoiding contamination
    - samples in the laboratory
- Selection of analytical method
    - target *versus* non-target analysis
    - selection of MS detector
- Performing analytical tests
    - quality criteria
    - internal standards
- Assessment of results
    - data management
    - interpretation of data
    - confirmation of data.

The different steps are discussed in the following sections.

Following this step-by-step approach, some challenges arise that need solutions. However, some topics actually cannot be solved, *e.g.* different ionization techniques for LC-MS in parallel in one run within sufficient time and sensitivity ranges, as the technical background is not yet fully elucidated – see also Section 10.7. These challenges and gaps in knowledge have to be understood by expert laboratories and solutions have to be provided.

Finally, the application of different MS techniques in the field of FCMs will change in the future as the focus might shift more and more to the food itself rather than testing simulants. The future perspectives are discussed in a separate section.

10.2 Consultancy

10.2.1 Defining the Analytical Procedure

When dealing with FCMs in a third-party (expert) laboratory, everything starts with an order or an idea from of a client, which might not be in any way clear at the beginning. Therefore, the discussion should start with initial relevant questions about the sample itself and what can be achieved with various analytical techniques, especially MS.

The first question deals with the nature of the samples to be investigated. For example, there is a huge difference in testing a plastic-based and a paper-based FCM. If the question can be clarified, an analysis can be performed in selected food matrices to verify early findings of extraction or migration studies. Therefore, the initial decision for FCMs themselves will be made between extraction of a material *versus* the migration approach.

10.2.1.1 Extraction of Food Contact Materials

When extraction of FCMs is selected as the handling procedure, it is necessary to consider what kind of solvent should be used. Often extraction is seen as a worst-case approach if the content of a substance in an FCM is to be determined. However, this depends on the nature of the solvent – is it destructive (the material will be completely dissolved), is it only able to create a swelling of the material or is it only a "washing" on the surface? These three different approaches later have a significant impact on the detection of individual substances using MS. To keep it simple, the more destructive the solvent is towards the material, the more matrix interferences will show up later in the MS detection, which may lead to significant problems in terms of sensitivity and selectivity, even when using high-resolution MS. At present no harmonized procedure has been established. However, an expert group within the Association of Independent German Laboratories (VUP – Verband unabhängiger Prüflabore) published a first guideline to "Screening Products for SVHC (Substances of Very High Concern) According to the REACH Regulation,"[2] which is actually being transferred into DIN 51012-1 (Draft), which can be taken as a first approach. Although the background of this guideline is with SVHC substances at higher concentrations as typically occur in FCMs, it gives a good overview of how to perform extractions of materials. As a starting procedure for extractions for most FCMs, the solvent mixture cyclohexene–ethyl acetate (1 : 1) can be used if GC-MS is proposed. This mixture has a good polarity range to cover extracted substances detectable by GC-MS. If LC-MS is proposed, acetonitrile might be the choice of solvent as it is applicable with most LC methods. However, the extraction procedures will have major impacts on the final measurement uncertainty, which is highly influenced by the choice of the extraction solvent. A problem in extraction studies is that no recovery procedures can be applied as the substances cannot be incorporated into the FCM itself.

Extraction studies are therefore suitable only if a full content analysis can be performed with suitable sensitivity. In many cases, *e.g.* extraction of plastic pellets, the pellets will only be "washed." Further, with the extraction of pellets it is difficult to obtain a link to a final product as the total surface area of the pellets cannot be correctly determined and the amount of pellets produced is not known, *e.g.* from a foil.

To summarize, extractions are a rapid means to obtain initial results, but their interpretation is limited as usually a total content analysis is not performed.

10.2.1.2 Migration of Food Contact Materials

If the decision is clearly in favour of migration studies, it is obvious that the migration conditions are important, especially as the selection of the simulant has a major impact on the MS technique to be used later. All polar simulants such as water and 10% and 20% ethanol cannot be used directly in GC-MS-based applications – additional extraction with a non-polar solvent, *e.g.* cyclohexene–ethyl acetate, is needed. Simulants such as 50% and 95% ethanol and also isooctane and/or heptane can be used for both LC-MS and GC-MS. Finally, the use of vegetable oils and Tenax® always needs an additional extraction run when using LC-MS and GC-MS. Therefore, it is obvious that for later migration studies the most significant contributor to the measurement uncertainty lies in the migration study itself, with uncertainties between ±20% (all simulants with the exception of oil) and 30% (only for oil). Also, with migration studies no exhaustive recovery studies can be performed – as in extraction studies – but as migration solutions typically have fewer interferences, the uncertainty is mainly influenced only by the migration itself; for further details on simulants, see Section 10.2.2.1.

10.2.1.3 Determination in Food

As clear results often cannot be achieved in extraction and migration studies, another option would be the direct determination of target substances stemming from FCMs in food matrices. As here the "other world" is affected, the whole range of analytical food chemistry can be applied. The great advantage of the analysis of food is the ability to run recovery experiments by spiking the food sample. However, the most challenging disadvantage is the huge number of food matrix-based interferences and, of course, the time required to develop new methods.

Nevertheless, Article 18, Section 6 of the European Commission Regulation (EU) 10/2011 on plastic materials and articles intended to come into contact with food clearly states: "The results of specific migration testing obtained in food shall prevail over the results obtained in food simulant. The results of specific migration testing obtained in food simulant shall prevail over the results obtained by screening approaches."[3] This statement clearly anticipates a programme for the future – screening of FCMs in food itself.

10.2.2 Legal Background

10.2.2.1 Defining Simulants

Once the decision has been made between extraction, migration and/or analysis directly in food, it is important to define which legal background should be applied when running the experiments. As already mentioned for extractions, no clear rules have been defined as the procedures might be material dependent. For direct food analysis, individual methods have to be applied. However, for simulation experiments, we see a kind of similarity with significant differences in time/temperature conditions. Table 10.1 provides an overview of some important simulation conditions in terms of simulants.

It can be clearly seen from Table 10.1 that there are both similarities and differences worldwide. The differences have a significant impact on the choice and the results of analytical methods, here MS. If a simulation is performed with 95% ethanol or Tenax® it does not necessarily have an impact on the MS method used, but it clearly has an impact on the results. These differences must be appreciated by an experienced laboratory, otherwise companies might believe that if a test is performed under US FDA conditions it

Table 10.1 Simulants used based on different legal requirements.

China[a]	EU[b]	USA[c]	Mercosur[d]	Japan[e]	South Korea[f]
—	—	Water	Water	Water	Water
—	—	8% ethanol	—	—	—
10% ethanol	10% ethanol	—	10% ethanol	—	—
4% acetic acid	3% acetic acid	—	3% acetic acid	4% acetic acid	4% acetic acid
20% ethanol	20% ethanol	—	>10% ethanol	20% ethanol	—
50% ethanol	50% ethanol	50% ethanol	50% ethanol	—	—
95% ethanol	95% ethanol	—	95% ethanol	—	—
—	Isooctane	—	Isooctane	—	—
—	—	Heptane		Heptane	Heptane
Vegetable oil	Vegetable oil	—	Vegetable oil	—	—
—	Tenax®	—	Tenax®	—	—

[a]Chinese Standard GB 31604.1-2016.[4]
[b]Regulation (EU) 10/2011.[3]
[c]US Food and Drug Administration (FDA) 21 CFR 175.300 and 176.170.[5]
[d]Mercosur: Argentina, Brazil, Paraguay, Uruguay and Venezuela. GMC Resolution No. 3/92.[6]
[e]Food Sanitation Act No. 233.[7]
[f]Standards and specifications for food utensils, containers and packages.[8]

would be sufficient for, *e.g.*, the EU market. As in fact there are currently no trends towards harmonization, this means that third-party (expert) laboratories and others must be fully aware of all these differences.

10.2.2.2 Defining Time/Temperature Conditions

Concerning time/temperature conditions, third-party (expert) laboratories and others are facing the same problem of non-harmonization, with similar influences on the subsequent assessment. In the various national regulations and subregulations cited in Table 10.1, the simulation conditions vary from 30 min or less at room temperature, through 2 h and more at boiling temperatures, to long-term conditions up to 10 days at 60 °C.

It is obvious that if the time/temperature conditions differ then different results may be obtained, even if the same MS technique is used. To avoid any comparisons of such unreliable data, in any final report it is mandatory to specify the conditions chosen. As this is not done in many so-called Declaration of Compliance documents, which are often a regulatory requirement, users of such documents must be very careful when drawing conclusions.

10.3 Sample Preparation/Sample Handling

10.3.1 Avoiding Contamination

With the availability of modern analytical techniques based on MS and the increasing concern about NIASs, contamination of FCMs is becoming a more and more important issue. The key is to avoid contaminants as they may lead to misleading interpretations of results. Knowing this, it is very surprising that clients continue to send samples wrapped in, *e.g.*, recycled paper or similar instead of choosing inert wrappings. On the one hand, there is a huge impact if results obtained by highly sophisticated analytical systems show non-conformity. On the other, the necessary careful handling of samples is neglected. Therefore, in Table 10.2 some basic rules are defined to help in avoiding contamination when dealing with samples for FCM analysis.

First, concerning extraction or migration studies that have to be performed, virgin packaging material that has never been in contact with any food product should be used for samples for chemical analysis. Any direct contact with other plastic materials, stickers, tapes, transport foam chips or markings with a pen or marker can spoil the analysis!

10.3.2 Samples in the Laboratory

When samples have been properly treated in clients' hand and shipped in good condition, it is then clearly the duty of the laboratory to avoid laboratory contaminations. As this is part of good laboratory practice and described

Table 10.2 Avoiding FCM contamination.

1. Wrapping material – Al foil

Use normal household aluminium foil only Do not use PET–Al–PE or paper–Al from your factory

2. Wrapping material – bottles

Use glass bottles with screw caps (glass bottles should be heated to 200 °C and immediately closed to avoid contamination)

3. Closing bottles

Close the glass bottle with aluminium foil directly after sampling Screw the cap tightly

4. Sampling

Do not use any plastic material for sampling (*e.g.* plastic spoons) The best sampling materials are metal or glass spoons that have previously been cleaned with acetone

5. Wrapping

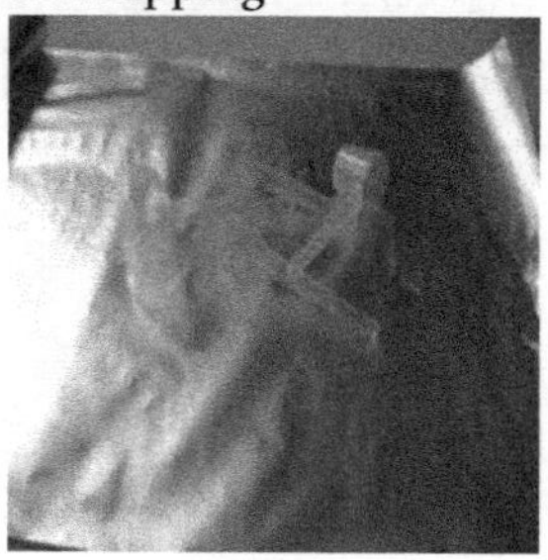

Wrap each sample separately in aluminium foil Do not write directly on the sample; instead, write on the wrapping (see below)

6. Foil samples

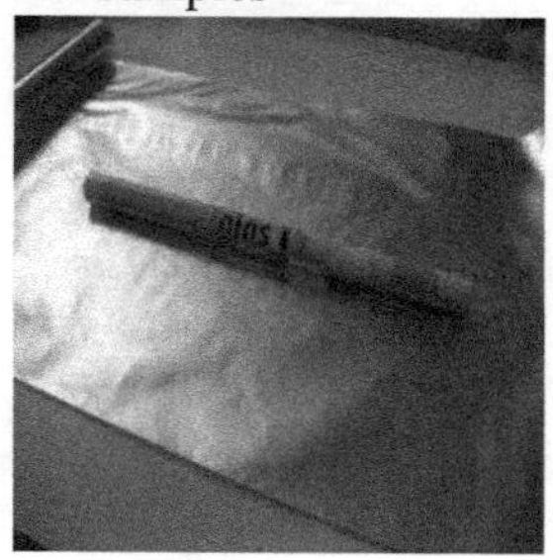

Keep samples of foil rolled-up (do not fold!) Supply at least 2–3 m of foil For large samples, cut the foil in half

(*continued*)

Table 10.2 (*continued*)

7. Labelling 	Wrap the aluminium foil tightly Label the sample with a sticker on the outside (aluminium foil)	8. Labelling 	Put the labels on the aluminium foil Do not label the sample directly on the material itself
9. Shipping 	When the samples are packed properly in aluminium foil, they can be shipped together in one box	10. Packing 	Only when the samples have been packed properly in aluminium foil can you foam chips be used to fill the box

several times elsewhere, it should be emphasized here that many items of laboratory equipment are made of plastic materials, which may be a possible source of contamination. Analysis of a laboratory blank (using the same equipment and solvents without sample) is mandatory. Special attention should also be paid to the laboratory labels used, *e.g.* a batch of thermopaper, which is used to create barcodes from a laboratory information management system, may contain bisphenol A or plasticizer such as a phthalate. Regular control of such typical laboratory aid equipment is also mandatory.

When the samples are ready for analysis, the predefined extraction and migration simulation procedures (see Section 10.2) can be applied.

10.4 Selection of Analytical Method

The selection of the analytical method clearly depends on the goal of the investigations. For some individual substances, methods exists that are based on classical detection systems such as GC-FID (flame ionization detection) or LC-UV detection. However, these methods are of limited use as they typically can be applied only to limited amounts of substances. Modern analytical strategies, however, try to cover as many substances as possible in a limited number of analytical runs. This can best be achieved with MS techniques. As shown in Figure 10.1, the methods used and proposed by ILSI[9] for *e.g.* NIAS analysis are mainly MS methods.

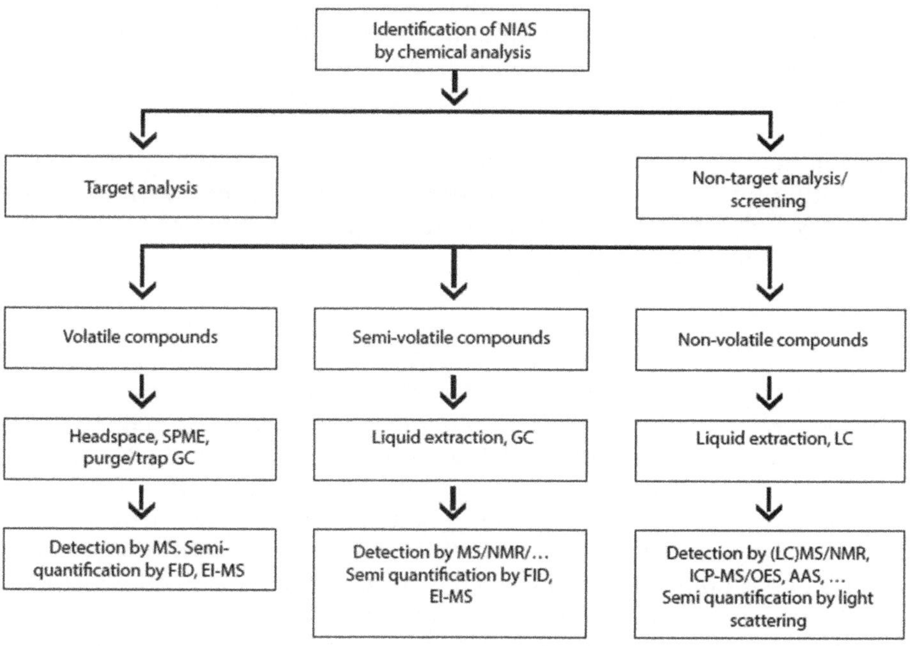

Figure 10.1 Chemical analysis methods to cover food contact materials.

10.4.1 Target *versus* Non-targeted Analysis

Nowadays there is a clear differentiation between target methods (single analyte *versus* multiple analytes) and non-target methods (general screening, NIAS identification).

10.4.1.1 Target Analysis

A few years ago, target analysis, especially the starting monomers for plastic-based FCMs, was the main goal in third-party (expert) laboratories. For substances with higher specific migration limits (SMLs), *e.g.* 7.5 mg kg^{-1} for terephthalic acid, a targeted method was used based on a classical LC-UV method. With the increasing introduction of mass spectrometers in many laboratories, those methods were replaced. As a next step, so-called multi-methods were implemented for the analysis of several materials, *e.g.* photoinitiators, acrylates, primary aromatic amines (PAAs), *etc.* With these multi-methods, laboratories were able to cover several analytes in parallel – the same trend as in classical food analysis of *e.g.* pesticides and/or veterinary drugs. The advantage is clear as the analytical effort and therefore costs could be minimized.

However, the target analysis mode became a clear competitor in so-called modelling software solutions. With this approach, computer programmers are able to predict a migration, especially for several plastics. Therefore, nowadays it can be concluded as a rule of thumb that if substances have SMLs higher than 1 mg kg^{-1}, the method of choice might be modelling. With target methods, the modelling solution might be confirmed. For substances at levels below 1 mg kg^{-1}, modelling might be also possible, but owing to production variations the uncertainty increases exponentially and confirmation is essential.

To summarize, target analysis is today a mixture of modelling and single- and/or multi-method approaches. Especially for substances with low SMLs (<1 mg kg^{-1}) or even non-listed substances, the non-target approach might be the better alternative.

10.4.1.2 Non-target Analysis

Non-target analysis became more and more popular with the increasing concern about NIASs and general contamination issues. With targeted methods, it is impossible to detect non-predefined substances analytically as they can be too numerous. This kind of task can only be performed with mass spectrometers with unit resolution or, even better, high-resolution instruments. However, non-targeted analysis has some weaknesses such as the continuing lack of harmonization and limited quantification – mainly the methods are semiquantitative and the interpretation of results is complex (see also Section 10.4.2.1). The advantage is also obvious, however – the maximum information can be achieved with a limited number of analytical runs. This is not only economically important, it is also nowadays essential to gain an overview of the NIAS situation.

The greatest challenge for non-targeted analysis in the near future is to establish better harmonized conditions to achieve better comparability between laboratories. Within ILSI Europe, some progress will be reported within the next few years as an expert group is working on such a harmonization.[10]

In addition, it is essential to develop better databases, as existing GC-MS databases are limited (mainly to known targets and less to NIASs) and LC-MS databases exist only in each individual laboratory. The future of non-targeted analysis depends on the willingness to set up such databases accessible to all interested laboratories.

10.4.2 Selection of Mass Spectrometric Detector

Just as typically the difference between target and non-target analysis has to be elucidated, so the choice of the most suitable MS detector also has to be made. For target analysis, classical quadrupole systems for coupled GC-MS and LC-MS are used; especially for LC-MS the triple quadrupole versions are the most favourable as they provide the highest capability for implementing multi-methods.

As quadrupole-based mass spectrometer are only able to operate in the unit mass resolution range, they are not very suitable for screening methods as typically used for the detection of NIASs. In some cases screening approaches can be based on quadrupole systems in the full-scan mode, but sensitivity might be a problem. This can be solved possibly by applying more concentration steps *e.g.* from the migration solution, but each concentration step may create artefacts or reaction products.

10.4.2.1 Unit-resolution MS

Unit-resolution mass spectrometers such as quadrupoles and ion traps are widely used for target analysis and have been extensively described. The advantages are clear as these systems offer high linearity and very good quantification possibilities, especially when used in the MS/MS mode by selecting single ions. Also, the lower cost compared with high-resolution mass spectrometers led to their adoption in many third-party (expert) laboratories. However, with the ongoing activity in the screening mode (NIAS detection) the disadvantages also become obvious. On the one hand quadrupole systems lose substantial sensitivity when operating in the scan mode. Although for GC-MS the typical databases are based on electron and chemical ionization (EI/CI) linked to the quadrupole technique and therefore a spectral library search is much easier, there is insufficient additional identification power as unit masses do not explain chemical structures adequately. The same is applicable with LC-MS systems. The highest sensitivity is achieved when operating in the MS/MS mode based on very few ion transitions, but the identification power, especially for unknown substances, is too low. In the scan mode, the sensitivity is also too low and also there are no or very limited databases available for mass spectral searches.

Based on these considerations, it can be concluded that unit-resolution mass spectrometers are most useful in the target analysis mode. Although it is possible to run them in a screening mode, the disadvantages are too numerous.

10.4.2.2 High-resolution MS

As addressed in the previous section, unit resolution is the method of choice when quantifying known (target) substances. However, in the area of FCMs non-target analysis is actually a challenge for third-party (expert) laboratories. For this purpose in the food and supply industries only high-resolution MS instruments such as time-of-flight (TOF) or Orbitrap systems are suitable. These systems offer a mass accuracy of less than 5 ppm, which may help in identifying known and unknown substances in the screening mode combined with high sensitivity. However, the greatest advantage and not just for third-party (expert) laboratories is the parallel measurement of all masses in one analytical run in contrast to predefined masses in quadrupole unit-resolution systems. This allows data to be created that can be reread again later in the case that new information becomes available. This can also be achieved with quadrupole systems operating in the full-scan mode, but the second major advantage is the possibility of elucidating structures by using exact masses. If in a more complex way TOF or Orbitrap systems are combined with an additional MS/MS capability such as Q-TOF or advanced Orbitrap instruments, the most powerful information tool for identification is then available. A disadvantage linked to high-resolution mass spectrometers is the slightly limited quantification power as the linearity is poorer compared with a unit-resolution quadrupole mass spectrometer. This disadvantage can be partly balanced by combining GC-MS with an FID detector and partly by combining LC-MS with a UV detector and/or an evaporative light scattering detector (ELSD) or by using a limited working concentration range.

In the following, some examples are highlighted to demonstrate the possibilities in routine work in a third-party (expert) laboratory.

10.4.2.3 GC-MS Example

Figure 10.2 shows a chromatogram of a 95% ethanol migration solution. Two signals were identified with very similar mass spectra. On comparing these mass spectra with a commercial MS library (unit resolution), more or less the same compound was identified: 4′-isopropylacetophenone ($C_{11}H_{14}O$) (CAS 645-13-6).

Even when comparing the high-resolution mass spectral profile against an MS library the same result was obtained. However, when using the exact

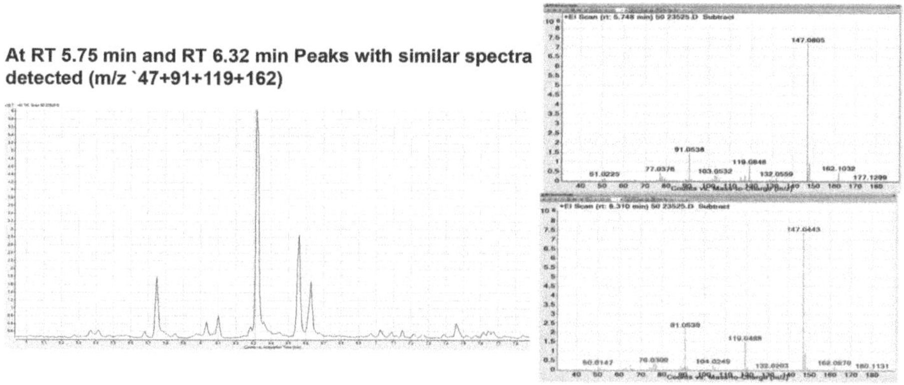

RT [min]	Summenformel	Masse (Soll) m/z
5.75	$C_{11}H_{13}O + H^+$	162.1032
6.21	$C_{10}H_{09}O_2 + H^+$	162.0670

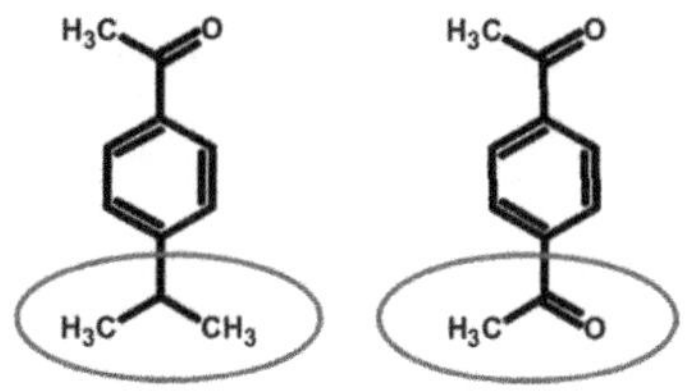

Figure 10.2 Example of high-resolution GC-MS.

mass of the molecular ion (the minor masses in the spectrum), a difference could be observed between the two signals: the mass difference is 0.0362 Da, which is significant with respect to being same compound. Extracting an empirical formula from the exact molecular ion, the difference can be explained: the second signal is due to 1,4-diacetylbenzene ($C_{10}H_{10}O_2$) (CAS 1009-61-6). This simple example demonstrates the power of the high-resolution MS technique.

10.4.2.4 LC-MS Examples

As already explained, when using LC-MS technology commercially available databases for MS library searches are lacking. Also, as shown in Figure 10.3, the LC traces might be not as clear as in GC – a clear substance location is not given.

On examining the LC results by checking each spectrum, there are two possible approaches:

1. One can search for known substances by filtering known exact masses out of the total spectrum. In Figure 10.3 the exact mass of 509.2363 Da matches a known acrylate used in printing inks.
2. The alternative is to extract a spectrum which is in the first instance unknown. Figure 10.4 shows an extracted exact mass of 353.1581 Da. Then either this mass is compared with formulation information supplied by several suppliers, or programs such as ChemSpider are

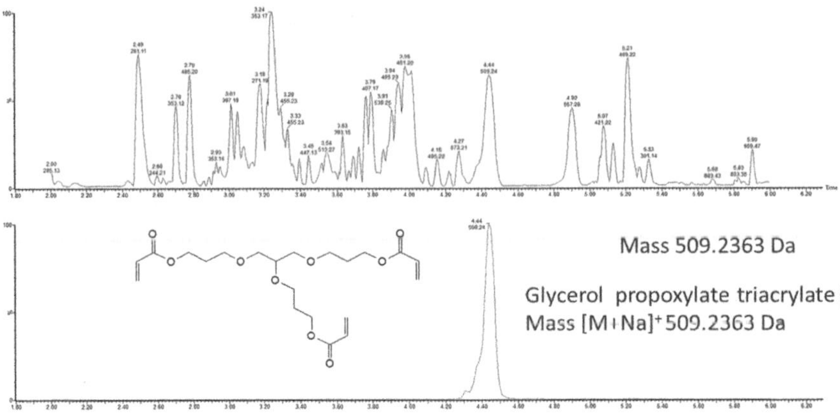

Figure 10.3 High-resolution mass spectrometry – known substance identification.

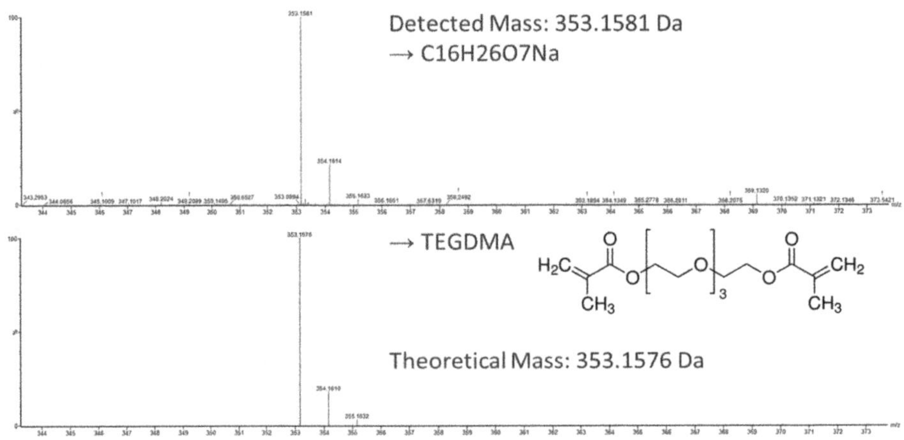

Figure 10.4 High-resolution mass spectrometry – unknown substance identification.

used. The closest fitting match given in Figure 10.4 yields a theoretical mass of 353.1576 Da. The mass difference is only 0.0005 Da (1.7 ppm). Therefore, the match with TEGDMA (triethylene glycol dimethacrylate) is considered perfect. Nevertheless, verification is needed as described in Section 10.5.

10.5 Performing Analytical Tests

Once the third-party (expert) laboratory has agreed with the client the terms of the testing conditions and the kind of MS analysis, the experiments can be performed. As already mentioned, for target analysis standard methods may be used, including quality parameters such as quality control samples, external calibration curves and/or using internal standards to achieve high precision. However, in non-target analysis it is not clear from the beginning what could be expected. Therefore, external calibration curves are not an option. The use of internal standards, however, is essential here, although no clear definition exists as to what kind of internal standards should be used. As no harmonized method for non-target analysis exists, there is a wide variety of internal standards. Figure 10.5 shows as an example a set of internal standards for GC-Q-TOF-MS-FID. Heptadecane is used for the FID signal, whereas labelled (deuterated) common contaminants in the FCM area (several phthalates) are used. By using several internal standards spread over the chromatogram, a high level of data safety is achieved.

The combination of GC-Q-TOF-MS and FID is a very good alternative with regard to detection capability, especially as within one system it combines high identification power with strong quantification possibilities.

For LC-MS non-target applications the situation is more limited than with GC-MS. Typically as an internal standard – used more as an injection and general performance control – substances such as urethane acrylate (CAS 63255-53-6) are used as they may provide a good compromise in terms of response factors.

Compared with GC-MS, LC-MS techniques show some significant disadvantages in non-target analysis. GC-MS generates ions always with the same energy level (70 eV), which leads to greater comparability of spectra even with unit- or high-resolution detection. In LC-MS, the ionization process is highly individual and any non-target analysis is either a compromise or must use several ionization techniques such as electrospray ionization (ESI) and atmospheric pressure chemical ionization (APCI) in negative and positive

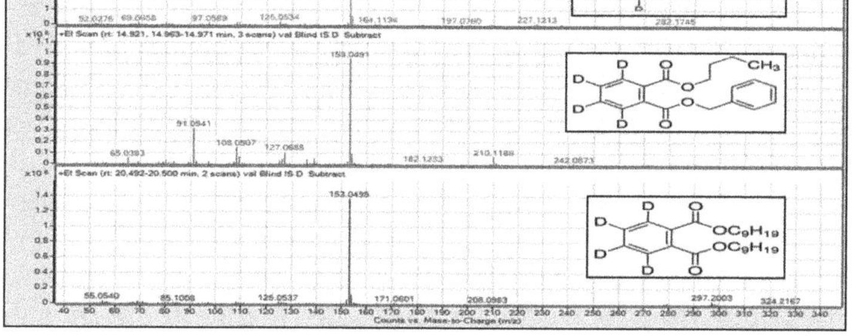

Heptadecane, RRT: 0.620

Di-n-butyl phthalate-D4 (D4-DBP), RRT: 0.756

Benzylbutyl phthalate-D4 (D4-BBP), RRT: 1.000

Di-n-nonyl phthalate-3,4,5,6-D4 (D4-DnNP), RRT: 1.369

Figure 10.5 Possible internal standards for GC-Q-TOF-MS-FID.

modes. Actually, most non-target analyses, *e.g.* LC-Q-TOF-MS, are performed in the ESI+ mode as most substances can be ionized in this way. However, as the focus is on only one ionization mode it is highly unsatisfactory. Data based on only one ionization technique have to be assessed very carefully as major information may be missed. This disadvantage of LC-MS compared with GC-MS explains why most laboratories actually perform only GC-MS-based non-targeted analysis.

From the above, it becomes clear that new developments in LC-MS technology must find a way to achieve less discrimination of substances in the screening mode. This means quicker ramping of the ionization energy in the ionization phase and much more robust switching between positive and negative modes. The current situation is not yet really satisfactory, which leads to limited use of LC-MS screenings. Nevertheless, the use of high-resolution LC-MS screening tools provides good additional insight into FCM behaviour.

10.6 Assessment of Results

The work of a third-party (expert) laboratory does not end with generating raw data/spectra, as this can nowadays be achieved with a high degree of automation. As shown with the examples in Section 10.4.2, the identification of screening results is time consuming and very complex. If, however, only target analysis has to be performed, all the criteria for quantitative method validation have to be applied. As indicated earlier, the greatest impact for the uncertainty determination lies in the extraction and migration experiments as here no spiking experiments can be performed.

10.6.1 Single/Multiple Analytes

When single and multiple analytes are clearly identified – rules for identification are given in Table 10.3, taken from an EU pesticide guideline[11] – then in either target or non-target analysis the quantification procedure is key. Within target analysis typically external calibration curves are the basis for quantification. In the case of highly matrix-impacted samples – mainly occurring by complete extractions destroying sample material – even matrix-based calibration curves may be a good alternative. Within non-target analysis, the quantification is based initially on a semiquantitative estimation *via* the internal standards used. If later, especially for initially unknown substances, analytical standards become commercially available then quantification procedure follows the same rules as for target analysis.

The quantification of known analytes is standard work for all third-party (expert) laboratories, but they face some limitations if oligomers or substance mixtures have to be identified and also quantified, as explained in the following sections.

Table 10.3 Mass spectrometric identification criteria. Reproduced from ref. 11, Copyright © 2006–2018 EU Reference Laboratories for Residues of Pesticides. All Rights Reserved.

MS detector/ characteristics	Typical systems (examples)	Acquisition	Requirements for identification	
			Minimum number of ions	Other
Unit mass resolution	Quadrupole, ion trap, TOF	Full scan, limited *m/z* range, SIM	3 ions	S/N ≥3[e]
MS/MS	Triple quadrupole, ion trap, Q-trap, Q-TOF, Q-Orbitrap	Selected or multiple reaction monitoring (SRM, MRM), mass resolution for precursor ion isolation equal to or better than unit mass resolution	2 product ions	Analyte peaks in the extracted ion chromatograms must fully overlap
Accurate mass measurement	High-resolution MS: (Q-)TOF, (Q-)Orbitrap, FT-ICR-MS, sector MS	Full scan, limited *m/z* range, SIM, fragmentation with or without precursor ion selection, or combinations thereof	2 ions with mass accuracy ≤5 ppm[a,b,c]	Ion ratio within ±30% (relative)of average of calibration standards from same sequence
		Combined single-stage MS and MS/MS with mass resolution for precursor ion isolation equal to or better than unit mass resolution	*2 ions*: 1 molecular ion, (de) protonated molecule or adduct ion with mass accuracy ≤5 ppm[a,c] *plus* 1 MS/MS product ion[d]	

[a]Preferably including the molecular ion, (de)protonated molecule or adduct ion.
[b]Including at least one fragment ion.
[c]<1 mDa for *m/z* <200.
[d]No specific requirement for mass accuracy.
[e]In case noise is absent, a signal should be present in at least five subsequent scans.

10.6.2 Oligomers

Oligomers are a common phenomenon in the area of FCM analysis. In addition to a lack of a regulatory background, reference standards are also mostly missing. In terms of MS identification and quantification there are several limitations, as summarized in the following four cases.

10.6.2.1 Structure of Known Oligomers Including Standards

If the structures of the identified oligomers within a non-targeted analysis are known and external standards are available (for each or for the majority of the oligomers), quantification is a challenge. As the distribution of the individual oligomers may vary from application to application, this might lead to an incorrect estimate in its quantification if a certain reference standard is used. In Figure 10.6, the distribution of Omnipol oligomers is shown. The relative ratio of each oligomer may vary from one technical mixture to the next depending on the supplier. The best procedure would be to quantify each oligomer with an external standard, build up the sum and compare with the SML if it exists. The comparison of one individual oligomer with a limit may lead to incorrect results.

10.6.2.2 Structure of Known Oligomers Without Standards for Each Oligomer

The second case is when reference standards are not available for all known oligomers, as shown in Figure 10.7.

For poly(ethylene glycol) bis(*p*-dimethylaminobenzoate) (CAS 71512-90-8), only one oligomer is commercially available. For quantification purposes, also here the best option would be to sum the areas of all oligomers used for quantification and compare with an external oligomer standard. The response factors of each oligomer may be different; however, if compared with one known oligomer there is no difference between technical mixtures as everything is integrated.

10.6.2.3 Structure of Known Oligomers With No Standards Available

The situation with the complete lack of commercially available oligomer standards is in practice fairly common. As shown in Figure 10.8 for poly(lactic acid) (PLA), no branched oligomers are available, but the cyclic compound – cyclic poly(lactic acid) nonamer – is available.

The best procedure here would be to quantify against a similar structure in the same way as before and sum all oligomers detectable. Here definitely only semiquantitative results will be obtained and this must be indicated when reporting.

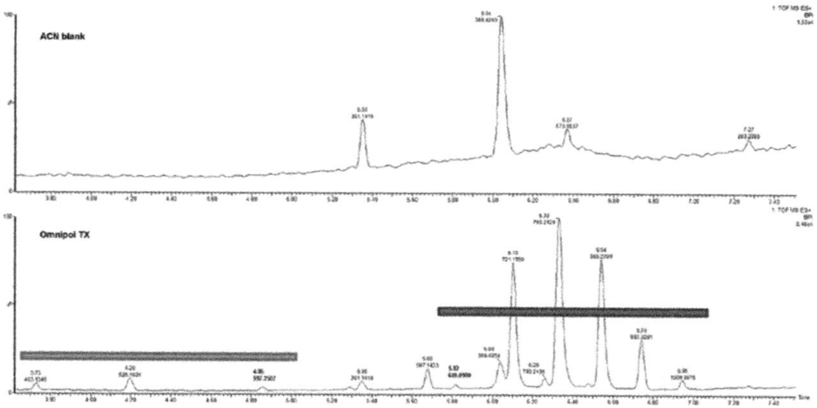

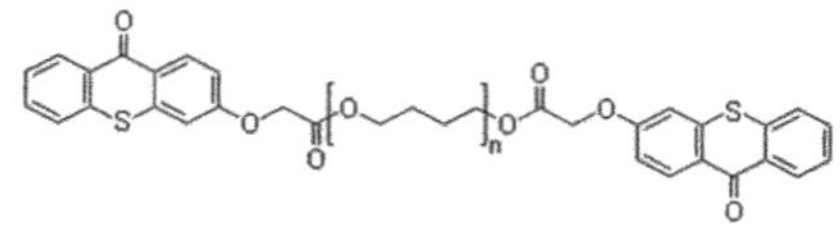

$(C_4H_8O)_n \cdot C_{30}H_{18}O_7S_2$

RT	H^+	Na^+	ID
5.82	627	649	Omnipol TX n=1
6.1	699	721	Omnipol TX n=2
6.33	771	793	Omnipol TX n=3
6.54	843	865	Omnipol TX n=4
6.74	915	937	Omnipol TX n=5
6.95	987	1009	Omnipol TX n=6
3.73	431	453	„ITX“ plus 2*72
4.2	503	525	„ITX“ plus 3*72
4.86	575	597	„ITX“ plus 4*72

Figure 10.6 Omnipol quantification using a reference for each oligomer.

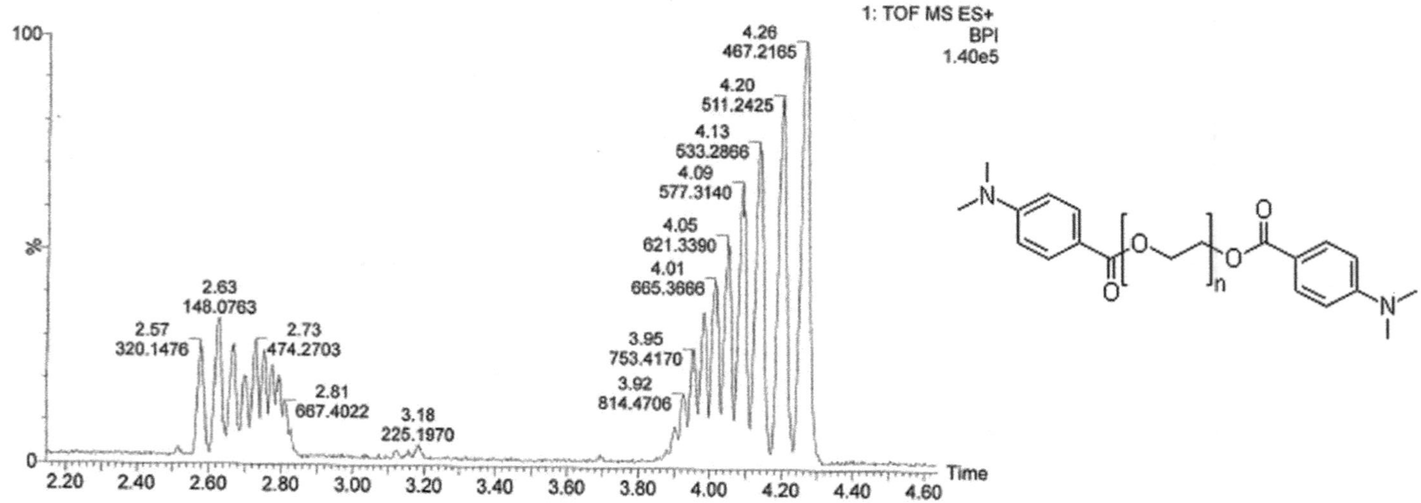

Figure 10.7 Omnipol ASA quantification using a limited number of references.

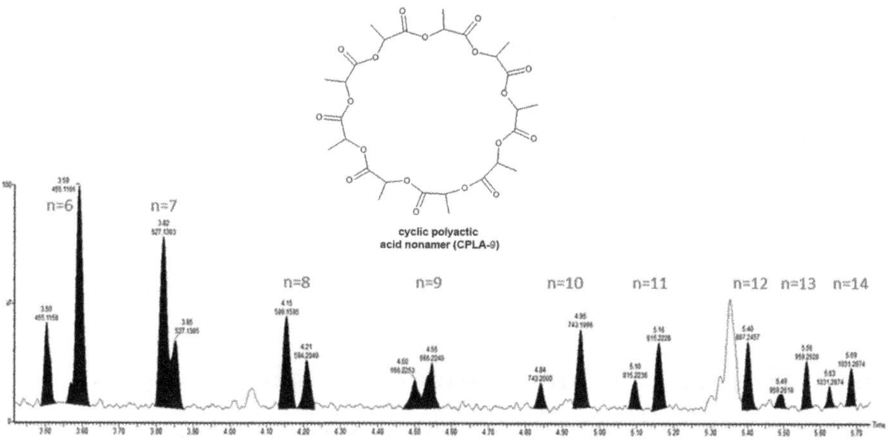

Figure 10.8 Quantification of poly(lactic acid). (PLA) oligomers.

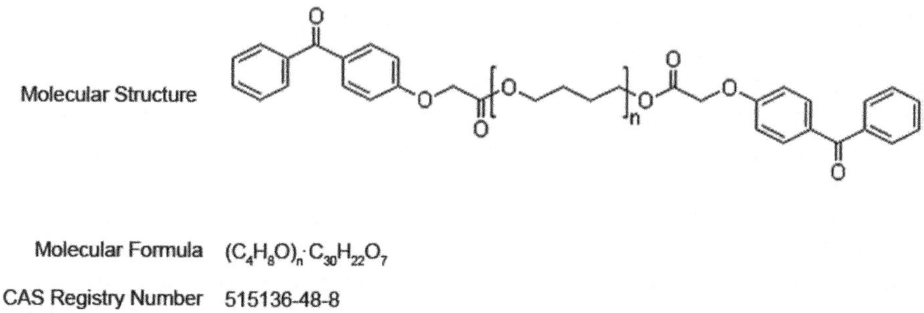

Figure 10.9 Structure of Poly(oxy-1,4-butanediyl), alpha-((4-benzoylphenoxy)acetyl)-omega-(((4-benzoylphenoxy)acetyl)oxy) (CAS No 515136-48-8).

10.6.2.4 Structure of Unknown Oligomers and No Standards Available

The need to detect oligomers clearly when no standard is available is also fairly common in modern non-targeted analysis. For quantification, only the approach *via* an internal standard as described in Section 10.5 can help in semiquantification. For that purpose, the sum area of all oligomers is also calculated. In Figure 10.9 and Table 10.4 a typical example of an oligomer for which no standard exists is presented. The quantification has to be performed as already described.

The above four examples of oligomer quantification show the overall scope to be covered in the targeted and non-target analysis of FCMs. They also illustrate the difficulties that third-party (expert) laboratories face. It is not only the identification but also the quantification that have problems to be solved.

Table 10.4 Omnipol BP oligomers.

Oligomer	*m/z* detected	Standard available
Oligomer 1	589.1841	No
Oligomer 2	661.2405	No
Oligomer 3	733.2991	No
Oligomer 4	805.3572	No
Oligomer 5	877.4137	No
Oligomer 6	949.4720	No
Oligomer 7	10 121.5320	No

10.7 Future of Food Contact Material Analysis

For both target and non-target analysis, MS, which is based on unit and/or high resolution, provides many opportunities in the analysis of FCMs. Most results are based on extraction or migration studies. On the one hand, extraction and migration studies help in understanding the mechanism of migration under worst-case conditions. However, such studies are always only simulations. Where real food samples are concerned, it not clear in any case whether the simulation studies actually represent the worst case. On the other hand, simulation situations may lead to a result such that a good FCM solution is not used for a final packaging application owing to some positive findings. An overestimation can have a high economic impact. This situation may arise more and more in the near future as NIASs will be detected "everywhere."

In the future, producing a material/food packaging without positive findings will no longer occur, especially if the sensitivity of non-target MS based screenings is increased. This brings up the question of whether all findings in simulation experiments can be confirmed in food itself, or in other words proof of simulation results in food is increasingly needed.

Targeted analysis of food products is already possible today, and MS provides the necessary tools. However, to establish the targets that need to be looked for in foods is still time consuming and possibly migrant substances are not reported within the value chain from the raw material supplier to the food manufacturer using FCMs. The next step could be to implement known targets in the HACCP (hazard analysis and critical control point) concept of the food manufacturer and to set up a monitoring programme. However, the willingness to perform such a self-control programme seems low. Therefore, a quick and reliable alternative for the food industry is needed in the future: non-target screening in the food itself. Figure 10.10 shows the possible future trend in FCM analysis.

However, there are some obstacles to this need for non-targeted screening in food. The food matrices are highly variable and the interferences that occur need very good profiling to differentiate their signals from signals that have their origin in the FCMs. Nevertheless, as food profiling will also be an increasing task in the near future driven by authenticity issues, these data

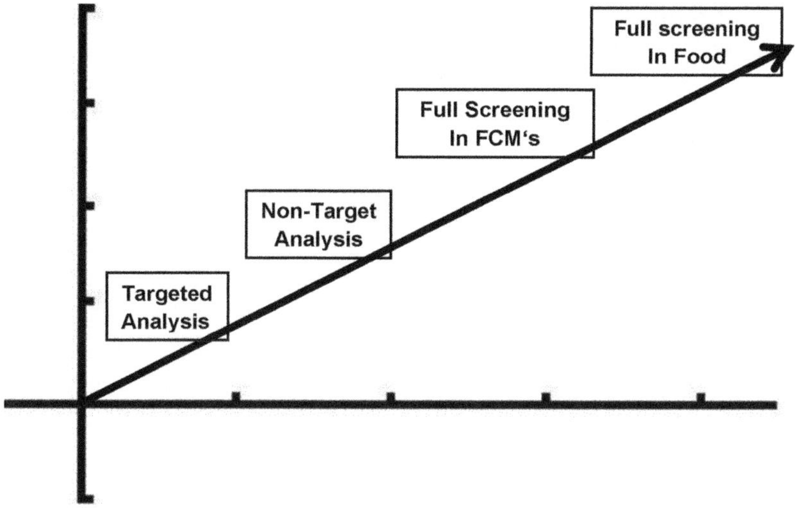

Figure 10.10 Future trend in food contact material analysis.

will help to differentiate which signals are directly food related and which signals are FCM related. To achieve this goal many new data are necessary and ultimately very good database solutions will also be needed. MS carried out in the non-targeted high-resolution mode will be one important tool alongside others to achieve the aim of coming closer to full screening in foods.

References

1. EN 13130 series – National norm committee.
2. *VUP – Verband unabhängiger Prüflabore published a First Guideline to "Screening Products for SVHC According to the REACH Regulation"*, 01.01.2013, http://www.vup.de/.
3. *COMMISSION REGULATION (EU) No 10/2011 of 14 January 2011 on plastic materials and articles intended to come into contact with food.*
4. Chinese Standard GB 4806.1-2016-National Food Safety Standard, *General Safety Requirements for Food Contact Materials and Products*, 19.10.2017.
5. US FDA: CFR 21-SUBCHAPTER B—FOOD FOR HUMAN CONSUMPTION, § 170ff.
6. Mercosur: GMC Resolution No. 3/92, *General Criteria for Packaging and Articles to Come into Contact with Foodstuffs: Terminology, General Criteria and Classification of Materials.*
7. *Japan: Food Sanitation Act No. 233*, 1947.
8. *South Korea: Standards and Specifications for Food Utensils, Containers and Packages*, 2013.
9. *ILSI: Guidance on Best Practices on the Risk Assessment of Non Intentionally Added Substances (NIAS) in Food Contact Materials and Articles*, July 2015.

10. ILSI, http://ilsi.eu/task-forces/food-safety/packaging-materials/.
11. *Guidance Document on Analytical Quality Control and Method Validation Procedures for Pesticides Residues Analysis in Food and Feed. SANTE/11945/2015, Supersedes SANCO/12571/2013 Implemented by 01/01/2016.*

Subject Index